GRADUATE STUDIES
TEXAS TECH UNIVERSITY

**Aspects of Avian Endocrinology:
Practical and Theoretical Implications**

Edited by
*C. G. Scanes, M. A. Ottinger, A. D. Kenny,
J. Balthazart, J. Cronshaw, and I. Chester Jones*

No. 26   October 1982

TEXAS TECH UNIVERSITY

Lauro F. Cavazos, President

Graduate Studies No. 26
411 pp.
22 October 1982
Paper, $29.95
Cloth, $59.95

Graduate Studies are numbered serially, paged separately, and published on an irregular basis under the auspices of the Dean of the Graduate School and Director of Academic Publications, and in cooperation with the International Center for Arid and Semi-Arid Land Studies. The preferred abbreviation for citing this series is Grad. Studies, Texas Tech Univ. Copies can be purchased on standing order or separately by title from the Texas Tech Press Sales Office, Texas Tech University Library, Lubbock, Texas 79409, U.S.A. Orders from individuals should be accompanied by remittance, and all payments should be made in U.S. currency, with check, money order, or bank draft drawn on a U.S. bank. Prices include normal handling and postage when mailed within the United States (add $2.00 per title for foreign postage; Texas residents must pay a 5 per cent sales tax on total purchase price). Institutions interested in exchanging publications should address the Exchange Librarian at Texas Tech University.

QL
698
.A79
1982

ISSN 0082-3198
ISBN 0-89672-102-7 (paper)
ISBN 0-89672-103-5 (cloth)

Texas Tech Press, Lubbock, Texas

1982

CONTENTS

SECTION VI: EFFECTS OF ENVIRONMENTAL POLLUTANTS ON
AVIAN ENDOCRINE SYSTEMS

PREFACE

Birds have held a fascination for man since before written historical records. They have been domesticated for food (in the Old World, chicken, goose, duck, and guinea fowl; in the New World, turkey), as hunting animals (hawks and kestrels), for racing (pigeon), and as pets or for show (parrots, canaries, and doves). At a more quixotic level, one can point to the importance of birds in mythology and ancient religions; as symbols, from the Imperial eagle to the dove of peace; and as analogies in colloquial English (for example, "wise as an owl").

The field of avian endocrinology can be traced back to the classical study of Berthold (1849). That study provided evidence for a testicular factor which affected secondary sex characteristics and behavior in birds. It is pertinent to ponder why birds have been used in many studies in the last hundred years. Practical reasons include availability, size, ease of maintenance, cost, and economic importance (particularly poultry and other domesticated species). Other important considerations include radiative evolution of a large number of species, patterns of migration, defined annual reproductive cycles, and defined behaviors.

The present volume took shape resulting from the endeavors of August Epple and Milton Stetson who organized the Second International Symposium on Avian Endocrinology (supported by grants from the National Institute of Health; United States Department of Agriculture; International Union of Biological Sciences; Merck, Sharp, and Dohme Research Laboratories, Rahway, New Jersey; Upjohn Company, Kalamazoo, Michigan; Smith, Kline, and French Laboratories, Philadelphia, Pennsylvania; and Arbor Acres Farm, Inc., Glastonbury, Connecticut). This meeting at Benalmadena, Spain, contained six round table conferences: 1) Pharmacological Techniques in Avian Endocrinology—Recent Progress and Future Applications; 2) Mechanisms in the Hormonal Control of Avian Behavior; 3) Nutritional Influences upon Growth, Reproduction, and Metabolism: Endocrine Effects; 4) Endocrinology of Calcium Metabolism and its Clinical Relevance; 5) Avian Osmoregulatory Mechanisms in Phylogenetic Perspective; and 6) Effects of Environmental Pollutants on Avian Endocrine Systems.

These areas are of obvious significance in endocrinology, particularly with respect to birds. An element of a common theme running through these topics was that they represented examples of basic biological research which had some practical or applied component. This ranged from the use of birds in bioassay (Kenny; and Hertelendy *et al.*), the interaction of hormones on growth or reproduction in domesticated species (Scanes *et al.*; El Halawani *et al.*; Harvey *et al.*; and Taylor *et al.*), the interaction of calcium (section IV) and other nutrients (section III) with the endocrinology of growth and reproduction, and chronic effects of oil pollution on the endocrinology of wild birds (section VI). In addition, there are sections on the endocrine in-

7

volvement in reproductive behavior (section II) and osmoregulation (section V). These biological characteristics are not only essential to the production of poultry species, but also to the survival of wild species, especially in these days of shrinking or changing habitats.

Because these topics have not received similar detail elsewhere, it was decided to put the conferences together as a single volume. It is hoped that this volume will be of interest to a wide variety of research workers including comparative endocrinologists, poultry scientists, ornithologists, and wildlife specialists.

LITERATURE CITED

BERTHOLD, A. A. 1849. Transplantation der Hoden. Arch. Anat. Physiol. U. Wiss. Med., 2:42-46.

SECTION I

PHARMACOLOGICAL TECHNIQUES IN AVIAN ENDOCRINOLOGY— RECENT PROGRESS AND FUTURE APPLICATIONS

INTRODUCTION

Colin G. Scanes

Department of Physiology, Rutgers—The State University, New Brunswick, New Jersey 08903 USA

A pharmacological approach can be applied to the endocrinology of birds (or mammals for that matter) for a number of distinct reasons. These include the use of drugs which are specific agonists, antagonists, synthesis blockers and the like for *in vivo* studies in physiology and *in vitro* studies in cell biology. In addition, this approach includes the development and application of both *in vivo* and *in vitro* biological assays.

The rationale for the investigator using an avian species obviously varies. The raison d'être for many avian studies is the regulation of reproduction and growth in poultry species. In recent years attention has been focused on the involvement of putative neurotransmitters in the central (hypothalamic) control of pituitary hormone secretion. These studies can employ the peripheral administration of a multitude of (largely) specific blockers of neurotransmitters synthesis/reuptake/degradation or receptor agonists or antagonists and the estimation of circulating concentrations of pituitary hormones. Examples of this approach are seen in the papers of Scanes and colleagues and El Halawani *et al.* These papers deal with the involvement of catecholamines and serotonin in the *in vivo* control of luteinizing hormone (LH), growth hormone, and prolactin secretion in gallinaceous birds. Neurotransmitters or analogues can also be administered centrally into the brain (for example, the ventricular system or specific nuclei in the hypothalami) and pituitary hormone secretion measured. An alternative approach involves the *in vitro* incubation of pituitary tissue alone or together with hypothalamic fragments in the presence of neurotransmitters or their agonists/antagonists. Harvey and his colleagues have employed these latter methods to examine the role of neurotransmitters in the control of prolactin secretion in the domestic fowl. Methallibure has a well-established antifertility action in both mammals and birds. Evidence is presented by Taylor *et al.* that methallibure depresses gonadotropin secretion at the hypothalamic level by inhibiting dopamine β-hydroxylase and hence norepinephrine synthesis.

Birds offer a unique model for biological research, including the use of avian species in bioassays. Examples of this include a widely used bioassay for parathyroid hormone (PTH). Kenny describes this system and the characterization of different forms of PTH. Granulosa cells from the chicken ovary offer a potentially very useful model for hormone action or simply as a bioassay for LH. In the present volume, Hertelendy *et al.* describe studies on the effects of the ammonium ion on granulosa cell progesterone production.

11

Reprinted from
ASPECTS OF AVIAN ENDOCRINOLOGY:
PRACTICAL AND THEORETICAL IMPLICATIONS (C. G. Scanes *et al.*, eds.)
Grad. Studies, Texas Tech Univ., 1982, 26:1-411.

BRAIN AMINES AND THE REGULATION OF ANTERIOR PITUITARY SECRETION IN THE DOMESTIC FOWL

COLIN G. SCANES, JAMSHID RABII, AND FRANCES C. BUONOMO

Department of Physiology, Rutgers—The State University, New Brunswick, New Jersey 08903 USA

An abundance of physiological and pharmacological evidence has firmly established the notion that brain amines (the catecholamines and serotonin) are intimately involved in the hypothalamic regulation of pituitary hormone release in mammals (Ganong, 1974; Mornex, *et al.*, 1976; Roser and Ganong, 1976; Meites, *et al.*, 1977; Müller *et al.*, 1977; Weiner and Ganong, 1978). In contrast to the thousands of papers published on this subject for mammalian species, only few reports exist on the aminergic control of pituitary hormone secretion in birds. This review focuses primarily on our efforts to study the participation of biogenic amines in the regulation of luteinizing hormone (LH), prolactin (PRL), and growth hormone (GH) release in the domestic fowl.

The neurosecretory neurons of the hypothalamus, by releasing their hormones into the hypophyseal portal capillaries, are largely responsible for controlling the secretion of the anterior pituitary gland. These hypothalamic hormones are generally referred to as releasing or release inhibiting hormones. In mammalian species studied, these appear to exert a stimulator influence (by a releasing hormone) over the release of LH, dual stimulating and inhibiting effects on that of GH, and inhibitory influence (inhibitory hormone) on the release of PRL. In birds, however, only LH and GH appear to follow the mammalian pattern, while PRL secretion is predominantly under stimulatory influence of the hypothalamus (Kragt and Meites, 1965; Nicoll, 1965; Schally and Kastin, 1972; Hall and Chadwick, 1976; Scanes *et al.*, 1977).

Our approach to the question of aminergic control of pituitary hormone secretion has been from a pharmacological standpoint using centrally active drugs presumed to inhibit the biosynthesis of the transmitter or to interact with the receptor sites (either as agonists of antagonists). It therefore is appropriate to review briefly the formation of the catecholamines and serotonin as well as the site of action of various drugs before describing the experimental results.

Biosynthesis of the Amines

The catecholamines (dopamine, norepinephrine, and epinephrine) are derived from the amino acid tyrosine.

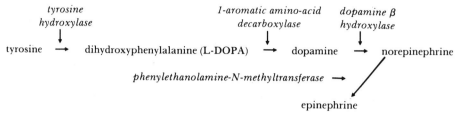

The conversion of tyrosine to dihydroxyphenylalanine, a reaction catalyzed by the enzyme tyrosine hydroxylase, is the rate limiting step in the biosynthesis of the catecholamines. The conversion of dopamine to norepinephrine is catalyzed by the enzyme dopamine-β-hydroxylase which is present only in tissues producing norepinephrine or epinephrine. The enzyme phenylethanolamine-N-methyltransferase is responsible for N-methylation of norepinephrine and its conversion to epinephrine. This enzyme appears to be present in relatively few neurons in the brain.

Serotonin is derived from tryptophan in the following way:

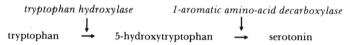

L-tryptophan first is converted to 5-hydroxytryptophan (5-HTP) catalyzed by tryptophan hydroxylase. 5-HTP is then rapidly decarboxylated by the enzyme l-aromatic amino acid decarboxylase to form serotonin (5-HT).

Both the catecholamines and serotonin are stored in synaptic vesicles, and, upon release into the synapse, interact with pre- and post-synaptic membrane receptors. The major mechanism of inactivation for these amines is reuptake into the nerve terminal and exposure to the catabolic enzyme monoamine oxidase.

Drugs Influencing the Catecholaminergic System

Inhibitors of catecholamine synthesis.—α-Methyl-p-tyrosine (α-MPT; Sigma Chemical Company, St. Louis, Missouri) inhibits the enzyme tyrosine hydroxylase, and blocks the synthesis of both dopamine and norepinephrine (Hanson, 1965; Udenfriend *et al.*, 1965). Diethyldithiocarbamate (DDC; Sigma Chemical Company, St. Louis, Missouri) inhibits the enzyme dopamine-β-hydrolase, and thus blocks the conversion of dopamine to norepinephrine (Johnson *et al.*, 1970, 1972).

Catecholamine depletors.—Reserpine (CIBA Pharmaceutical Company, Summit, New Jersey) interferes with the intraneuronal storage of catecholamines and serotonin. The released amine is not able to exert an effect, however, unless the enzyme monoamine oxidase is inhibited (Shore, 1962; Kopin and Gordan, 1963; Glowinski and Axelrod, 1965).

Adrenergic Receptor Antagonists.—Phenoxybenzamine (Smith, Kline, and French, Philadelphia, Pennsylvania) is a long lasting α-adrenergic receptor blocker (Nickerson and Gump, 1949; Furchgott, 1954). Propranolol (Sigma Chemical Company, St. Louis, Missouri) is a β_1 and β_2 adrenergic receptor antagonist (Black *et al.*, 1965; Dunlop and Shanks, 1968; Black and Prichard, 1973).

Dopaminergic Receptor Antagonists.—Pimozide (McNeil Laboratories, Ft. Washington, Pennsylvania) is a selective dopamine receptor blocker (Andén *et al.*, 1970).

Adrenergic Receptor Agonists.—Phenylephrine (Sigma Chemical Company, St. Louis, Missouri) is an α-adrenergic receptor stimulator (Meyers *et al.*, 1974), and isoproterenol (Sigma Chemical Company, St. Louis, Missouri) is a β-adrenergic receptor stimulator (Furchgott, 1960).

It should be noted that α receptors can be classified into two types: α_1 is post-synaptic, and α_2 is predominantly pre-synaptic (Hoffman and Lefkowitz, 1980). Norepinephrine and epinephrine interact equally well with both α_1 and α_2 receptors. Phenylephrine is predominantly an α_1-agonist, whereas clonidine is specific for α_2 receptors. With respect to the antagonists, phenoxybenzamine tends to be specific for α_1 receptors, yohimbine for α_2 receptors, whereas phentolamine blocks both α_1 and α_2 receptors (Hoffman and Lefkowitz, 1980).

Dopaminergic Receptor Agonists.—Apomorphine (Merck, Sharpe, and Dohme, Rahway, New Jersey) is a selective dopamine receptor stimulator (Andén *et al.*, 1967; Ernst, 1967).

Precursors of the catecholamines.—Dihydroxyphenylalanine (L-DOPA, Sigma Chemical Company, St. Louis, Missouri) is the immediate precursor of dopamine, and thus leads to an increase in the synthesis of both dopamine and norepinephrine (Horneykiewicz, 1966; Bartholini and Pletscher, 1968) Dihydroxyphenylserine (L-DOPS, Sigma Chemical Company, St. Louis, Missouri) is converted without dopamine intervention directly to norepinephrine by direct decarboxylation (Creveling and Daly, 1971; Puig *et al.*, 1974.

Drugs Influencing the Serotonergic System

Inhibitors of Serotonin Biosynthesis.—*p*-Chlorophenylalanine (pCPA; Sigma Chemcial Company, St. Louis, Missouri) is most widely used to inhibit both the *in vitro* and *in vivo* synthesis of serotonin by the brain. pCPA inhibits the enzyme tryptophan hydroxylase, thus blocking the conversion of tryptophan to 5-hydroxytryptophan (Koe and Weissman, 1966). pCPA, however, possesses several other actions which should be kept in mind when interpreting the observed effects. At high doses, pCPA is capable of inducing catecholamine-depletion, reducing the norepinephrine turnover rate, and reducing DA synthesis (Stolk *et al.*, 1969; Miller *et al.*, 1970; Tagliamonte *et al.*, 1973).

Serotonin Depletors.—Reserpine (see section under Catecholamine Depletors).

Serotonin Receptor Antagonists.—Methysergide (Sandoz Pharmaceutical Company, East Hanover, New Jersey), a derivative of lysergic acid, is a competitive blocker of serotonin receptors (Goodman and Gilman, 1980). At higher doses, methysergide also has been shown to interfere with dopaminergic neurotransmission (Lambert and MacLeod, 1978). Cyproheptadine (Merck, Sharpe, and Dohme, Rahway, New Jersey) is another serotonin receptor blocker that has been shown to have weak anticholinergic activity (Stone *et al.*, 1961). SQ-10631 (E. R. Squibb and Sons, Princeton, New Jersey) is also a serotonin receptor antagonist.

Serotonin Receptor Agonists.—2-(1-Piperazinyl) quinoline maleate (Quipazine, Miles Research Products, Elkhart, Indiana) is a serotonin receptor stimulator (Hong *et al.*, 1969).

Precursors of Serotonin.—5-Hydroxytryptophan (5-HTP, Sigma Chemical Company, St. Louis, Missouri) is the immediate precursor of serotonin (Moir and Eccleston, 1968). It should be kept in mind that large doses of 5-HTP can lead to interference with catecholamine transport, storage, and metabolism (Johnson *et al.*, 1968).

Other Drugs

Cholinergic Antagonists.—Atropine (Sigma Chemical Company, St. Louis, Missouri) is an anticholinergic drug that blocks muscarinic receptors (Bonnet and Bremer, 1952; Phillis and Tebecis, 1967).

Cholinergic Agonists.—Pilocarpine (Sigma Chemical Company, St. Louis, Missouri) is a cholinomimetic drug which stimulates muscarinic receptors (Goodman and Gilman, 1980).

MATERIALS AND METHODS

The experiments were performed on young (six-week-old) and adult male domestic fowl (strain White Leghorn). The young animals were maintained on a 16 hours light:8 hours dark (16L:8D) lighting regimen, whereas the adults were exposed to 14L:10D. The drugs and their respective vehicles (for control groups) were administered intraperitoneally. Blood samples were obtained either by venipuncture from the brachial vein (adult birds) or following decapitation (young birds).

The plasma concentration of each hormone in the samples was determined by homologous radioimmunoassay (LH, Follett *et al.*, 1972; PRL, Scanes *et al.*, 1976; Harvey and Scanes, 1977).

AMINERGIC ROLE IN LUTEINIZING HORMONE (LH) SECRETION

There is overwhelming evidence in mammals that release of luteinizing hormone releasing hormone (LHRH), and hence LH, is provoked by catecholamines (reviewed by Weiner and Ganong, 1978). This evidence includes results from the intraventricular injection of neurotransmitters (Sawyer, 1952; Sawyer, 1979), from pharmacological studies using biogenic amine synthesis blockers (Kalra *et al.*, 1972; Kalra and McCann, 1974; Drouva and Gallo, 1976), and neurotransmitter agonists and antagonists (Sawyer and Radford, 1978; Weick, 1978). Furthermore, there is evidence that

TABLE 1.—*Effect of neurotransmitter synthesis and granular reuptake inhibitors and antagonists on the plasma LH concentration in six-week-old and adult male chickens.*

	Time (hours)	Plasma LH as % of vehicle ± (N) SE	
		ADULT MALE	YOUNG MALE
Aminergic Synthesis Blockers			
p-Chlorophenylalanine (250 mg/kg)	48	42 ± (25) 3***	47 ± (14) 6***
Vehicle (aud saline)	48	100 ± (19) 12	100 ± (10) 17
α-Methyl-p-tyrosine (250 mg/kg)	1	51 ± (6) 12	65 ± (10) 4
Vehicle	1	100 ± (6) 14	100 ± (10) 17
Diethyldithiocarbamic acid (400 mg/kg)	1	77 ± (8) 5*	52 ± (16) 7***
Vehicle	1	100 ± (8) 10	100 ± (15) 8
Granular Reuptake Inhibitor			
Reserpine (25 mg/kg)	4	34 ± (5) 4	64 ± (10) 6*
Vehicle	4	100 ± (5) 20	100 ± (10) 17
Reserpine (25 mg/kg)	24		45 ± (10) 4**
Vehicle	24		100 ± (10) 17
Antagonists			
Pimozide (2.5 mg/kg)	1	78 ± (6) 12	80 ± ((19) 11
Vehicle	1	100 ± (6) 23	100 ± (17) 14
Phenoxybenzamine (20 mg/kg)	1	67 ± (7) 5*	65 ± (12) 9*
Vehicle	1	100 ± (7) 10	100 ± (11) 15
Propranolol (2 mg/kg)	1	89 ± (8) 12	110 ± (18) 16
Vehicle	1	100 ± (8) 14	100 ± (18) 14
Methysergide (2.5 mg/kg)	1	86 ± (6) 18	61 ± (21) 7
Vehicle	1	100 ± (6) 16	100 ± (20) 14
Atropine (10 mg/kg)	1		89 ± (18) 8
Vehicle	1		100 ± (18) 14

*$P < 0.05$, **$P < 0.01$, ***$P < 0.001$ that treatment is different from respective vehicle-injected control based on Student t-test.

serotonin has both positive and negative influences on the hypothalamic control of LH secretion in mammals (Kamberi *et al.*, 1970; Porter *et al.*, 1971/1972; Gallo and Moberg, 1977).

In birds, however, there are relatively few published reports on the effect of catecholamines on reproductive function (El Halawani and Burke, 1975; El Halawani *et al.*, 1978) and on LH release (Davis and Follett, 1974; Sharp, 1975).

Effect of Inhibition of Catecholamine Biosynthesis and Storage

The administration of the catecholamine synthesis blocker α-MPT (250 mg/kg dissolved in saline) to birds of both ages led to a significant reduction in circulating LH levels within one hour after injection (Table 1). Specific blockage of norepinephrine synthesis by DDC (400 mg/kg dissolved in saline) also reduced plasma LH levels by one hour after injection (Table 1). Similarly, the granular reuptake blocker reserpine (25 mg/kg dissolved in 80% acetone) depressed the circulating LH concentration within four hours in both young and adult birds. In the young birds this decline in plasma LH was still evident 24 hours following reserpine administration (Table 1).

TABLE 2.—*Effect of catecholamine precursors and agonists on LH release in young chicks pretreated with α-MPT (250 mg/kg).*

Treatment	Plasma LH as % of vehicle ± (N) SE
Precursors†	
L-DOPA (100 mg/kg)	173 ± (8)10**
DL-DOPS (100 mg/kg)	202 ± (8)17***
Vehicle	100 ± (8)19
Agonists‡	
Apomorphine (2 mg/kg) (DA)	148 ± (8) 9*
Phenylephrine (0.1 mg/kg) (α-adrenergic)	145 ± (8)12*
Isoproterenol (0.1 mg/kg) (β-adrenergic)	129 ± (8) 8
Vehicle	100 ± (8) 9

*P < 0.05; **P < 0.001, compared with vehicle-injected by ANOVA.
†L-DOPA and DL -DOPS were injected three hours after α-MPT and two hours before taking blood sample.
‡Agonists were administered 1/2 hour after α-MPT and three hours before blood removal.

Effect of Catecholamine Receptor Inhibition

The administration of the α-adrenergic receptor blocker phenoxybenzamine (20 mg/kg dissolved in acidified saline) was effective in reducing the plasma levels of LH in birds of both ages; the effect was significant within one hour after drug injection (Table 1).

The β-adrenergic receptor blocker propranolol (2 mg/kg dissolved in saline), on the other hand, did not significantly alter plasma LH levels in either the young or adult birds (Table 1).

Administration of the dopaminergic receptor blocker pimozide (2.5 mg/kg dissolved in 2% tartaric acid) had no apparent influence on plasma LH levels one hour after injection in both age groups (Table 1).

Effect of Catecholaminergic Receptor Stimulation

Injection of the α-adrenergic receptor agonist phenylephrine (0.1 mg/kg dissolved in saline) in young animals pretreated with α-MPT significantly elevated circulating LH concentration. Similarly, the dopaminergic receptor agonist apomorphine (2 mg/kg dissolved in saline) elevated plasma LH levels in the male chicks pretreated with α-MPT. The β-adrenergic receptor agonist isoproterenol (0.1 mg/kg dissolved in saline), on the other hand, did not lead to a significant rise in circulating LH levels in the α-MPT treated chicks (Table 2).

Effect of Catecholamine Precursors

In animals whose catecholamine synthesis is inhibited with α-MPT, the administration of either L-DOPA (100 mg/kg dissolved in saline) or L-DOPS (100 mg/kg dissolved in saline) led to an increase in plasma LH levels two hours following administration of the precursor (Table 2). The one hour LH levels in these animals are, however, lower than those in the vehicle injected controls. The reason for this apparent biphasic effect is not clear.

Effect of Inhibition of Serotonin Biosynthesis

The administration of the tryptophan hydroxylase inhibitor pCPA (250 mg/kg dissolved in acidified saline) to young and adult males significantly depressed the plasma concentrations of LH at 48 hours following injection (Table 1).

Effect of Serotonergic Receptor Inhibition

Injection of methysergide (2.5 mg/kg dissolved in 20% ethanol) led to inhibition of LH secretion in young birds by one hour and in adult animals at one hour following administration (Table 1). Similarly, another serotonin antagonist cyproheptadine (5 mg/kg dissolved in 20% ethanol) reduced plasma LH levels in the young males by one hour after injection.

General Discussion of Aminergic Role in Luteinizing Hormone Secretion

The secretion of LH by the avian pituitary is regulated by the hypothalamus. Early evidence came from the observation of gonadal atrophy following transplantation of the pituitary to a site away from the sella turcica or after formation of electrolytic lesions in the hypothalamus (reviewed by Dodd et al., 1971). More recently, Davies and Follett (1974) demonstrated that LH release in Japanese quail could be provoked by electrical stimulation of specific areas in the hypothalamus. Data on the involvement of biogenic amines in hypothalamic regulation of LH secretion in birds have been scarce. There is some evidence that catecholamines are essential for reproductive function (Ferrando and Nalbandov, 1969; Kao and Nalbandov, 1972; El Halawani and Burke, 1975; El Halawani et al., 1978) and maintenance of plasma LH levels (Davies and Follett, 1974; Sharp, 1975). From experiments described in this paper, it is evident that the biogenic amines are intimately involved in the regulation of LH secretion in male domestic fowl. Centrally active compounds, presumed to be synthesis blockers of catecholamines and of serotonin, are all capable of depressing circulating LH levels in both adult and sexually immature chickens. This reduction in LH release occurred with inhibitors of catecholamine synthesis (by α-MPT), norepinephrine synthesis (by DDC), and serotonin synthesis (by pCPA). Furthermore, the amine reuptake inhibitor, reserpine, which affects all three neurotransmitters, is effective in lowering the circulating levels of LH in both age groups.

The effect of inhibition of catecholamine synthesis can be at least partially overcome by the administration of precursors (DOPA and DOPS), and by α-adrenergic (phenylephrine) and dopaminergic (apomorphine) agonists. These observations suggest that there is a major involvement of α-adrenergic transmission in the central LH secretion in this species. Further evidence comes from our observations of the inhibitory influence of α-adrenergic blockade by phenoxybenzamine on LH release and a lack of effect of β-adrenergic blockade by propranolol. The dopaminergic receptor antagonist pimozide produced a non-significant reduction in plasma LH levels in both young and adult birds. This is similar to the proposed α-adrenergic role for norepinephrine in the hypothalamic control of LH secretion in mammals (see for example, Weiner and Ganong, 1978). For instance, the proestrus surge of LH in the rat can be inhibited by catecholamine depletion (Kalra and McCann, 1974) and α-adrenergic blockade (Gnodde and Schulling, 1976; Jackson, 1977). Furthermore, it can be provoked by amine precursors and α-adrenergic agonists which overcome the inhibition by nonadrenergic synthesis blockers (Kalra et al., 1973; Kalra and McCann, 1973, 1974). The present data also are in agreement with observations made by others using female birds. For example, it has been observed that α-adrenergic antagonists inhibit ovulation (Ferrando and Nalbandov, 1969; Kao and Nalbandov, 1972). In addition, reserpine administration depresses the circulating concentrations of LH in adult female chickens, as well as in adult males and castrated males (Sharp, 1975). Furthermore, the photoperiodically induced increase in LH release and testes weight in the Japanese quail can be ablated by the administration of reserpine or α-MPT (Davis and Follett, 1974; El Halawani and Burke, 1975; El Halawani et al., 1978).

Our studies with the serotonin related drugs suggest a stimulatory role in the control of LH secretion in the male domestic fowl. An inhibition of serotonin synthesis (by pCPA) results in a decrease of plasma LH levels. Similarly, serotonin receptor blockade (by methysergide and cyproheptadine) is accompanied by an inhibition of circulating LH concentrations. This suggestion is in apparent contrast to the observation on pCPA effects in the Japanese quail (El Halawani and Burke, 1975). In mammals, on the other hand, there is evidence for serotonin having both stimulatory and inhibitory roles in the control of LH release (see for example, Gallo and Moberg, 1977).

AMINERGIC ROLE IN THE SECRETION OF GROWTH HORMONE (GH) AND PROLACTIN (PRL)

Pituitary secretion of GH in mammals is regulated by the central nervous system through a dual system of control based on an inhibiting (somatostatin GHRIH) and a stimulating (GHRF) factor (Brazeau et al., 1973; Reichlin et al., 1973). The latter is not fully characterized. In the case of PRL, it is generally believed that its secretion in mammals is mainly under a tonic inhibitory influence of the central nervous system by prolactin inhibitory factor (PIF) (Everett, 1954). The chemical characterization of PIF has not yet been reported. In spite of the vast physiological evidence supporting the view that PRL secretion (by the mammalian anterior pituitary) is under tonic inhibitory hypothalamic control, a number of reports have suggested the additional existence of a stimulatory component, prolactin releasing factor (PRF) in regulation of PRL (Meites et al., 1960; Mishkinsky et al., 1968).

It is widely recognized that in turn, the GH and PRL regulatory systems are controlled by complex networks of neurotransmitter and neuromodulator pathways, which represent the final link between the central nervous system and the neurosecretory hypophysiotropic hormones (Müller et al., 1977; Weiner and Ganong, 1978).

Similarly, in birds, there is evidence for both stimulatory and inhibitory hypothalamic control of avian GH secretion (Schally and Kastin, 1972; Hall and Chadwick, 1976; Scanes et al., 1977; Harvey et al., 1978a). In contrast to mammals, however, the avian hypothalamus seems to exert a predominantly stimulatory influence on the release of pituitary PRL (Ma and Nalbandov, 1961; Kragt and Meites, 1965; Nicoll, 1965; Harvey et al., 1978a, 1978b). Very little information is available on the particulars of the biogenic amines in the regulation of avian GH and PRL secretion.

Effect of Inhibition of Catecholamine Biosynthesis

Administration of the tyrosine hydroxylase inhibitor, α-MPT (250 mg/kg, dissolved in saline) was accompanied by increases in the circulating levels of both GH and PRL (Tables 3 and 4). DDC treatment (a dopamine β-hydroxylase inhibitor) resulted in a decrease (76%) of plasma levels of GH but had little effect on PRL levels.

Effect of Receptor Agonists and Antagonists on Catecholamines

The circulating concentration of PRL was little affected by antagonists to neurotransmitters. Small but significant increases in the circulating PRL levels were observed after the administration of pimozide and propranolol (see Table 3). The plasma concentration of GH was, however, depressed consistently by phenoxybenzamine and by pimozide (30 minutes following administration). It was unaffected by phentolamine or atropine but tended to be elevated by propranolol (see Table 4). α and β agonists (phenylephrine, isoproterenol, and epinephrine) all significantly decreased the circulating level of GH (Table 4).

Effect of Catecholamine Granular Reuptake Inhibitor

Reserpine administration was accompanied by a large decrease in the plasma concentration of GH and by a large increase in that of PRL (Table 4).

General Discussion of Involvement of Catecholamines in GH and PRL Secretion

There is evidence that catecholamines are involved in the control of both GH and PRL secretion. Additionally, a case can be made for dopamine having some inhibiting effect on PRL secretion. This is supported by observed increases in PRL following administration of both α-methyltyrosine (which inhibits the synthesis of dopamine and norepinephrine) and pimozide

TABLE 3.—*Effect of catecholamine antagonists, synthesis blockers, and reuptake inhibitors on circulating concentrations of PRL in young chickens.*

Drug	Time	PRL as % of vehicles treatment ± (N) SE	
Synthesis Blockers			
α-Methyl-*p*-tyrosine (250 mg/kg)	1 hr.	130 ± (25) 8	
Vehicle	1 hr.	100 ± (25) 5	$P < 0.01$
Diethyldithiocarbamate (400 mg/kg)	1 hr.	86 ± (8) 9	
Vehicle	1 hr.	100 ± (7) 6	
Antagonist			
Pimozide (2.5 mg/kg)	30 min.	121 ± (8) 4	
Vehicle	30 min.	100 ± (8) 3	$P < 0.01$
Pimozide (2.5 mg/kg)	1 hr.	94 ± (19) 5	
Vehicle	1 hr.	100 ± (16) 7	
Phenoxybenzamine (20 mg/kg)	1 hr.	88 ± (12) 5	
Vehicle	1 hr.	100 ± (11) 14	
Atropine (10 mg/kg)	1 hr.	110 ± (18) 8	
Propranolol (2 mg/kg)	1 hr.	139 ± (18) 11	
Vehicle	1 hr.	100 ± (18) 4	$P < 0.005$
Reuptake Inhibitor[a]			
Reserpine (25 mg/kg)	4 hr.	303 ± (7) 48	
Vehicle	4 hr.	100 ± (7) 14	$P < 0.001$

[a]Calculated from Harvey *et al.*, 1978.

(dopamine receptor blocker). However, diethyldithiocarbamate, which also inhibits synthesis of norepinephrine, produces no effect. The situation with GH is not clear, although there is substantial evidence that catecholamines affect the plasma concentration of GH. The high circulating levels of GH found in young chickens are depressed by epinephrine acting via β receptors (Harvey and Scanes, 1978). This may be due to a peripheral action; for instance systemically administered epinephrine might increase the clearance rate of GH. The data presented at least may be suggestive of norepinephrine acting at a hypothalamic level, via α receptors, to stimulate GH release. This is supported by observed decreases in circulating GH concentrations observed following α-adrenergic receptor blockade by phenoxybenzamine, inhibition of granule reuptake by reserpine, and suppression of norepinephrine synthesis by diethyldithiocarbamate. The lack of clarity in the suggested pattern of hypothalamic control is understandable, particularly in view of the dual releasing and release inhibitory hypothalamic hormone control of GH and the possibility of peripheral effects of the catecholamines.

Effect of Inhibition of Serotonin Biosynthesis

Inhibition of serotonin synthesis by pCPA (250 mg/kg dissolved in acidified saline) does not appear to have any influence on the tonic levels of either PRL or GH by 48 hours (Table 5). It may be noted, however, that there is a trend for plasma GH levels to increase after administration of pCPA, though not significantly, in these young birds.

TABLE 4.—*Effect of catecholamine agonists, antagonists, synthesis blockers and reuptake inhibitors on circulating concentrations of GH in young chickens.*

Drug	Time	GH as a % of vehicle treatment $\pm$ (N) SE	
Synthesis Blocker			
α-Methyl-*p*-tyrosine (250 mg/kg)	1 hr.	182 ± (17) 26	
Vehicle	1 hr.	100 ± (17) 12	$P < 0.005$
Diethyldithiocarbamate (400 mg/kg)	1 hr.	24 ± (7) 16	
Vehicle	1 hr.	100 ± (7) 14	$P < 0.001$
Agonists[a]			
Phenylephrine (α) (1 mg/kg)	20 min.	54 ± (5) 9	
Isoproterenol (β) (1 mg/kg)	20 min.	36 ± (5) 6	
Epinephrine (1 mg/kg)	20 min.	42 ± (5) 9	$P < 0.05$
Vehicle	20 min.	100 ± (5) 5	
Antagonists			
Pimozide (2.5 mg/kg)	30 min.	61 ± (8) 5	
Vehicle	30 min.	100 ± (8) 18	$P < 0.05$
Pimozide (2.5 mg/kg)	1 hr.	114 ± (19) 18	
Vehicle	1 hr.	100 ± (16) 15	
Phenoxybenzamine (20 mg/kg)	30 min.	35 ± (7) 2	
Vehicle	30 min.	100 ± (10) 23	$P < 0.05$
Phenoxybenzamine (20 mg/kg)	1 hr.	42 ± (12) 4	
Vehicle	1 hr.	100 ± (11) 14	$P < 0.001$
Phentolamine[a] (1 mg/kg)	20 min.	127 ± (5) 26	
Propranolol[a] (1 mg/kg)	20 min.	149 ± (5) 23	
Vehicle[a]	20 min.	100 ± (5) 5	$P < 0.05$
Atropine (10 mg/kg)	1 hr.	185 ± (18) 45	
Propranolol (2 mg/kg)	1 hr.	143 ± (18) 23	
Vehicle	1 hr.	100 ± (18) 12	
Reuptake Inhibitors[b]			
Reserpine (25 mg/kg)	4 hr.	17 ± (7) 5	
Vehicle	4 hr.	100 ± (7) 16	$P < 0.001$

[a]Calculated from Harvey and Scanes (1978)
[b]Calculated from Harvey et al., (1978).

Effect of Serotonin, Serotonin Precursors, and Serotonergic Receptor Stimulation

Peripheral administration of serotonin itself (10 mg/kg dissolved in saline) is effective in lowering blood levels of GH by two hours after injection (Fig. 1a), while showing no apparent influence on circulatory levels of PRL (Fig. 2a). 5-HTP, the immediate precursor of serotonin (100 mg/kg dissolved in acidified saline) lowered plasma GH (Fig. 1a) and increased plasma PRL (Fig. 2a) levels at one and two hours following administration. Similarly, the serotonergic receptor agonist quipazine (30 mg/kg dissolved in saline) reduced circulating levels of GH (Fig. 1a) while elevating the plasma levels of PRL (Fig. 2a) both one and two hours after injection.

In animals whose serotonin biosynthesis has been blocked by pCPA (250 mg/kg), serotonin, 5-HTP, and quipazine were still effective in lowering the

TABLE 5.—*Effect of serotonin antagonists and synthesis inhibitors on plasma levels of GH and PRL in male chickens.*

Treatment	Time	Plasma GH		Plasma PRL	
		as % of vehicle ± (N) SE			
Synthesis Blocker					
p-Chlorophenylalanine (250 mg/kg)	48 hr.	243 ± (8)	51*	109 ± (8)	8
Vehicle (acid saline)	48 hr.	100 ± (8)	16	100 ± (8)	16
Antagonists					
Methysergide (0.5 mg/kg)	1 hr.	414 ± (8)	98**	59 ± (8)	5***
SQ 10631 (2.5 mg/kg)	1 hr.	271 ± (10)	53	73 ± (10)	5**
Cyproheptadine (5 mg/kg)	1 hr.	329 ± (7)	100*	55 ± (7)	5***
Vehicle (20% ethanol)	1 hr.	100 ± (9)	23	100 ± (9)	9

*P < 0.05; **P < 0.01; ***P < 0.001 compared with vehicle control.

plasma concentrations of GH (Fig. 1b). In these animals, however, only quipazine appeared to have retained its ability to elevate the circulating levels of PRL (Fig. 2b).

Effect of Serotonergic Receptor Blockade

The administration of the serotonin antagonist methysergide caused a significant increase in plasma GH levels while decreasing those of PRL. Similarly, another serotonin antagonist, SQ-10631, elevated circulating GH and lowered circulating PRL concentrations (Table 5). A third receptor blocker, cyproheptadine, is equally effective in stimulating GH secretion while inhibiting the release of PRL (Table 5).

General Discussion of the Role of Serotonin in GH and PRL Secretion

A large body of information has implicated serotonin in the regulation of pituitary GH and PRL secretion in mammalian species (see for example, Weiner and Ganong, 1978). Serotonin is widely distributed in the hypothalamus, with high concentrations found in the suprachiasmatic, arcuate and premammilary nuclei (Saavedra *et al.*, 1974). Serotonin is also found in the median eminence.

Our experiments with young male birds clearly indicate an intimate relationship between the activity of the serotonergic neuronal system and the pituitary secretion of GH and PRL in this species. The blockade of the serotonin biosynthesis by pCPA results in an elevation of circulating GH levels. This pCPA effect on GH secretion can be reversed by the repletion of serotonin levels with 5-HTP, as well as by stimulation of serotonin receptors with quipazine.

Both 5-HTP and quipazine are also capable of reducing plasma concentrations of GH in intact chicks, supporting the notion of an inhibitory role of serotonin in the regulation of GH in birds. In mammalian species on the other hand, 5-HTP has been shown to increase the plasma levels of GH.

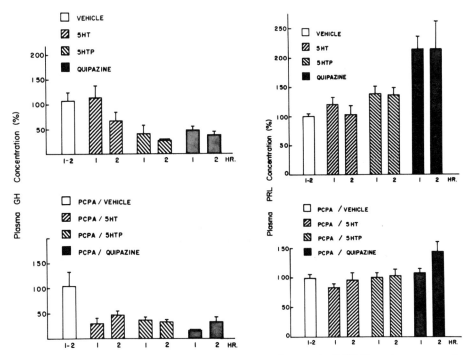

FIG. 1 (left).—Effect of serotonin (5-HT), a precursor of serotonin (5-HTP), and a serotonergic agonist (quipazine) on circulating GH concentrations in (a) control, and (b) p-chlorophenylalanine (pCPA) pretreated young male chickens. Plasma GH concentrations are expressed as a percentage of the mean in the control and the pCPA pretreated chicks. The vertical bars represent the standard error of the mean. Plasma GH concentrations were depressed ($P<0.05$) 2 hours following 5-HT, 1 and 2 hours following 5-HTP and quipazine administration, and at both 1 and 2 hours following 5-HT, 5-HTP, and quipazine in pCPA pretreated chicks.

FIG. 2 (right).—Effect of serotonin (5-HT), a precursor of serotonin (5-HTP) and a serotonergic agonist (quipazine) on circulating PRL concentrations in (a) control and (b) p-chlorophenylalanine (pCPA) pretreated young male chickens. Plasma PRL levels are expressed as a percentage of the mean in the control and the pCPA pretreated chicks. The vertical bars represent the standard error of the mean. Plasma PRL concentrations were elevated ($P < 0.05$) 1 hour following 5-HT, 1 and 2 hours following 5-HTP and quipazine, and 2 hours following quipazine in pCPA pretreated chicks.

Species studied in this regard include dogs (Weiner and Ganong, 1978), cats (Ruch *et al.*, 1977), monkeys (Jacoby *et al.*, 1974) and man (Nakai and Imura, 1974; Nakai *et al.*, 1974).

pCPA had no apparent influence on plasma PRL levels in birds. The effect of this drug on PRL secretion in mammals has also led to unpredictable results. pCPA had no influence on tonic levels of PRL in ovariectomized or orchidectomized rats (Donoso *et al.*, 1971; Gallo *et al.*, 1975) but inhibited PRL secretion in adult males (Gil-Ad *et al.*, 1976).

It should be realized that the same dose of pCPA as in the aforementioned reports has been shown to result in a large (greater than 60%) depletion of brain serotonin levels in rats (Gallo *et al.*, 1975) and quail (El Halawani *et al.*, 1978). It may turn out that the remaining serotonin is sufficient to maintain tonic PRL release in the chicks in the present study.

Systemic injection of serotonin is effective in lowering plasma GH levels without an apparent alteration in circulating PRL concentrations. A similar lack of effect of systemic serotonin injections on PRL has been observed in rats (Lu and Meites, 1973), whereas the intraventricular administration of this neurotransmitter is capable of stimulating PRL release in this rodent (Kamberi *et al.*, 1971). The lack of influence of systemic serotonin in birds might reflect not only the ability of the amine to reach its proper target site within the blood-brain barriers, but also variables such as sex, dose of the neurotransmitter, and time of blood removal.

Administration of 5-HTP, on the other hand, was effective in elevating plasma levels of PRL in intact birds and in birds pretreated with pCPA. 5-HTP has also been reported to result in increased PRL levels in rats (Chen and Meites, 1975), mice (Larson *et al.*, 1977), and man (MacIndoe and Turkington, 1973; Kato *et al.*, 1974). The serotonin receptor agonist quipazine leads to a marked elevation of circulating PRL levels, thus supporting a stimulatory role for serotonin.

The three serotonin antagonists, methysergide, SQ-10631, and cyproheptadine, all are effective in reducing plasma PRL levels while elevating the circulatory levels of GH in the male chicks. These observations further support the notion that serotonin acts as a stimulatory neurotransmitter in regulation of tonic PRL secretion while being inhibitory to the control of GH release in the male domestic fowl. The apparent stimulatory involvement of serotonin with PRL secretion in the fowl is similar to the case in mammalian species where this biogenic amine is suspected also of playing an excitatory role in the release of PRL. With GH, however, the situation seems to differ from that in mammals where serotonin is supposed to play a stimulatory role in the regulation of this hormone. Our experiments strongly suggest an inhibitory role for this indolamine in the control of avian GH secretion.

CONCLUSIONS

A number of amines, in particular dopamine, norepinephrine, and serotonin, appear to function as neurotransmitters involved in the regulation of the secretion of anterior pituitary hormones. Neurons containing these amines in their terminals innervate the hypothalamus and function in the transfer of neural information to the hormone-secreting hypothalamic neurons. These hypothalamic neurons, in turn, end on the primary capillary plexus of the hypophysial portal system and constitute the final common pathway for neuroendocrine regulation.

An enormous quantity of data has been reported in the literature pertaining to the participation of these biogenic amines in the regulation of the anterior pituitary hormones in mammalian species. Only recently reports have begun to emerge concerning the aminergic control of avian pituitary hormone secretion. Our efforts in studying the hypothalamic regulation of LH, GH, and PRL in the male domestic fowl have been reviewed, and the data clearly implicate an intricate relationship between the various neurotransmitters and these three hormones.

LITERATURE CITED

ANDÉN, N. E., A. RUBENSON, K. FUXE, AND T. HOKFELT. 1967. Evidence for dopamine receptor stimulation by apomorphine. J. Pharm. Pharmacol., 19:627-629.

ANDÉN, N. E., S. G. BUTCHER, H. CORRODI, K. FUXE, AND V. UNGERSTADT. 1970. Receptor activity and turnover of dopamine and noradrenaline after neuroleptics. Eur. J. Pharm., 11:303-314.

BARTHOLINI, G., AND A. PLETSCHER. 1968. Cerebral accumulation and metabolism of C^{14}-DOPA after selection inhibition of peripheral decarboxylase. J. Pharm. Exp. Therap., 161:14-20.

BLACK, J. W., AND B. M. C. PRICHARD. 1973. Activation and blockade of β-adrenoceptors in common cardiac disorders. Brit. Med. Bull., 29:163-167.

BLACK, J. W., W. A. M DUNCAN, AND R. G. SHANKS. 1965. Comparison of some properties of pronethalol and propanolol. Brit. J. Pharm. Chemoth., 25:577-591.

BONNET, V., AND F. BREMER. 1952. Les potentials synaptiques et la transmission de nerveuse centrâle. Arch. Int. Physiol., 60:33-93.

BRAZEAU, P., W. VALE, R. BURGUS, N. LING, M. BUTCHEE, J. RIEVIER, AND R. GUILLEMIN. 1973. Hypathalamic polypeptides that inhibit the secretion of immunoreactive pituitary growth hormone. Science, 179:423-425.

CHEN, A. J., AND J. MEITES. 1975. Effect of biogenic amines and TRH on release of prolactin and TSH in the rat. Endocrinology, 96:10-14.

CREVELING, C. R., AND J. W. DALY. 1971. The application of biochemical techniques in the search of drugs affecting biogenic amines. Part B, pp. 335-441, in Biogenic amines and physiological membranes in drug therapy (J. Biel and G. Abood, eds.), Dekker, New York, vii+525 pp.

DAVIES, D. T., AND B. K. FOLLETT. 1974. The effect of intraventricular administration of 6-hydroxydopamine on photo-induced testicular growth in the Japanese quail. J. Endocrinol., 60:277-283.

DODD, J. M., B. K. FOLLETT, AND P. J. SHARP. 1971. Endocrine aspects of the hypothalamus in sub-mammalian vertebrates. Pp. 144-223, in Advances in comparative physiology and biochemistry (O. Lowenstein, ed.), Academic Press, New York and London, 4:xii+410 pp.

DONOSO, A. O., W. BISHOP, C. P. FAWCETT, L. KRULICH, AND S. M. McCANN. 1971. Effects of drugs that modify brain monamine concentrations on plasma gonadotropin and prolactin levels in the rat. Endocrinology, 89:774-784.

DROUVA, S. V., AND R. V. GALLO. 1976. Catecholamine involvement in episodic luteinizing hormone release in adult ovariectomized rats. Endocrinology, 99:651-658.

DUNLOP, D., AND R. G. SHANKS. 1968. Selective blockade of adrenoceptive beta receptors in the heart. Brit. J. Pharm. Chemoth., 32:201-218.

EL HALAWANI, M. E., AND W. H. BURKE. 1975. Role of catecholamines in photoperiodically-induced gonadal development in Coturnix quail. Biol. Reprod., 13:603-609.

EL HALAWANI, M. E., W. H. BURKE, AND L. A. OGREN. 1978. Effects of drugs that modify brain monoamine concentration on photoperiodically-induced testicular growth in Coturnix quail (Coturnix coturnix japonica). Biol. Reprod., 18:198-203.

ERNST, A. M. 1967. Mode of action of apomorphine and dexamphetamine on gnawing compulsion in rats. Psychopharmacol., 10:216-323.

EVERETT, J. W. 1954. Luteotrophic function of autografts of the rat hypophysis. Endocrinology, 54:685-690.

FERRANDO, G., AND A. NALBANDOV. 1969. Direct effect of the ovary of the adrenergic blocking drug dibenzyline. Endocrinology, 85:38-42.

FOLLETT, B. K., C. G. SCANES, AND F. J. CUNNINGHAM. 1972. A radioimmunoassay for avian luteinizing hormone. J. Endocrinol., 52:359-378.

FURCHGOTT, R. F. 1954. Dibenamine blockade in strips of rabbit aorta and its use in differentiating receptors. J. Pharmacol., 111:265-284.

———. 1960. Receptors for sympathomimetic amines. Pp. 246-252, in Adrenergic mechanisms (J. R. Vane, G. E. W. Wolstenholme, and M. O'Connor, eds.), Little, Brown, and Company, Boston, and J. and A. Churchill, London, xx+632 pp.

GALLO, R. V., AND G. P. MOBERG. 1977. Serotonin mediated inhibition of episodic luteinizing hormone release during electrical stimulation of the arcuate nucleus in ovariectomized rats. Endocrinology, 100:945-954.

GALLO, R. G., J. RABII, AND G. P. MOBERG. 1975. Effect of methysergide, a blocker of serotonin receptors, on plasma prolactin levels in lactating and ovarioectomized rats. Endocrinology, 97:1096-1105.

GANONG, W. F. 1974. The role of catecholamines and acetylcholine in the regulation of endocrine function. Life Sci., 15:1401-1414.

GIL-AD, I. F., F. ZAMBOTTI, M. O. CARRUBA, L. VINCENTINI, AND E. E. MÜLLER. 1976. Stimulatory role of brain serotoninergic system on prolactin secretion in the male rat. Proc. Soc. Exp. Biol. Med., 151:512-518.

GLOWINSKI, J., AND J. AXELROD. 1965. Effect of drugs on the uptake, release, and metabolism of H^3-norepinephrine in the rat brain. J. Pharm. Exp. Therap., 149:43-49.

GNODDE, H. P., AND G. A. SCHULLING. 1976. Involvement of catecholaminergic and cholinergic mechanisms in the pulsatile release of LH in the longterm ovariectomized rat. Neuroendocrinol., 20:212-223.

GOODMAN AND GILMAN'S, The Pharmacological Basics of Therapeutics. 1980. 6th Edition. Macmillan Publishing Company, New York, Toronto, London, xvi+1843 pp.

HALL, T. R., AND A. CHADWICK. 1976. Effect of growth hormone inhibiting factor (somatostatin) on the release of growth hormone and prolactin from pituitaries of the domestic fowl in vitro. J. Endocrinol., 68:163-164.

HANSON, L. C. F. 1965. The disruption of conditioned avoidance response following selective depletion of brain catecholamines. Psychopharmacol., 8:100-110.

HARVEY, S., AND C. G. SCANES. 1977. Purification and radioimmunoassay of chicken growth hormone. J. Endocrinol., 73:321-329.

———. 1978. Effect of adrenaline and adrenergic active drugs on growth hormone secretion in immature cockerels. Experientia, 34:1096-1097.

HARVEY, S., C. G. SCANES, A. CHADWICK, AND N. J. BOLTON. 1978a. Effect of reserpine on plasma concentrations of growth hormone, and prolactin in the domestic fowl. J. Endocrinol., 79:153-154.

———. 1978b. The effect of thyrotropin-releasing hormone (TRH) and somatostatin (GHRIH) on growth hormone and prolactin secretion in vitro and in vivo in the domestic fowl (Gallus domesticus). Neuroendocrinol., 26:249-260.

———. 1979a. In vitro stimulation of chicken pituitary growth hormone and prolactin secretion by chicken hypothalamic extract. Experientia, 15:694-695.

HARVEY, S., C. G. SCANES, A. CHADWICK, G. BORDER, AND N. J. BOLTON. 1979b. Effect of chicken hypothalamus on prolactin and growth hormone secretion in male chickens. J. Endocrinol., 82:193-197.

HOFFMAN, B. B., AND R. J. LEFKOWITZ. 1980. Alpha-adrenergic receptor subtypes. New England J. Med., 302:1390-1396.

HONG, E., L. F. SANCILLO, R. VARGAS, AND E. PARDO. 1969. Similarities between the pharmacological actions of quipazine and serotonin. Eur. J. Pharm., 6:274-280.

HORNEYKIEWICZ, O. 1966. Dopamine (3-hydroxytyramine) and brain function. Pharm. Rev., 18:925-964.

JACKSON, G. L. 1977. Effect of adrenergic blocking drugs in secretion of luteinizing hormone in the ovariectomized ewe. Biol. Reprod., 16:543-548.

JACOBY, J. H., M. GREENSTEIN, J. F. SASSIN, AND E. D. WEITZMAN. 1974. The effect of monoamine precursors on the release of growth hormone in the Rhesus monkey. Neuroendocrinol., 14:95-112.

JOHNSON, G. A., E. G. KIM, P. A. PLATZ, AND M. M. MICKELSON. 1968. Comparative aspects of tyrosine hydroxylase and tryptophan hydroxylase inhibition: arterenones and dihydroxyphenylacetamide. Biochem. Pharm., 17:403-410.

JOHNSON, G. A., S. J. BOUKMA, AND E. G. KIM. 1970. In vivo inhibition of dopamine β-hydroxylase by 1-phenyl-3-(2 thiazoyl)-2-thiourea. J. Pharm. Exp. Therap., 171:80-87.

JOHNSON, G. A., E. G. KIM, AND S. J. BOUKMA. 1972. 5-Hydroxyindole levels in rat brain after inhibition of dopamine β-hydroxylase. J. Pharm. Exp. Therap., 180:539-546.

KALRA, S. P., AND S. M. MCCANN. 1973. Effect of drugs modifying catecholamine synthesis on LH release induced by preoptic stimulation in the rat. Endocrinology, 93:356-362.

———. 1974. Effects of drugs modifying catecholamine synthesis on plasma LH and ovulation in the rat. Neuroendocrinol., 15:79-91.

KALRA, P. S., S. P. KALRA, L. KRULICH, C. P. FAWCETT, AND S. M. MCCANN. 1973. Involvement of norepinephrine rise in transmission of the stimulatory influence of progesterone on gonadotropin release. Endocrinology, 90:1168-1176.

KAMBERI, I. A., R. S. MICAL, AND J. C. PORTER. 1970. Effect of anterior pituitary perfusion and intraventricular injection of catecholamines and indolamines on LH release. Endocrinology, 87:1-12.

———. 1971. Effect of anterior pituitary perfusion and intraventricular injection of catecholamines on prolactin release. Endocrinology, 88:1012-1020.

KAO, L. W. L., AND A. NALBANDOV. 1972. The effect of antiadrenergic drugs on ovulation in hens. Endocrinology, 90:1343-1349.

KATO, Y., Y. NAHAI, H. IMURA, K. CHIHARA, AND S. OHGO. 1974. Effect of 5-hydroxytryptophan (5-HTP) on plasma prolactin levels in man. J. Clin. Endocrinol. Metab., 38:695-697.

KOE, B. K., AND A. WEISSMAN. 1966. p-Chlorophenylalanine: a specific depletor of brain serotonin. J. Pharm. Exp. Therap., 154:499-516.

KOPIN, I. J., AND E. K. GORDON. 1963. Metabolism of administered and drug-released norepinephrine-7-H^3 in the rat. J. Pharm. Exp. Therap., 140:207-216.

KRAGT, C. L., AND J. MEITES. 1965. Stimulation of pigeon pituitary prolactin release by pigeon hypothalamic extract in vitro. Endocrinology, 76:1169-1176.

LAMBERT, S. W. J., AND R. M. MACLEOD. 1978. The interaction of the serotonergic and dopaminergic systems on prolactin secretion in the rat: the mechanism of action of the "specific" serotonin receptor antagonist, methysergide. Endocrinology, 103:287-295.

LARSON, B. A., Y. N. SINHA, AND W. P. VANDERLAAN. 1977. Effect of 5-hydroxytryptophan on prolactin in the mouse. J. Endocrinol., 74:153-154.

LU, K. H., AND J. MEITES. 1973. Effects of serotonin precursors and melatonin on serum prolactin release in rats. Endocrinology, 93:152-155.

MA, R. C. S., AND A. V. NALBANDOV. 1961. The transplanted hypophysis: Discussion. Pp. 306-312, in Advances in neuroendocrinology (A. V. Nalbandov, ed.), Univ. Illinois Press, Urbana, xii+525 pp.

MACINDOE, J. H., AND R. W. TURKINGTON. 1973. Stimulation of human prolactin secretion by intravenous infusion of L-tryptophan. J. Clin. Invest., 52:1972-1978.

MEITES, J., P. K. TALWALKER, AND C. S. NICOLL. 1960. Initiation of lactation in rats with hypothalamic or cerebral tissue. Proc. Soc. Exp. Biol. Med., 103:298-300.

MEITES, J., J. SIMPKINS, J. BRUNI, AND J. ADVIS. 1977. Role of biogenic amines in control of anterior pituitary hormones. IRCS Med. Sci., 5:1-7.

MEYERS, F. H., E. JAWETZ, AND A. Goldfi. 1974. Review of medical pharmacology. Large Medical Publications, Los Altos, California, 762 pp.

MILLER, E. P., R. H. COX, W. R. SNODGRASS, AND R. P. MAICKEL. 1970. Comparative effects of p-chlorophenylalanine, p-chloroamphetamine and p-chloro-N-methylamphetamine on rat brain norepinephrine, serotonin, and 5-hydroxyindol-3-acetic acid. Biochem. Pharm., 19:435-442.

MISHKINSKY, J., K. KHAZEN, AND F. G. SULMAN. 1968. Prolactin releasing activity of the hypothalamus in post-partum rats. Endocrinology, 82:611-613.

MOIR, A. T. B., AND D. ECCLESTON. 1968. The effects of precursor loading in the cerebral metabolism of 5-hydroxyindoles. J. Neurochem., 15:1093-1108.

MORNEX, R., D. JORDAN, AND A PICKERING. 1976. Contrôle neuroaminergique des antéhypophysaires sécrétions. Pathol. Biol., 24:555-574.

MÜLLER, E. E., G. NISTICO, AND V. SCAPAGNINI. 1977. Neurotransmitters and anterior pituitary function. Academic Press, New York, xi+435 pp.

NAKAI, Y., AND H. IMURA. 1974. Effect of adrenergic blocking agents on plasma growth hormone responses to l-5-hydroxytryptophan (5-HTP) in man. Endocrinol. Jap., 21:493-497.

NAKAI, Y., H. IMURA, H. SAKURAI, H. KIRAHACHI, AND T. YOSHEMI. 1974. Effect of cyproheptadine on human growth hormone secretion. J. Clin. Endocrinol., 38:446-449.

NICKERSON, M., AND W. S. GUMP. 1949. The chemical basis for adrenergic blocking activity in compounds related to dibenamine. J. Pharm. Exp. Therap., 97:25-47.

NICOLL, C. S. 1965. Neural regulation of adenohypophyseal prolactin secretion in tetrapods: indications from in vitro studies. J. Exp. Zool., 158:203-210.

NG, L. K. Y., T. N. CHASE, R. W. COLBURN, AND I. J. KOPIN. 1972. L-Dopa in Parkinsonism. Neurology, 22:688-696.

PHYLLIS, J. W., AND A. K. TEBECIS. 1967. The effects of topically applied cholinomimetic drugs on the isolated spinal cord of the toad. Comp. Biochem. Physiol., 23:541-552.

PORTER, J. C., R. S. MICAL, AND O. M. CRAMER. 1971/1972. Effect of serotonin and other indoles on the release of LH, FSH and prolactin. Gynecol. Invest., 2:13-33.

PUIG, M., G. BARTHOLINI, AND A. PLESTSCHER. 1974. Formation of noradrenaline in the rat brain from the four stereoisomers of 3, 4-dihydroxyphenylserine. Naunyn Schmiedberg's Arch. Pharmacol., 281:443-446.

REICHLIN, S., R. SAPERSTEIN, I. M. D. JACKSON, A. E. BOYD, AND Y. PATEL. 1973. Hypothalamic hormones. Ann. Rev. Physiol., 38:389-424.

ROSE, J. C., AND W. F. GANONG. 1976. Neurotransmitter regulation of the pituitary gland secretion. Pp. 87-123, in Current developments in psychopharmacology (W. F. Essman and L. Valzelli, eds.), Spectrum Press, Holliswood, New York, 267 pp.

RUCH, W., R. C. MIXTURE, R. M. RUSSELL, J. F. GARCIA, AND C. G. GALE. 1977. Aminergic and thermoregulatory mechanisms in hypothalamic regulation of growth hormone in cats. Amer. J. Physiol., 223:E61-E69.

SAAVEDRA, J. M., M. PALKOVITS, M. J. BROWNSTEIN, AND J. AXELROD. 1974. Serotonin distribution in the nuclei of the rat hypothalamus and preoptic region. Brain Research, 77:157-165.

SAWYER, C. H. 1952. Stimulation of ovulation in the rabbit by the intraventricular injection of epinephrine or norepinephrine. Anat. Rec., 112:385.

———. 1979. Brain amines and pituitary gonadotropin secretion. Can. J. Physiol. Pharm., 57:667-680.

SAWYER, C. H., AND H. M. RADFORD. 1978. Effects of intraventricular injections of norepinephrine on brain-pituitary-ovarian function in the rabbit. Brain Res., 146:83-93.

SCANES, C. G., A. CHADWICK, AND N. J. BOLTON. 1976. Radioimmunoassay of prolactin in the plasma of the domestic fowl. Gen. Comp. Endocrinol., 30:12-20.

SCANES, C. G., S. HARVEY, A. CHADWICK, AND N. J. BOLTON. 1977. Studies on the hypothalamic control of growth hormone and prolactin secretion in the chicken. J. Endocrinol., 73:10P.

SCHALLY, A. V., AND A. J. KASTIN. 1972. Hypothalamic releasing and inhibiting hormones. Gen. Comp. Endocrinol., Suppl. 3:176-185.

SHARP, P. J.   1975.   The effect of reserpine and plasma LH concentrations in intact and gonadectomised domestic fowl.   Brit. Poul. Sci., 16:79-82.

SHORE, P. A.   1962.   Release of serotonin and catecholamines by drugs.   Pharm. Rev., 14:531-
550.

STOLK, J. M., J. BARCHAS, W. DEMENT, AND S. SCHANBERG.   1969.   Brain catecholamine metabolism following p-chlorophenylalanine treatment.   Pharmacologist, 11:258.

STONE, C. L., H. C. WENGER, C. T. LUDDEN, J. M. STRAVORSKI, AND C. A. ROSS.   1961.   Anti-
serotinin-antihistimine properties of cyproheptadine.   J. Pharmacol. Exp. Therap.,
131:73-84.

TAGLIAMONTE, A., P. TAGLIAMONTE, G. V. CORSINI, G. P. MEREU, AND G. L. GESSA.   1973.   Decreased conversion of tyrosine to catecholamines in the brain of rats treated with p-
chlorophenylalanine.   J. Pharm. Pharmacol., 25:101-108.

UDENFRIEND, S., P. ZALTMAN-NIRENBERG, AND T. NAGULSU.   1965.   Inhibitors of purified beef
adrenal tyrosine hydroxylase.   Biochem. Pharm., 14:837-845.

WEICK, R. F.   1978.   Acute effects of adrenergic blocking drugs and neuroleptic agents on pulsatile discharges of lutenizing hormone in the ovariectomized rat.   Neuroendocrinol.,
26:108-117.

WEINER, R. I., AND W. F. GANONG.   1978.   The role of brain amines and histamine in regulation of anterior pituitary secretion.   Physiol. Rev., 58:905-976.

Reprinted from
ASPECTS OF AVIAN ENDOCRINOLOGY:
PRACTICAL AND THEORETICAL IMPLICATIONS (C. G. Scanes *et al.*, eds.)
Grad. Studies, Texas Tech Univ., 1982, 26:1-411.

NEUROPHARMACOLOGICAL ASPECTS OF NEURAL REGULATION OF AVIAN ENDOCRINE FUNCTION

M. E. EL HALAWANI, W. H. BURKE, P. T. DENNISON, AND J. L. SILSBY

*Department of Animal Science, University of Minnesota, St. Paul,
Minnesota 55108 USA*

Brain monoamines have recently become the subject of several reports concerning their role in the regulation of avian anterior pituitary hormone secretion. The effect of neurotransmitters upon the hypothalamic neurohormones which control anterior pituitary secretion has mainly been studied using pharmacological tools. Pharmacological modulation of dopamine (DA), norepinephrine (NE), epinephrine (E), and serotonin (5-HT) can affect pituitary hormone release. However, interpretation of their physiological role in neuroendocrine regulation causes some difficulties. Shifts in central biogenic amine turnover associated with alterations in hormonal release could either be the cause or the result of changes in the central neuroendocrine mechanism. Moreover, pharmacological manipulation of monoaminergic neurons could affect the pituitary in many ways. It might control the transfer of sensory inputs or affect the biosynthesis and release of releasing factors.

The specificity of the pharmacological tools available for such studies is not entirely satisfactory, and the drugs which are used affect amine metabolism in the entire nervous system. This makes it almost impossible to distinguish primary neuroendocrine effects from indirect ones elicited by overall changes in neural activity.

Before using such drugs, it is important that the investigator: 1) be aware of their specificity, as very few drugs are specific for the desired monoamine; 2) make sure that the drug acts in the central nervous system; and 3) obtain the same experimental results with more than one drug before reaching firm conclusions.

With these reservations in mind, we will be discussing the proposed monoamines involved in avian neuroendocrine function: DA, NE, E, and 5-HT. In this presentation, drugs affecting pituitary hormone release are grouped into categories according to their pharmacological function. We have limited our discussion to those drugs which are used in avian studies.

Drugs can modulate monoaminergic activity by several mechanisms: 1) inhibition of synthesis; 2) cell damage leading to neuronal degeneration; 3) receptor antagonists and agonists; 4) inhibition of storage mechanism; or 5) inhibition of uptake mechanism.

Pharmacologically Induced Changes in Catecholaminergic Transmission

Effects of Inhibitors of Catecholamine Synthesis

Biosynthesis of DA, NE, and E from tyrosine in adrenergic neurons includes four enzymatic steps: 1) tyrosine to dihyroxyphenylalanine (DOPA); 2) DOPA to DA; 3) DA to NE; and 4) NE to E. This pathway can readily be disrupted at any of the steps with available specific drugs. Tyrosine hydroxylase, which catalyzes the conversion of tyrosine to DOPA, is the rate limiting enzyme in the biosynthesis of catecholamines. Depletion of DA and NE can then be induced by inhibition of this enzyme by the administration of α-methyl-p-tyrosine (MPT) (Spector *et al.*, 1965). This drug effectively reduces brain DA, NE, and E and blocks the photoperiodically-induced testicular growth in Japanese quail (El Halawani and Burke, 1975).

Further experimentation indicated the catecholamine involved in regulation of testicular growth appears to be NE and not DA. Diethyldithiocarbamate (DDC) blocks the conversion of DA to NE by inhibiting dopamine-β-hydroxylase (Carlsson *et al.*, 1966), and results in reduced levels of NE and increased levels of DA. This treatment suppresses testicular weight. The involvement of noradrenergic neurons is also supported by the observation that preferential restoration of normal NE levels with dihydroxyphenylserine (DOPS) after administration of dopamine-β-hydroxylase inhibitor restores gonadal growth.

The suppression of testicular growth in response to photostimulation by catecholamine-depleting drugs appears to be mediated by gonadotropins. Reduction in brain catecholamine levels achieved by treatment with MPT is effective in blocking luteinizing hormone (LH) release in response to photostimulation in quail (El Halawani *et al.*, 1980*b*). Selective inhibition of NE with a more potent and apparently specific DA-β-hydroxylase inhibitor, bis(4-methyl-1-hymopiperazinylthicarbonyl)-disulfide (FLA-63) (Corrodi *et al.*, 1970), shows that inhibition of the enzyme completely blocks the photoperiodically stimulated rise in serum LH in quail. Further support for the concept that a functional noradrenergic system is essential for LH release comes from a study in which the rise in serum LH levels following castration of turkeys is significantly reduced by MPT administration (El Halawani *et al.*, 1980*c*).

Although these experiments do not rule out dopaminergic involvement, they indicate that noradrenergic mechanisms play a predominant role in regulating LH release and gonadal development.

Effect of Receptor-blocking Drugs

A different approach, used in the early experiments by van Tienhoven *et al.* (1954), is the pharmacological blockade of post-synaptic receptors. Blockade of α-adrenergic receptors by phenoxybenzamine decreases LH levels in cockerels, while stimulation of these receptors by phenylephrine overcomes the DDC-reducing effect of plasma LH levels (Rabii *et al.*, 1980). Release of

TABLE 1.—*Effects of reserpine on hypothalamic catecholamine concentrations and serum LH levels on juvenile male turkeys. Results are expressed as mean ± one SE from six to eight birds.*

Treatment	LH ng/ml	Catecholamine (ng/g)		
		DA	NE	E
Saline control	2.28 ± 0.16	2291 ± 221	3166 ± 125	725 ± 123
Reserpine	1.65 ± 0.21	391 ± 64	713 ± 157	136 ± 29

LH is similarly stimulated by the DA agonist apomorphine. The results of these experiments further confirm the theory of a noradrenergic involvement in LH regulation and provide evidence for a stimulatory effect of DA. However, intraventricularly infused DA has been shown to inhibit LH release in the chicken (Hibbs *et al.*, 1977), and the post-castration rise in serum LH is associated with a reduced hypothalamic DA turnover rate in the turkey (El Halawani *et al.*, 1980*c*).

We must be careful in our interpretation of results obtained from experiments using receptor blockers. Amine specificity of central receptors towards endogenous transmitters has not been well established. Moreover, drugs such as phenoxybenzamine can block the reuptake of catecholamines and might therefore increase the available amount of catecholamines at receptor sites.

Effects of Drugs Causing Degeneration of Catecholaminergic Fiber System

The prototype for this mechanism is 6-hydroxydopamine (6-OHDA) which must be applied locally in order to act in the central nervous system (Ungerstedt, 1971; Calas *et al.*, 1974). Infusion of this drug into the third ventricle of quail suppresses testicular growth in response to photostimulation and reduces testicular weight in previously photostimulated birds (Assenmacher and Biossin, 1972; Calas, 1975). 6-hydroxydopamine has also been intraventricularly injected and the effect of such injection on pituitary LH release has been evaluated. Davies and Follett (1974) have observed an inhibition of the photoperiodically induced rise in plasma LH to that of non-photostimulated following infusion of 6-OHDA in quail.

Effects of Drugs Blocking Amine-storage Mechanism

These drugs cause marked depletion of both catecholamines and indolamines by blocking the incorporation of the amines into the storage sites. Reserpine is the prototype of this group (Carlsson, 1965). Fluorescence histochemistry reveals a complete disappearance of fluorescence (monoamine depletion) from catecholaminergic fibers in the median eminence of the duck hypothalamus 18 hours after reserpine treatment (Calas *et al.*, 1974). Biochemical analysis of turkey hypothalami after treatment with a single injection of reserpine (2 mg/kg; i.p.) shows a significant reduction in levels of DA, NE, and E (Table 1).

Reserpine has been shown to cause gonadal atrophy in several avian species (Hagen and Wallace, 1961; Brown and Mewaldt, 1967) and to inhibit the

TABLE 2.—*Effects of* p-*chlorophenylalanine on serum LH levels (ng/ml) of six sexually mature female turkeys. Results are expressed as mean ± one* SE.

Treatment	Time after treatment (hr)					
	0	1	2	3	24	48
Saline	1.7 ± 0.1	1.8 ± 0.2	1.9 ± 0.2	1.7 ± 0.2	1.9 ± 0.2	2.2 ± 0.3
p-Chlorophenylalanine	1.8 ± 0.2	1.8 ± 0.2	1.9 ± 0.2	1.9 ± 0.2	1.3 ± 0.3	4.1 ± 0.5

photogonadal response in the duck (Tixier-Vidal and Assenmacher, 1962). Gonadal atrophy following reserpine treatment might be related to a reduced gonadotropin release. Indeed, Sharp (1975) observed an inhibition of post-castration release of LH following reserpine treatment in the chicken. In addition, turkeys treated with reserpine (2 mg/kg, i.p.) had lower LH levels than saline treated birds (Table 1).

Pharmacologically Induced Changes in Serotonergic Transmission

Effects of Inhibitors of Serotonin (5-HT) Synthesis

5-Hydroxytryptophan (5-HTP) is the intermediate step in the synthesis of serotonin from the amino acid tryptophan. Tryptophan hydroxylase, which catalyzes the conversion of tryptophan to 5-HTP, is the rate limiting enzyme in the indolamine pathway. Consequently, levels of 5-HT decline markedly following inhibition of this enzyme by administration of *p*-chlorophenylalanine (pCPA) (Koe and Weissman, 1966). This drug has been shown to augment photoinduced gonadal growth in quail (El Halawani *et al.*, 1978). Clarification of the specificity of serotonergic mechanisms involves the use of 5-HTP, the precursor of 5-HT. Restoration of brain 5-HT levels by administration of 5-HTP in pCPA treated quail results in testicular weights similar to those found in the control group.

In view of the pCPA augmented effect of testicular weight, an experiment was conducted to determine if a single i.p. injection of pCPA (200 mg/kg) could elicit LH release in female turkeys. Prior to pCPA treatment, serum LH levels of saline treated birds and pCPA treated turkeys were comparable (1.7 ng/ml versus 1.8 ng/ml). Serum LH was found to be twice as high in pCPA treated turkeys as in saline treated controls 48 hours after injection (Table 2).

The effect of pCPA on circulating prolactin (PRL) levels has also been investigated. Whereas pCPA does not have an effect on basal PRL levels of juvenile male chickens (Buonomo *et al.*, 1980), it inhibits the nesting-induced PRL increase in female turkeys (El Halawani *et al.*, 1980a). When 5-HTP, the immediate precursor of 5-HT, is injected in pCPA pretreated turkeys it increases PRL levels in a dose-related manner.

Effects of Serotonergic Receptors Blockers

Further indications of the involvement of serotonergic mechanisms in PRL and growth hormone (GH) release were obtained when the levels of

TABLE 3—*Effects of tranylcypromine sulfate on brain monoamine concentrations, serum LH levels and testicular weights of seven sexually mature male turkeys. Values given are mean ± one* SE.

Treatment	Testis weight (g)	LH ng/ml	monoamines (ng/g)			
			DA	NE	E	5-HT
Saline	27.6 ± 1.3	35.08 ± 3.38	336 ± 12	481 ± 27	74 ± 3	1410 ± 67
Tranylcypromine	15.6 ± 2.3	19.6 ± 2.00	416 ± 19	873 ± 43	98 ± 5	2251 ± 121

these hormones were measured following treatment with methysergide, cyproheptadine and 2-chloro-2-3-dimethylaminopropylthiocinnamanilde (SQ-10361; Buonomo *et al.*, 1980). These serotonergic blockers were effective in reducing plasma PRL levels and increasing circulating GH levels of young cockerels.

Effects of Drugs Causing Degeneration of Serotonergic Fibers

5,6-Dihydroxytryptamine (5,6-DHT) is typical of these drugs (Baumgarten *et al.*, 1972). As with 6-OHDA, 5,6-DHT must be applied in the central nervous system to be effective. Calas (1975) and Calas *et al.* (1974) has reported 1) a degeneration of serotonergic fibers in the median eminence of the duck hypothalamus, and 2) an augmentation in photostimulated testicular growth of quail following 5,6-DHT treatment. In addition, 5,6-DHT has been found to inhibit the testicular regression induced by exposure of quail to nonstimulatory photoperiod.

Effects of Pharmacologically-induced Increase in Endogenous 5-HT Levels

Monoamine oxidase (MAO) appears to be the major enzyme by which endogenous monoamines are degraded in the turkey (El Halawani and Waibel, 1975). The enzyme occurs in high levels in the hypothalami of birds (Aprison *et al.*, 1964), especially the median eminence (Follett *et al.*, 1966). Inhibition of this enzyme by tranylcypromine causes testicular weight depression of young cockerels (Forbes, 1970). Similar results were obtained in mature male turkeys. Daily injection of tranylcypromine sulfate (5 mg/kg; i.p.) for 6 days increased brain DA, NE, E, and 5-HT concentrations. This was associated with significant reductions in serum LH levels and testicular weights (Table 3). Data obtained by selective blockade of either catecholamines or 5-HT synthesis prior to MAO inhibition suggest that the LH level and testicular weight depressing effects of monoamine oxidase inhibitor is due to an increase in 5-HT levels (El Halawani *et al.*, 1978). The observation that testicular growth is augmented by decreased 5-HT levels resulting from inhibition of 5-HT synthesis by pCPA further substantiates this interpretation.

The Use of Synthesis and Degradation Inhibitors for the
Determination of Monoamine Turnover Rates

Several methods have been developed for the determination of monoamine turnover rates (Neff, 1972). Two of these methods employ drugs that interfere with central monoamine metabolic pathways.

El Halawani and Burke (1976) determined DA, NE, E, and 5-HT turnover rates in female turkeys at different stages of their reproductive cycle. They accomplished this by determining the increase in concentrations of these four monoamines after blocking catecholamine degradation by inhibiting activity of monoamine oxidase (Green and Erickson, 1960; Tozer *et al.*, 1966). The results showed a substantial increase in NE turnover rate following light stimulation. In addition, El Halawani and Burke (1976) found an increase in 5-HT turnover rate in turkeys that were nesting. In another study by El Halawani *et al.* (1980*c*), turnover rate of catecholamines was determined from the rate of decline of the amine after inhibition of tyrosine hydroxylase by α-MPT (Brodie *et al.*, 1966). The results showed that following castration, DA turnover rate decreased while NE and E turnover rates increased. These changes in catecholamine turnover rate were accompanied by an enhanced LH release, suggesting that LH release is related to central catecholaminergic mechanisms.

Monoamine Interactions

It is likely that the monoamine systems implicated in the control of avian hormone release and reproduction interact with one another. For example, NE depletion by DA-β-hydroxylase inhibitors or 6-OHDA (which reduces LH release and gonadal development), causes an increase in 5-HT turnover in the brain (Johnson *et al.*, 1972; Blondaux *et al.*, 1973). In addition, DOPA can displace and deplete 5-HT from nerve terminals (Carlsson, 1975), and this effect might contribute to the reinstatement by DOPA of normal gonadal growth in MPT pretreated birds. Thus, in addition to activating NE systems, it is possible that photostimulation could also activate ascending NE neurons that inhibit 5-HT cells. The reduced 5-HT activity could then be a second step in enhancing LH release.

SUMMARY

The secretion of avian pituitary hormones is regulated in large part by neural circuits in the hypothalamus and adjacent portions of the brain. Some of the principal transmitters in these circuits are dopamine (DA), norepinephrine (NE), epinephrine (E), and serotonin (5-HT). It is thus not surprising that the concentrations and turnover rates of DA, NE, E, and 5-HT in the hypothalamus were found to change during the reproductive cycle. In addition, pharmacological modulations of these amines were also found to affect pituitary hormone release.

Literature Cited

APRISON, M. H., R. TAKAHASHI, AND T. L. FOLKERTH. 1964. Biochemistry of the avian central nervous system-I. The 5-hydroxytryptophan decarboxylasemonoamine oxidase and cholineacetylase-acetylocholinestrase systems in several discrete areas of the pigeon brain. J. Neurochem., 11:341-350.

ASSENMACHER, I., AND J. BOISSIN. 1972. Circadian endocrine and related rhythms in birds. Gen. Comp. Endocrinol., Suppl., 3:489-498.

BAUMGARTEN, H. G., A. BJORKLUND, A. F. HOLSTEIN, AND A. NOBIN. 1972. Chemical degeneration of indolamine axons in rat brain by 5,6-dihydroxytryptamine; an ultrastructural study. Z. Zellforsch., 129:256-271.

BLONDAUX, J., A. SORDET, G. CHOUVET, M. JOUVET, AND J. F. PUJOL. 1973. Modification du metabolisme de la serotonine (5-HT) cérébrale induite chez le rat par administration de 6-hydroxdopamine. Brain Res., 50:101-114.

BRODIE, B. B., E. COSTA, A. BLABAC, N. H. NEFF, AND H. H. SMOOKLER. 1966. Application of steady-state kinetics to the estimation of synthesis rate and turnover time of tissue catecholamines. J. Pharm. Exp. Therap., 154:493-498.

BROWN, I. L., AND L. R. MEWALDT. 1967. Effects of reserpine on the white-crowned sparrow (Zonotrichia leucophrys gambelli). Brit. J. Pharm. Chemoth., 30:251-257.

BUONOMO, F., J. RABII, AND C. G. SCANES. 1980. The involvement of serotonin (5-HT) in the hypothalamic regulation of prolactin (PRL) and growth hormone (GH) secretion in the domestic fowl. Fed. Proc., 39:375.

CARLSSON, A. 1965. Drugs which block the storage of 5-hydroxtryptamine and related amines. Pp. 529-592, in Handbook of Experimental Pharmacology. Vol. 19, (V. Erspamer, ed.), Springer-Verlag, Berlin xx+928 pp.

———. 1975. Monoamine precursors and analogues. Pharm. Therap., B, 1:382-392.

CARLSSON, A., M. LINDQUIST, K. FUXE, AND T. HOKFELT. 1966. Histochemical and biochemical effects of diethyldithiocarbamate on tissue catecholamines. J. Pharm. Pharmacol., 18:60-62.

CORRODI, H., K. FUXE, B. HAMBERGER, AND A. LJUNGDAHL. 1970. Studies on central and peripheral noradrenaline neurons using a new dopamine-β-hydroxylase inhibitor. Eur. J. Pharm., 12:145-155.

DAVIES, D. T., AND B. K. FOLLETT. 1974. The effect of intraventricular administration of 6-hydroxydopamine on photo-induced testicular growth in Japanese quail. J. Endocrinol., 60:277-283.

EL HALAWANI, M. E., AND W. II. BURKE. 1975. Role of catecholamines in photoperiodically-induced gonadal development in Coturnix quail. Biol. Rep., 13:603-609.

———. 1976. Brain monoamine metabolism of turkey hens in various stages of their reproductive life cycle. Biol. Rep., 15:254-259.

EL HALAWANI, M. E. BURKE, L. A. OGREN, AND P. E. WAIBEL. 1975. The relative importance of monoamine oxidase and catechol-O-methyl transferase on the physiologic response to administered norepinephrine in the turkey. Comp. Biochem. Physiol., 52:35-39.

EL HALWANI, M. E., W. H. BURKE, AND L. A. OGREN. 1978. Effects of drugs that modify brain monoamine concentrations of photoperiodically-induced testicular growth in Coturnix quail (Coturnix coturnix japonica). Biol. Rep., 18:198-203.

EL HALAWANI, M. E., W. H. BURKE, and P. T. DENNISON. 1980a. Effects of p-chlorophenylalanine on the rise in serum prolactin associated with nesting in broody turkeys. Biol. Rep., 23:815-819.

———. 1980b. Involvement of catecholaminergic mechanisms in the photoperiodically-induced rise in serum luteinizing hormone of Japanese quail (Coturnix coturnix japonica). Gen. Comp. Endocrinol., 41:14-21.

———. 1980c. Age-dependent changes in hypothalamic catecholamine turnover rate following castration in turkeys. Gen. Comp. Endocrinol., 42:290-296.

FOLLETT, B. K., H. KOBAYASHI, AND D. S. FARNER. 1966. The distribution of monoamine oxidase and acetylcholinesterase in the hypothalamus and its relation to the hypothala-
mohypophysial neurosecretory system in the white-crowned sparrow, *Zonotrichia leucophrys gambelii*. Z. Zellforsch. Mikrosk. Anat., 75:57-65.
FORBES, W. R. 1970. The effects of the monoamine oxidase inhibitor tranylcypromine on the
testis weight of the cockerel. J. Endocrinol., 47:387-388.
GREEN, H., AND R. W. ERICKSON. 1960. Effect of trans-2-phenylcyclopropylamine upon norepinephrine concentration and monoamine oxidase activity of rat brain. J. Pharm.
Exp. Therap., 129:237-242.
HAGEN, P., AND A. C. WALLACE. 1961. Effect of reserpine on growth and sexual development
of chickens. Brit. J. Pharm. Chemoth., 17:267-275.
HIBBS, M., R. T. GALDWELL, AND F. J. CUNNINGHAM. 1977. Effect of the intraventricular
administration of a dopamine on plasma concentration of luteinizing hormone in the
domestic fowl. J. Endocrinol., 75:43.
JOHNSON, F. A., E. G. KIM, AND S. J. BOUHMA. 1972. 5-Hydroxyindole levels in rat brain after
inhibition of dopamine-β-hydroxylase. J. Pharm. Exp. Therap., 180:539-546.
KOE, K. B., AND A. WEISSMAN. 1966. *p*-Chlorophenylalanine: A specific depletor of brain serotonin. J. Pharm. Exp. Therap., 154:499-516.
NEFF, N. H. 1972. Methods for estimating the *in vivo* rate of formation of norepinephrine
and serotonin in nerve tissue. Pp. 604-640, *in* The thyroid and biogenic amines (J. E.
Rall and J. J. Kopin, eds.), Amsterdam, North Holland, 687 pp.
RABII, J., F. BUONOMO, AND C. G. SCANES. 1980. Studies on the hypothalamic regulation of
luteinizing hormone secretion in the domestic fowl. Pp. 25-32, *in* Recent advances of
avian endocrinology (G. Pethes, P. Peczely, and P. Rudas, eds.), Pergamon Press,
Oxford, xvii + 469 pp.
Sharp, P. J. 1975. The effect of reserpine on plasma LH concentration in intact and gonadec-
tomized domestic fowl. Brit. Poult. Sci., 16:79-82.
SPECTOR, S., A. SJOERDSMA, AND S. UDENFRIEND. 1965. Blockade of endogenous norepinephrine synthesis by α-methyl-tyrosine, an inhibitor of tyrosine hydroxylase. J. Pharm.
Exp. Therap. 147:86-95.
TIXIER-VIDAL, A., AND I. ASSENMACHER. 1962. Effects d'un traitement isolé ou combiné à la
reserpine et à la lumiere permanente, sur la prehypophyse du canard mâle. C. R. Soc.
Biol. (Paris), 156:37-43.
TOZER, T. N., N. H. NEFF, AND B. B. BRODIE. 1966. Application of steady-state kinetics to the
synthesis rate and turnover time of serotonin in the brain of normal and reserpine-
treated rats. J. Pharm. Exp. Therap., 153:177-182.
UNGERSTEDT, U. 1971. Histochemical studies on the effect of intracerebral and intraventricular injection of 6-hydroxydopamine on monoamine neurons in the rat brain. Pp. 101-
126, *in* 6-Hydroxydopamine and catecholamine neurons (T. Malmfors and H.
Thoenen, eds.), Amsterdam, North, Holland 386 pp.
VAN TIENHOVEN, A., A. V. NALBANDOV, AND H. W. NORTON. 1954. Effect of dibenamine on
progesterone induced and "spontaneous" ovulation in the hen. Endocrinology,
54:605-611.

Reprinted from
ASPECTS OF AVIAN ENDOCRINOLOGY:
PRACTICAL AND THEORETICAL IMPLICATIONS (C. G. Scanes *et al.*, eds.)
Grad. Studies, Texas Tech Univ., 1982, 26:1-411.

NEUROENDOCRINE CONTROL OF PROLACTIN SECRETION

S. HARVEY[1], A. CHADWICK[2], G. BORDER[2],
C. G. SCANES[3], AND J. G. PHILLIPS[1]

[1]*The Wolfson Institute, University of Hull, Hull HU6 7RX, England*; [2]*Department of Pure and Applied Zoology, University of Leeds, Leeds LS2 9JT, England*; [3]*Department of Physiology, Rutgers—The State University, New Brunswick, New Jersey 08903 USA*

The hypothalamic control of prolactin secretion in mammals has been extensively investigated and frequently reviewed (for example, Nicoll and Fiorindo, 1969; Meites, 1972; Meites and Clemens, 1972; Schally *et al.*, 1974a; MacLeod, 1976; Clemens and Meites, 1977; Weiner and Ganong, 1978; Franks, 1979; Ganong, 1980), and considerable physiological and pharmacological evidence has demonstrated that the mammalian hypothalamus exerts a predominately inhibitory role on prolactin secretion. In marked contrast, it has been established that prolactin secretion in birds is not under tonic inhibitory control and that the avian hypothalamus contains predominately prolactin-releasing activity (Meites and Nicoll, 1966). This species difference in prolactin physiology might reflect differences in the nature and type of the hypothalamic factors that control prolactin release, but, as yet, elucidation of the hypophysiotropic factors involved still must be achieved. Attempts to purify the peptide releasing factors thought to be responsible from mammalian (see Clemens and Meites, 1977) and avian (Hall, 1975) hypothalami so far have proved unsuccessful; indeed, the existence of specific inhibiting and stimulating peptide releasing factors (in accordance with the classical concept of dual hypothalamic control) is in doubt. It is now well recognized that much, if not all, of the prolactin release inhibiting activity of the mammalian hypothalamus is due to its catecholamine content, particularly dopamine (Shaar and Clemens, 1974; MacLeod and Lehmeyer, 1974; Kordon and Enjalbert, 1980), although other neurotransmitters, putative neurotransmitters, and neurohormones in the hypothalamus are capable, in some circumstances, of exerting direct and indirect effects of pituitary prolactin release (Horrobin, 1975; Weiner and Ganong, 1978; Ganong, 1980; Muller *et al.*, 1980). The effects of catecholamines and other neurotransmitters on avian prolactin secretion have not, however, been assessed, and the possibility remains that differences in aminergic pathways or dopamine action may be partly responsible for the difference between birds and mammals in the hypothalamic control of prolactin secretion. The neuroendocrine control of avian prolactin secretion has therefore been investigated in the present study.

MATERIALS AND METHODS

Preparation of Hypothalamic Extract

An acid (0.1M hydrochloric acid) extract was prepared from hypothalami obtained from eight-week-old broiler fowl using the method of Follett (1970). The extract was neutralized before use and diluted with Medium 199 (Wellcome Laboratories). The extract showed an immunoreactive prolactin potency of approximately 650-700 ng per hypothalamic equivalent. Hypothalamic extract (HE) free from dopamine and other biogenic amines was prepared by the method of Enjalbert et al., (1977) by adsorption on finely ground alumina. This technique removes more than 99.5 per cent of the amines present in the extract.

In vitro Experiments

Pituitaries and whole hypothalami were dissected from the heads of freshly killed eight to 10-week-old broiler fowl and incubated separately or together in one milliliter Medium 199 at 37°C in a shaking water bath under an atmosphere of 95 per cent O_2 and 5 per cent CO_2 as previously described (Border and Chadwick, 1977; Harvey et al., 1979a). Following a one-hour preincubation period the medium was discarded and the test substances were added in one milliliter of fresh medium. Five hours later the tissues were removed, the pituitaries weighed and the media stored deep frozen at −20°C prior to analysis. In all pituitary-hypothalamus incubation experiments, each incubation consisted of a longitudinally bissected hemipituitary and a single whole hypothalamus. In all experiments in which HE was used, a standard dose of extract derived from one hypothalamus equivalent was employed throughout.

In vivo Experiments

In some preliminary experiments, biogenic amines and other neurotransmitters and drugs were injected (1 μg per bird in a volume of 0.1 ml) intracisternally into the region of the third ventricle. This route was chosen to circumvent the blood-brain barrier which would otherwise introduce difficulties in interpretation of the results. Blood samples were obtained from each four week-old cockerel by brachial vein venipuncture before and 30 minutes after injection. Following centrifugation and separation the plasma samples were stored deep frozen pending analysis.

Drug Treatments

All drug solutions used were freshly prepared on the day of use and were applied to the incubation media or injected in vivo in a volume of 0.1 milliliter. L-DOPA, dopamine, noradrenaline, histamine, serotonin, and gamma-aminobutyric acid (GABA) (all from Sigma Chemical Company) were dissolved in 1 per cent ascorbic acid. Melatonin and L-DOPS (Sigma Chemical Company) were moistened with 70 per cent ethanol before dilution in 1 per cent ascorbic acid. CB154 (Sandoz Ltd.) was moistened with 70 per cent ethanol and, after the addition of an equivalent weight of tartaric acid, was suspended in saline. Haloperidol (Jansen Pharmaceuticals) was dissolved in 1N HCl, neutralized, and diluted with 0.9 per cent saline. Prostaglandin E_2 (Sigma Chemical Company) was dissolved in 70 per cent ethanol and diluted with 0.9 per cent saline. All other drugs were dissolved and diluted in 0.9 per cent saline. In each *in vitro* experiment the controls were incubated with the appropriate vehicle.

Determination of Prolactin

The concentrations of prolactin in the plasmas and incubation media were determined by a specific homologous radioimmunoassay for chicken prolactin (Scanes *et al.*, 1976). All the samples from the same experiment were assayed together to eliminate interassay variation. The assay has an intra-assay coefficient of variation of 4.2 per cent and has a minimum detectable dose of 2.0 ng/ml. Statistical differences in the results were determined by Student's paired and unpaired *t*-test where appropriate. Variability in the basal and stimulated level of prolactin release from one incubation experiment to another was not due to interassay variation but to differences in the time taken after death (between 50 and 250 minutes) to transport the chicken heads from a local packing station to the laboratory and to dissect the pituitaries and hypothalami prior to incubation. Variability also might have been due to the physiological state of the bird prior to death.

RESULTS

In vitro *Experiments*

Influence of Hypothalamic Amines

The effect of hypothalamic amines on prolactin secretion was determined by comparing the prolactin response of incubated hemipituitaries to normal hypothalami and HE with the response to alumina treated HE and hypothalami that had been preincubated with finely ground alumina (1 mg/ml) to absorb or inactivate the biogenic amines. Pituitary prolactin release (expressed per mg of pituitary tissue) was greatly increased by incubation with the untreated hypothalamus (3.1-fold, $P < 0.001$) and the untreated extract (16.1-fold, $P < 0.001$ as shown in Fig. 1a and b. The degree of stimulation induced by incubation with the alumina pretreated hypothalamus and alumina treated HE was significantly less than that by their untreated controls ($P < 0.001$ in both cases), although in both cases prolactin release was increased over basal levels (0.35, $P < 0.05$, and 5.4-fold, $P < 0.001$, respectively).

To demonstrate that this loss of hypothalamic releasing activity was probably due to the loss of biogenic amines, hemipituitaries were incubated alone or in the presence of finely ground alumina (1 mg/ml) with dopamine, noradrenaline, serotonin, and histamine (all at doses of 20 ng/ml).

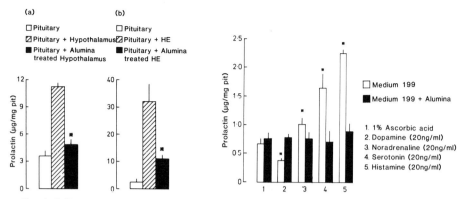

FIG. 1 (left).—**a**, Prolactin release (μg/mg pituitary) from incubated chicken hemipituitaries (open histogram) in the presence of a single whole chicken hypothalamus (shaded histogram) and alumina-treated hypothalami (solid histogram). Mean $\pm$ SE ($N = 8$). **b**, Prolactin release from incubated chicken hemipituitaries (open histogram) in the presence of chicken hypothalamic extract (HE 1 hypothalamic equivalent: shaded histogram) and alumina-treated HE (solid histogram). Mean $\pm$ SE ($N = 8$). Significant ($P < 0.001$) inhibition of hypothalami or HE induced prolactin secretion is indicated by the asterisk.

FIG. 2 (right).—Prolactin release (μg/mg pituitary) from chicken hemipituitaries incubated alone (open histograms) or in the presence of finely ground alumina (1 mg/ml, solid histograms) with biogenic amines (dopamine, noradrenaline, serotonin, and histamine, all at 20 ng/ml). Mean $\pm$ SE ($N = 8$). Prolactin release significantly ($P < 0.05$) different from that in the controls (1% ascorbic acid) is indicated by the asterisks.

Although dopamine significantly lowered ($P < 0.05$) prolactin release, noradrenaline, serotonin, and histamine all markedly increased ($P < 0.01$) prolactin secretion (Fig. 2). The ability of alumina to adsorb these amines was clearly demonstrated by the fact that coincubation with alumina completely blocked the inhibitory effect of dopamine and the stimulatory effect of the other three amines. Alumina incubation itself had no effect on pituitary prolactin release.

In view of the fact that these preliminary experiments suggested that these biogenic amines have direct effects on pituitary prolactin release (in addition to any effect they might have on inhibiting or stimulating the release of any peptide releasing factors in the intact hypothalamus), the influence of hypothalamic amines on prolactin secretion was further investigated.

Influence of Catecholamines

Hemipituitaries were incubated alone or in the presence of HE with noradrenaline (80 ng/ml). Prolactin release was increased by incubation with HE ($P < 0.001$) and with noradrenaline ($P < 0.05$), as shown in Fig. 3. When hemipituitaries were coincubated with noradrenaline the stimulation of prolactin release was greater ($P < 0.05$) than that induced by either noradrenaline or HE when incubated separately. Moreover the degree of stimulation by the coincubation of noradrenaline and HE (11.8-fold) was greater than that (7.3-fold) due to their additive effects. A stimulatory role of norad-

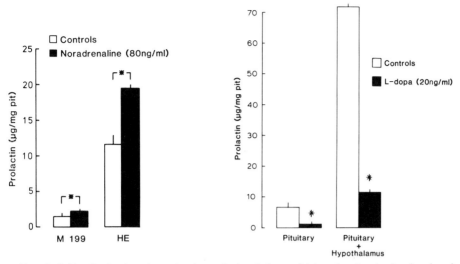

FIG. 3 (left).—Prolactin release (μg/mg pituitary) from chicken hemiptuitaries incubated alone or with hypothalamic extract (HE) (open histograms) in the presence of noradrenaline (80 ng/ml, solid histograms). Mean ± sᴇ (N = 8). Prolactin release significantly (P < 0.05) different from the respective controls is indicated by the asterisks.

FIG. 4 (right).—Prolactin release (μg/mg pituitary) from chicken hemipituitaries incubated alone or with whole hypothalami (open histograms) in the presence of L-DOPA (20 ng/ml, solid histograms). Mean ± sᴇ (N = 8). Prolactin release significantly (P < 0.001) different from the respective controls is indicated by the asterisks.

renaline in prolactin secretion was further demonstrated by the fact that L-DOPS (a noradrenaline precursor) also increased the release of prolactin from incubated hemipituitary glands (values given are mean ± sᴇ, sample size): controls, 1.28 ± 0.17 μg prolactin/mg pituitary, 8; treatment with L-DOPS at 20 ng/ml, 1.80 ± 0.22 μg/mg pituitary, 8; P < 0.05.

In contrast to the stimulatory effect of noradrenaline on basal and HE induced prolactin secretion, both L-DOPA (a dopamine precursor) and dopamine were found to suppress prolactin release and the releasing effect of the hypothalamus. The incubation of hemipituitaries with 20 ng/ml of L-DOPA strikingly lowered (P < 0.001) basal prolactin release (Fig. 4), and its coincubation with single whole hypothalami completely blocked its prolactin-releasing activity, although the level of prolactin release was higher (P < 0.001) than that induced by L-DOPA itself. Dopamine, at doses of 5 (P < 0.05), 20 (P < 0.01), and 80 (P < 0.001) ng/ml also reduced the basal release of prolactin from incubated hemipituitaries in a dose-related way (Fig. 5). The possibility that dopamine might also inhibit prolactin secretion at the hypothalamic level was also suggested by the fact that coincubating the same doses of dopamine with hypothalamic tissue suppressed the prolactin releasing activity of the tissue (Fig. 5). This possibility was emphasized by the fact that a stimulatory (P < 0.05) effect of HE on prolactin release was still discernible in the presence of 5 and 20 ng/ml of dopa-

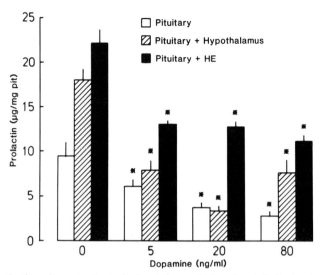

FIG. 5.—Prolactin release (μg/mg pituitary) from chicken hemipituitaries incubated alone (open histograms) or with whole hypothalami (shaded histograms) or with hypothalamic extract (HE, solid histograms) in the presence of different doses of dopamine. Mean ± SE (N = 8). Prolactin release significantly ($P < 0.01$) different from the respective controls is indicated by the asterisks.

mine (in contrast to its effect in intact tissue), although the degree of stimulation was less ($P < 0.01$) than that induced by HE itself. Dopamine at 80 ng/ml, however, blocked the effect of HE.

The potent inhibitory effect of dopamine on prolactin secretion and the presence of dopamine receptors on the pituitary was further shown by the ability of 20 ng/ml apomorphine (a specific dopamine receptor agonist) to reduce the basal level of prolactin release from incubated hemipituitary glands by the same degree as the same dose of dopamine (Fig. 6a). A nonspecific dopamine agonist, CB154, had no effect on basal prolactin release when incubated with hemipituitaries at doses of 5, 20, and 80 ng/ml, but when coincubated with whole hypothalami, these doses greatly reduced ($P <$ 0.01) the releasing activity of the hypothalamus, although in each case the level of prolactin release remained higher ($P < 0.01$) than that of the control pituitaries (Fig. 6b). The marked inhibitory effect of dopamine on prolactin secretion and its direct pituitary action was also indicated by showing that coincubating haloperidol and pimozide (specific dopamine receptor antagonists) with dopamine (all at 20 ng/ml) completely blocked the decrease ($P <$ 0.001) in prolactin release induced by incubation with dopamine alone (Fig. 7). Furthermore, evidence of a dopaminergic pathway in the hypothalamic control of prolactin secretion was suggested by the fact that a dopamine synthesis blocker, α-methyl DOPA, stimulated ($P < 0.01$) both the basal and hypothalamus induced prolactin release at a dose of 20 ng/ml and increased ($P < 0.01$) the pituitary prolactin response to the coincubated hypothalami at a dose of 10 ng/ml., as shown in Fig. 8.

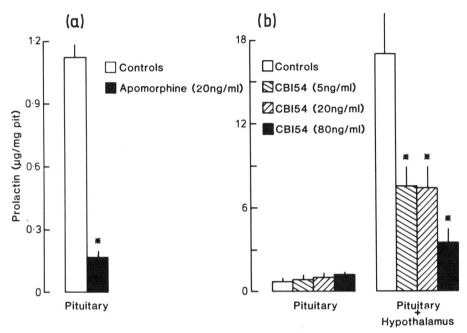

FIG. 6.—**a**, Prolactin release (μg/mg pituitary) from chicken hemipituitaries incubated alone (open histogram) or with apomorphine (20 ng/ml, solid histogram). Mean ± SE ($N = 8$). prolactin release significantly ($P < 0.001$) different from the controls is indicated by the asterisk. **b**, Prolactin release (μg/mg pituitary) from chicken pituitaries incubated alone or with whole hypothalami (open histograms) in the presence of doses of CB154. Mean ± SE ($N = 8$). Prolactin release significantly ($P < 0.01$) different from the respective controls is indicated by the asterisks.

Influence of Indoleamines

The earlier suggestion of serotoninergic involvement in the control of prolactin secretion and the ability of serotonin to directly enhance pituitary prolactin release was subsequently confirmed in another series of experiments. Serotonin itself, at dose of 20 ng/ml, greatly increased ($P < 0.05$) basal prolactin release (Fig. 9a). Coincubating this dose of serotonin with whole hypothalami resulted in a much greater ($P < 0.01$) stimulation of prolactin release than that induced by either serotonin or the hypothalamus in separate incubations and this increase in response was much more than an additive effect (Fig. 9a). Moreover, although L-tryptophan (a serotonin precursor) had no direct effect itself on basal prolactin release (at doses of 20 and 80 ng/ml), its coincubation with whole hypothalami again resulted in a greater increase ($P < 0.05$) in prolactin secretion than that due to the hypothalamic tissue itself (Fig. 9b). Further involvement of serotoninergic pathways in the control of prolactin secretion was also shown by the demonstration of a pronounced ($P < 0.01$) inhibitory effect of melatonin on the potent releasing activity of HE and whole hypothalami (Fig. 9c). In addition, methysergide (a serotonin receptor antagonist) reduced the releasing activity of coincubated hypothalami by 33 per cent and markedly reduced ($P < 0.01$) the basal level

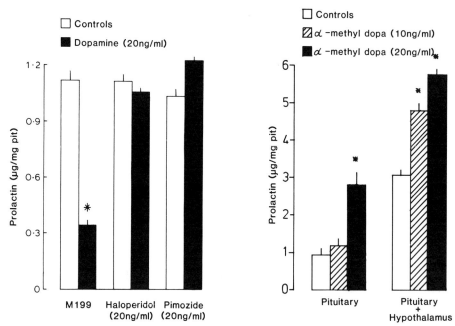

Fig. 7 (left).—Prolactin release (μg/mg pituitary) from chicken hemipituitaries incubated alone (open histograms) or with dopamine (20 ng/ml, solid histograms) in the presence of haloperidol (20 ng/ml) or pimozide (20 ng/ml). Mean ± SE (N = 8). Prolactin release significantly (P < 0.001) different from the controls is indicated by the asterisk.

Fig. 8 (right).—Prolactin release (μg/mg pituitary) from chicken hemipituitaries incubated alone or with whole hypothalmi (open histograms) in the presence of 10 ng/ml (shaded histograms) and 20 ng/ml (solid histograms) of α-methyl DOPA. Mean ± SE (N = 8). Prolactin release significantly (P < 0.01) different from the respective controls is indicated by the asterisks.

of prolactin release (controls, 1.29 ± 0.18 μg/mg pituitary, 8; methysergide treated 0.13 ± 0.05 μg/mg pituitary, 8).

Influence of Histamine

Potent and direct prolactin releasing activity was also demonstrated by 5 and 20 ng/ml of histamine, which significantly (P < 0.05) increased basal prolactin release from incubated hemipituitaries (Fig. 10). Moreover, both doses of histamine also greatly increased (P < 0.001) the stimulation (P < 0.001) of prolactin release induced by single whole hypothalami; in both cases the stimulation was greater than that expected from a combination of their separate effects. The possibility that this was due to a stimulatory effect of histamine on the release of hypothalamic releasing factors was suggested by the fact that coincubating histamine with HE had no effect on the HE response (Fig. 10).

Influence of Acetyl Choline

A stimulatory cholinergic pathway in the control of prolactin secretion was suggested by the finding that incubating hemipituitaries (which had

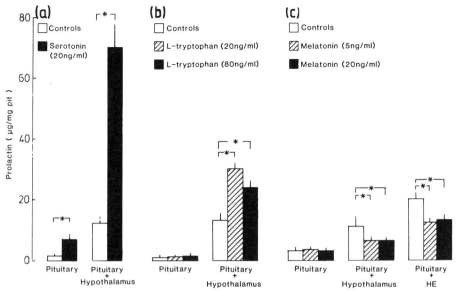

FIG. 9.—**a**, Prolactin release (μg/mg pituitary) from chicken hemipituitaries incubated alone or with whole hypothalami (open histograms) in the presence of serotonin (20 ng/ml, solid histograms). Mean $\pm$ SE ($N = 8$). Prolactin release significantly ($P < 0.01$) different from the respective controls is indicated by the asterisks. **b**, Prolactin release (μg/mg pituitary) from chicken hemipituitaries incubated alone or with whole hypothalami (open histograms) in the presence of 20 ng/ml (shaded histograms) or 80 ng/ml (solid histograms) of L-tryptophan. Mean $\pm$ SE ($N = 8$). Prolactin release significantly ($P < 0.01$) different from the controls is indicated by the asterisks. **c**, Prolactin release (μg/mg pituitary) from chicken hemipituitaries incubated alone or with whole hypothalami or hypothalamic extract (HE, open histograms) in the presence of 5 ng/ml (shaded histograms) or 20 ng/ml (solid histograms) of melatonin. Mean $\pm$ SE ($N = 8$). Prolactin release significantly ($P < 0.05$) different from the respective controls is indicated by the asterisks.

been preincubated with 20 ng/ml physostigmine for 30 minutes to prevent acetyl choline metabolism) with 5 ng/ml acetyl choline increased ($P < 0.05$) basal prolactin release (controls, 1.19 ± 0.23 μg/mg pituitary, 8; acetyl choline treated, 1.95 ± 0.29 μg/mg, 8). Moreover, following the initial preincubation period the level of prolactin released by hemipituitaries incubated with 20 ng/ml atropine (a cholinergic antagonist) and single whole hypothalami (10.00 ± 0.97 μg/mg pituitary, 10) was less ($P < 0.02$) than that released by pituitaries incubated only with hypothalami (15.58 ± 1.12 μg/mg pituitary, 10). Atropine itself had no effect on basal prolactin release.

Influence of Gamma Amino Butyric Acid (GABA)

A direct inhibitory effect of GABA on pituitary prolactin release was shown by the ability of 20 ng/ml GABA to lower ($P < 0.02$) the basal level of prolactin release from incubated hemipituitary glands (controls, 2.04 ± 0.27 μg/mg pituitary, 10; GABA treated 0.78 ± 0.22 μg/mg pituitary, 10). In addition, the possibility that GABA might have an effect on prolactin secretion via the hypothalamus was suggested by the reduction ($P < 0.02$) in the

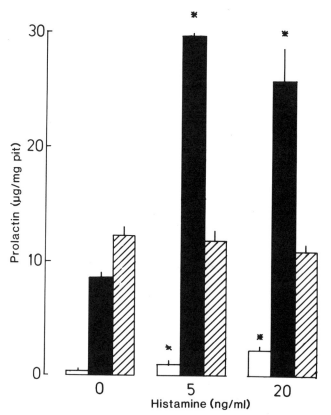

FIG. 10.—Prolactin release (μg/mg pituitary) from chicken pituitaries incubated alone (open histograms) or with whole hypothalami (solid histograms) or hypothalamic extract (HE, shaded histograms) in the presence of different doses of histamine. Mean ± SE ($N = 8$). Prolactin release significantly ($P < 0.01$) different from the respective controls is indicated by the asterisks.

releasing activity of the hypothalamus when the same dose of GABA was incubated with single whole hypothalami (hypothalami incubation, 5.23 ± 0.13 μg/mg pituitary, 8; GABA and hypothalami coincubation, 2.58 ± 0.63 μg/mg pituitary, 10).

Influence of Prostaglandin E_2

Potent prolactin releasing activity was also exerted by prostaglandin E_2 when incubated at a dose of 20 ng/ml with hemipituitaries, alone and in the presence of single whole hypothalami (hemipituitaries 0.66 ± 0.06 μg/mg pituitary, 10; prostaglandin treated, 1.14 ± 0.12 μg/mg pituitary, 10; $P < 0.01$; hemipituitaries incubated with hypothalami 2.51 ± 0.62 μg/mg pituitary, 10; prostaglandin treated 7.22 ± 1.24 μg/mg pituitary, 10; $P < 0.01$). The degree of stimulation induced by coincubating prostaglandin E_2 with the hypothalamus was 2.19-fold greater than that expected from its additive effects.

TABLE 1.—*Effect of intracisternal injections of various amines, hormones and drugs on plasma prolactin concentration in anesthetized (with Equithesin, Gandal, 1969) 4-week-old cockerels.*

Treatment (1 μg/bird)	Plasma prolactin (ng/ml + SE, $N = 8$)	
	Before injection	30 minutes after injection
1% Ascorbic Acid (Vehicle)	60.4 ± 8.5	63.7 ± 6.6
Serotonin	36.4 ± 5.2	140.5 ± 45.8*
L-tryptophan	69.8 ± 6.0	90.7 ± 7.1*
Melatonin	74.2 ± 21.0	75.9 ± 18.5
Noradrenaline	41.0 ± 5.1	72.9 ± 11.0*
L-DOPS	39.8 ± 3.8	152.7 ± 75.8*
L-DOPA	68.9 ± 16.6	67.1 ± 22.3
α-Methyl DOPA	19.5 ± 4.6	152.5 ± 48.4*
Pimozide	22.7 ± 1.9	75.2 ± 17.2*
CB154	63.0 ± 8.4	125.6 ± 13.3*
Histamine	59.6 ± 5.7	80.9 ± 11.2*
GABA	79.2 ± 16.4	14.4 ± 4.4*
Carbachol	109.7 ± 44.5	265.5 ± 45.6*

*Significantly different from pretreatment level, $P < 0.05$ (paired t-test).

In vivo *Experiments*

The influence of biogenic amines and other neurotransmitters and drugs on prolactin secretion was also assessed in some preliminary *in vivo* experiments in immature cockerels. Although the basal level of plasma prolactin varied from one experiment to another, the intracisternal injection of noradrenaline, L-DOPS, L-tryptophan, serotonin, α-methyl DOPA, pimozide, and histamine, all increased plasma prolactin levels 30 minutes after injection, while GABA lowered the prolactin concentration (Table 1), confirming the *in vitro* results. Carbachol, a cholinergic agonist, also stimulated prolactin secretion, as did CB 154. At the dose used, L-DOPA and melatonin had no inhibitory effect on prolactin secretion.

DISCUSSION

The results of the present study clearly demonstrate the potent prolactin-releasing activity of the avian hypothalamus. Considerable evidence from six different species of birds has now been accumulated which demonstrates that, unlike fish, amphibia and mammals (see Ensor, 1978, for review), the avian hypothalamus promotes rather than inhibits the release of prolactin. This was first suggested by Ma and Nalbandov (1963) who showed that auto-transplanted pigeon pituitaries had reduced prolactin secretion. Tixier-Vidal *et al.* (1968) and Tixier-Vidal and Gourdji (1972) subsequently demonstrated that this was a result of prolactin cell regression. Consequently, pigeons bearing transplanted pituitaries have unstimulated crop sacs (Bayle and Assenmacher, 1965; Hall *et al.*, 1975; Knight and Chadwick, 1975; Hall and Chadwick, 1979a). Similarly, in the absence of hypothalamic stimulation, the prolactin cells from pigeon and quail pituitaries are quiescent *in vitro* (Tixier-Vidal, 1967), and pituitary glands from pigeons (Gala and Reece, 1965; Nicoll, 1965; Gourdji, 1970; Hall *et al.*, 1975), quail (Gourdji, 1970; Tixier-Vidal and Gourdji, 1972), tricolored blackbird (Nicoll, 1965; Nicoll *et al.*, 1970), and chickens (Bolton *et al.*, 1976; Hall *et al.*, 1975) secrete very little prolactin when incubated *in vitro*. The existence of a prolactin releasing factor (PRF) in the avian hypothalamus has also been demonstrated by Assenmacher and Tixier-Vidal (1964), who reported that sectioning of the pituitary portal vessels in the duck resulted in atrophy of the prolactin cells. The stimulatory influence of hypothalamic extracts on *in vitro* prolactin secretion has also been shown many times in pigeons (Kragt and Meites, 1965; Hall and Chadwick, 1975, 1979a, 1981; Hall *et al.*, 1975, 1976; Chadwick, 1980), quail (Kragt, 1966; Tixier-Vidal and Gourdji, 1972), tricolored blackbirds (Nicoll, 1965), turkeys (Chen *et al.*, 1968), ducks (Gourdji and Tixier-Vidal, 1966), and chickens (Hall *et al.*, 1975, 1976; Bolton *et al.*, 1976; Chadwick *et al.*, 1977, 1977, 1978a, 1978b; Scanes *et al.*, 1977; Border and Chadwick, 1977; Hall and Chadwick, 1979b; Harvey *et al.*, 1979a). Pigeon and chicken HE have also been shown to induce crop sac proliferation in pigeons (Knight and Chadwick, 1975; Hall and Chadwick, 1979a), while chicken HE also increases plasma prolactin levels in fowl (Harvey *et al.*, 1979b). Acid extracts of other brain tissues have no effect on prolactin secretion (Nicoll *et al.*, 1970; Hall and Chadwick, 1979a; Harvey *et al.*, 1979b).

The hypothalamic control of pituitary hormone secretion has classically been thought to be mediated by specific inhibiting or stimulating polypeptide hypophysiotropic releasing factors secreted by the peptidergic neurones of the hypothalamus (Harris, 1955). However, although peptide releasing factors for thyrotropin stimulating hormone (TSH), luteinizing hormone (LH) and growth hormone (GH) have been isolated and purified from mammalian hypothalami (see Schally and Kastin, 1972; Schally *et al.*, 1974a; Arimura and Schally, 1979), as yet hypothalamic fractions with PRF-activity have only been isolated (for example, Dular *et al.*, 1974) and not purified (Kordon and Enjalbert, 1980). Nevertheless there is a large body of opinion that considers that thyrotropin releasing hormone (TRH) may also be the PRF of the hypothalamus. Indeed there is overwhelming evidence that TRH stimulates prolactin release in mammalian species under many physiological, pathological and pharmacological conditions (see Horrobin, 1975, 1976; Clemens and Meites, 1977; and Ensor, 1978, for reviews), and Rivier and Vale (1974) concluded that all the PRF activity in their ovine HE was attributable to TRH. However, the belief that TRH may be the PRF of the hypothalamus (Bowers *et al.*, 1971) has also been disputed, as several physiological states (such as suckling, insulin hypoglycemia, and ether stress) exist when TSH and prolactin are not released together; furthermore the stimulatory action of TRH on prolactin but not TSH secretion can be blocked by L-DOPA (Clemens and Meites, 1977). Moreover, it has been demonstrated that PRF activity is present in hypothalamic extracts which are free from

TRH (Valverde *et al.*, 1972; Machlin *et al.*, 1974; Boyd *et al.*, 1976; Szabo and Frohman, 1976). TRH is therefore unlikely to be the sole PRF or be responsible for all the stimulatory effects of hypothalamic extracts, but as prolactin secretion in mammals is evidently under complex hypothalamic control, it may be one of several stimulating and inhibiting releasing factors that plays a physiological role in the regulation of prolactin secretion.

In birds, TRH is also unlikely to be responsible for all the PRF activity of the avian hypothalamus as it has no consistent effect on *in vivo* prolactin secretion (Bolton *et al.*, 1974; Chadwick and Hall, 1975; Scanes *et al.*, 1977; Harvey *et al.*, 1978a; Harvey, Chadwick, and Sterling, unpubl. observ.) despite consistently increasing prolactin release from chicken and pigeon pituitaries *in vitro* (Hall *et al.*, 1975; Hall and Chadwick, 1974, 1976; Scanes *et al.*, 1977; Harvey *et al.*, 1978a; Hall, 1980). Furthermore the TRH content of the chicken hypothalamus (approximately 0.4-0.5 ng, Jackson and Reichlin, 1974) is too low to account for the *in vitro* PRF activity of the hypothalamus. The possibility that the prolactin releasing activity of the avian hypothalamus may be due solely or in part to a specific polypeptide PRF therefore remains to be investigated. In either event, the results of the present study clearly show that catecholamines and other neurotransmitters have direct and possibly indirect (via the hypothalamus) stimulatory effects on pituitary prolactin secretion. Indeed, it would appear that at least half of the prolactin releasing activity of the avian hypothalamus may be due to its amine content, particularly noradrenaline, serotonin, and histamine (Figs. 1 and 2).

Hypothalamic monoamines are known to play a vital role in the control of prolactin secretion in mammals, although under normal circumstances only dopamine and noradrenaline have direct effects on pituitary prolactin release (reviewed by Weiner and Ganong, 1978). In mammals, hypothalamic monoamines usually affect adenohypophysial function by acting as stimulatory and/or inhibitory neurotransmitters in the transfer of neural information from the internal and external environment to the hypothalamic neurones which secrete stimulatory or inhibitory hypophysiotrophic releasing factors directly into portal circulation. However, because of the multisynaptic nature of the brain input pathways, the non-specific nature of some amine receptors and pharmacological agents and because all brain amines may have multiple effects and can be metabolized into each other (Weiner and Ganong, 1978; Hökfelt *et al.*, 1980; Ganong, 1980), elucidation of the aminergic pathways and roles of the amines involved in the control of prolactin secretion has proved difficult. Nevertheless, the existence of adrenergic or noradrenergic, serotonergic, and cholinergic pathways in the neuroendocrine control of prolactin secretion has been demonstrated in mammals and in birds (this study).

In birds, the direct effect of noradrenaline on pituitary prolactin release is different from that in mammals, in which it inhibits prolactin secretion (MacLeod, 1969; Birge *et al.*, 1970; Koch *et al.*, 1970; MacLeod and Lehymayer, 1974; Shaar and Clemens, 1974; Schally *et al.*, 1974*b*, 1976). Pharmacological evidence, however, suggests that the action of noradrenaline in mammals is not mediated through noradrenergic or adrenergic receptors but through dopamine receptors (reviewed by Clemens and Meites, 1977). In contrast, as dopamine has an inhibitory effect on prolactin secretion in the fowl, it is most unlikely that the pituitary action of noradrenaline is mediated through dopamine receptors in this species. Corroborative evidence of a direct adrenergic system in the control of avian prolactin secretion was provided by the fact that L-DOPS directly increased pituitary prolactin release *in vitro* and by the demonstration of increased plasma prolactin levels *in vivo* following noradrenaline or L-DOPS administration. L-DOPS has also been found to promote *in vivo* prolactin release in rats (Donoso *et al.*, 1971) and considerable pharmacological evidence derived from the use of adrenergic active drugs and hypothalamic lesions has revealed a stimulatory noradrenergic and adrenergic pathway involved in the mediation of estrogen-induced and stress-induced prolactin release (for example, Fenske and Wuttke, 1976; Subramanian and Gala, 1976; Langlelier and McCann, 1977), although it is not thought to be involved in basal (Weiner, 1973) or suckling-induced (Carr *et al.*, 1977) prolactin secretion. Whether or not adrenaline itself affects prolactin secretion in birds is not known; the preliminary data so far obtained has been inconclusive (Border, Harvey, and Chadwick, unpubl. observ.).

The difference between birds and mammals in the *in vitro* effects of noradrenaline may, therefore, partly explain the difference between them in the pituitary prolactin response to hypothalamic extracts. This difference is probably also due to differences in the effects of other amines (particularly histamine and serotonin) and neurotransmitters (such as acetyl choline) which are likely to be present in normal HE preparations (Brownstein *et al.*, 1974, 1976; Zimmerman, 1976). For instance, while acetyl choline and cholinergic drugs reduce prolactin secretion in mammals (reviewed by Horrobin, 1976; Clemens and Meites, 1977), acetyl choline directly stimulated basal prolactin release from the fowl pituitary *in vitro*, atropine (a cholinergic antagonist) inhibited the releasing activity of the incubated hypothalamus, and carbachol (a cholinergic agonist) increased prolactin secretion *in vivo*. Similarly, whereas serotonin was a potent stimulator of basal prolactin release *in vitro* in this study, serotonin has no effect on prolactin secretion when added to *in vitro* cultures of mammalian pituitaries (Birge *et al.*, 1970) or when injected into hypophyseal portal vessels (Kamberi *et al.*, 1971). Histamine was also found to have a potent direct effect on pituitary prolactin release in the present study in contrast with its inability to affect prolactin release following intrapituitary injection (Libertun and McCann, 1976) or *in vitro* incubation with mammalian pituitaries (Rivier and Vale, 1977).

In the present study, the coincubation of whole hypothalami and pituitaries with serotonin or histamine showed a synergistic effect on prolactin secretion, suggesting that in addition to their direct effects, these drugs may have stimulated the release of hypothalamic releasing factors or inhibited the release of inhibiting factors. Hall (1980) also found that 5-hydroxytryptamine (5-HT) stimulated hypothalamus induced prolactin release (as measured by densitometery following polyacrylemide gel electrophoresis) from pigeon pituitaries *in vitro*, although in this study 5-HT had no direct effect itself. The fact that L-tryptophan also had a synergistic effect with the hypothalamus on prolactin release but had no effect itself on basal prolactin release *in vitro* suggests that this stimulation of prolactin release resulted from an indirect or direct effect of its metabolism to serotonin. Presumably the increase in prolactin secretion *in vivo* following L-tryptophan administration was also due to the conversion to serotonin.

Serotonin and L-tryptophan also stimulated prolactin secretion *in vivo* in the present study. Similarly, the intraventricular treatment of pigeons with drugs enhancing serotonin transmission has also been found to produce a marked stimulation of prolactin secretion, as judged by crop sac proliferation; this effect is blocked by methysergide pretreatment (Nistico *et al.*, 1980). In the present study, methysergide and melatonin also had an inhibitory effect on hypothalamus-induced prolactin secretion *in vitro*; similar effects have also been observed in pigeons (Hall, 1980). The fact that methysergide decreased basal prolactin release from the incubated pituitary in the present study probably resulted from the dopamine-agonist property of this drug. The inhibitory effect of melatonin on serotoninergic pathways is thought to be due to its competition with serotonin for serotonin receptors (Smythe and Lazarus, 1973).

In mammals, serotonin-containing neurons are also thought to indirectly stimulate prolactin secretion under certain conditions (such as during suckling or during oestrogen surges) but have no effect on tonic secretion (reviewed by Weiner and Ganong, 1978). The action of this is presumably via the hypothalamus, but whether it acts by stimulating PRF or inhibiting PIF (Prolactin Inhibiting Factor) is not known. In mammals, histamine also affects prolactin secretion *in vivo* but the effects are inconsistent; stimulatory (Donoso and Bannza, 1976; Liberton and McCann, 1976) and inhibitory (Arakelian and Liberton, 1977; Carlson and Ippoliti, 1977) effects have been reported. These effects are thought to be mediated through H_1- and H_2-type receptors, respectively, and their physiological role, if any, in the control of prolactin secretion is not known. Histamine is found in very high concentrations in the hypothalamus (Brownstein *et al.*, 1976; Zimmerman, 1976), but it is not clear whether it is located in nerve cells (and involved in neurotransmission) or in hypothalamic mast cells.

Another difference between birds and mammals in the neuroendocrine control of prolactin secretion was also shown in the present study through the influence of prostaglandin E_2, which directly stimulated prolactin release from fowl pituitaries. Intra-pituitary injections (Harms *et al.*, 1973) and *in vitro* pituitary incubations (Sunberg *et al.*, 1975) with PGE_2 and other prostaglandins had no effect on prolactin secretion in rats. Prostaglandins of the

E series do have a stimulatory effect on prolactin secretion *in vivo* in mammals (Ojeda *et al.*, 1974; Sato *et al.*, 1974; Davis *et al.*, 1977), suggesting an extra pituitary action which is probably at the hypothalamus and effective by inhibiting PIF release (Labrie *et al.*, 1976). Prostaglandins E_1, E_2 and $F_2\alpha$ are also potent stimulators of *in vivo* prolactin secretion in birds (Scanes *et al.*, 1978; Harvey and Scanes, 1981), and the results of the present study demonstrate that PGE_2 is also effective at the hypothalamic level; coincubation of PGE_2 with the hypothalamus had a greater than additive effect on pituitary prolactin release.

In the absence of intact hypothalamic tissue, a synergistic effect of noradrenaline on pituitary prolactin secretion was also observed in the present study when coincubated with HE. Border and Chadwick (1977) have previously reported a synergistic effect of serotonin and HE on *in vitro* prolactin release. The reason for this phenomenon is unknown, but might suggest that these amines have the ability to enhance or respectively reduce the sensitivity of the prolactin cells to the PRF(s) or PIF(s) of the extract. Alterations in pituitary sensitivity to hypothalamic releasing factors do occur, as exemplified by the changes in sensitivity of the avian (Davies and Bicknell, 1976; Davis *et al.*, 1976; Davies and Follett, 1980) and mammalian (reviewed by Labrie *et al.*, 1976) pituitary to luteinizing hormone releasing hormone (LHRH) induced by changes in photoperiod and gonadal steroid feedback. If biogenic amines can affect pituitary sensitivity, then the synergism between them and the intact hypothalamus acting to increase prolactin release simply might not be due to increased secretion of stimulatory releasing factors or suppressed release of inhibiting factors. Similarly, if this hypothesis is correct, the fact that melatonin not only depressed the releasing activity of the intact hypothalamus but reduced the stimulatory effect of HE (despite having no direct effect itself) may be due to melatonin inducing changes in pituitary sensitivity to the releasing factors of the extract. The possibility that biogenic amines might be able to affect the sensitivity of the avian pituitary to hypothalamic releasing factors therefore requires investigation.

In the present study, dopamine was also found to inhibit the releasing activity of the hypothalamus, but it also had a very potent inhibitory influence on basal prolactin release. Thus, although it is possible that dopamine inhibited prolactin secretion by hypothalamic action or changes in pituitary sensitivity, this effect might be due solely to its direct effect on the pituitary.

The involvement of an inhibitory dopaminergic system in the hypothalamic control of prolactin secretion was conclusively demonstrated in the present study. Dopamine inhibited *in vitro* prolactin release directly as did its precursor (L-DOPA) and receptor agonist (apomorphine). The inhibitory effect of dopamine was blocked by its receptor antagonists, haloperidol and pimozide; another dopamine antagonist (α-methyl DOPA) increased basal and hypothalamus induced prolactin release. Moreover, while L-DOPA had no effect on *in vivo* prolactin secretion, the *in vivo* effects at the dose used of pimozide and α-methyl DOPA corroborated with their *in vitro* effects. CB154, a less specific dopamine agonist, also suppressed the releasing activity of the hypothalamus, possibly by action at the hypothalamic level, but increased prolactin secretion *in vivo*. Dopamine itself has been shown to depress *in vivo* prolactin secretion in the fowl following intraperitoneal injection of a pharmacological dose (12 mg) into immature cockerels (Border and Chadwick, unpubl. observ.).

Further evidence of an inhibitory role of dopamine in the control of avian prolactin secretion has recently been reported by Hall (1980), who found that dopamine suppressed TRH and hypothalami-induced prolactin secretion from pigeon pituitaries *in vitro*; these effects were blocked by pimozide. Nistico *et al.*, (1979) also reported that neuroleptic drugs (reserpine, haloperidol, and sulpiride) given systemically or intraventricularly produce marked crop sac and pituitary prolactin cell stimulation. These results suggest that drugs able to block dopamine receptors increase prolactin secretion as a result of an indirect or direct effect; indeed, Harvey *et al.*, (1978*b*) have shown that systemic administration of reserpine to domestic fowl produces a marked increase in plasma prolactin concentrations. These findings contradict the earlier belief that such drugs were ineffective in birds (Calas, 1977) and are very similar to the effects of dopamine and dopaminergic drugs which have been observed in mammals (reviewed by Clemens and Meites, 1977; Weiner and Ganong, 1978). In mammals, a vast literature on the inhibitory role of dopamine on prolactin secretion now exists since Kanematsu *et al.*, (1963) first showed that reserpine injections induced lactation and decreased pituitary prolactin content in rabbits and since Van Maanen and Smelik (1968) first proposed that dopamine might be the PIF of mammalian hypothalamic extracts. Since then, highly purified hypothalamic fractions with PIF activity are known to contain significant amounts of catecholamines (Schally *et al.*, 1974*b*), and treatment of the extracts with monoamine oxidase (MAO, an enzyme that catalyzes the metabolism of catecholamines) or alumina (which extracts catecholamines) removes the PIF activity. Alumina extraction does not, however, remove all the PIF activity of synaptosomal preparations of HE (Enjalbert *et al.*, 1977). Furthermore, some purified fractions that contain PIF activity do not contain catecholamines (Greibrokk *et al.*, 1974; Schally *et al.*, 1977).

The PIF activity in these catecholamine-stripped fractions is thought to be due to GABA, which has been reported to directly inhibit *in vitro* prolactin release and to suppress *in vivo* prolactin secretion (Schally *et al.*, 1977), although these effects are controversial (Mioduszewski *et al.*, 1976; Pass and Ondo, 1977). Nevertheless, neither dopamine nor GABA can completely account for the PIF activity of the median eminence of the hypothalamus; other factors, possibly peptides, are undoubtedly involved (Kordon and Enjalbert, 1980).

The role of GABA in the control of prolactin secretion is unknown, but for various reasons (reviewed by Weiner and Ganong, 1978), including its low PIF potency, it is not thought to be a physiological PIF but might be involved in neural regulation of prolactin secretion. In the present study, GABA was also found to inhibit prolactin secretion *in vitro*, possibly by effects at the pituitary level.

Whether or not GABA plays a physiological role in the avian hypothalamus as a PIF remains to be investigated, but the fact that GABA and dopamine pharmacologically inhibit prolactin secretion in birds suggests that prolactin release in this species may be under dual hypothalamic control. Evidence for PIF activity in the avian hypothalamus has previously been derived from the facts that 1) pigeon HE inhibited prolactin release from rat pituitary glands (Nicoll *et al.*, 1970), 2) cultured duck pituitary glands are able to autonomously secrete prolactin and synthesize it *de novo* (Tixier-Vidal and Gourdji, 1965, 1972; Tixier-Vidal and Picart, 1967), 3) the degree of crop sac stimulation in pigeons injected with chicken HE was far less than that expected from HE-induced changes in prolactin cell morphology (Knight and Chadwick, 1975), and 4) rat HE inhibits the *in vitro* stimulation of prolactin release induced by chicken HE (Bolton *et al.*, 1976). The reduction in the releasing activity of the chicken hypothalamus following catecholamine removal suggests, however, that under normal conditions, that PRF activity of noradrenaline, serotonin, and histamine is greater than the PIF activity of L-DOPA and dopamine. The results of this study, therefore, suggest that the effects of these amines on prolactin secretion are partly responsible for the fact that the avian hypothalamus contains predominantly PRF activity while the mammalian hypothalamus contains predominantly PIF activity.

Summary

The avian hypothalamus, unlike the mammalian hypothalamus, contains predominantly prolactin releasing activity. Biogenic amines and neurotransmitters in the hypothalamus, particularly noradrenaline, serotonin, histamine, acetyl choline, and prostaglandin E_2, are partly responsible for this releasing activity as a result of their direct pituitary action and/or hypothalamic stimulation. The possibility of prolactin secretion in birds being under dual hypothalamic control is suggested by an inhibitory effect of L-DOPA and dopamine on prolactin release. The difference between mammals and birds in the hypothalamic control of prolactin secretion is partly due to differences in aminergic effects on prolactin secretion.

Acknowledgments

This work was supported in part by grants from the Agricultural Research Council (AG 24/102) and the Science Research Council (GR/B47782). G. B. was in receipt of a SRC studentship during the course of this work.

LITERATURE CITED

ARAKELIAN, M. C., AND C. LIBERTON. 1977. H_1 and H_2 histamine receptor participation in the brain control of prolactin secretion in lactating rats. Endocrinology, 100:890-895.

ASSENMACHER, I., AND A. TIXIER-VIDAL. 1964. Répercussions de la section veines portes hypophysaires sur la préhypophyse du canard pekin male, entier ou castré. Arch. Anat. Microse. Morphol. Exp., 53:83-108.

BAYLE, J. D., AND I. ASSENMACHER. 1965. Absence de stimulation du jabot du pigeon après autogreffe hypophysaires. C. R. Acad. Sci., Ser. D (Paris), 261:5667-5670.

BIRGE, C. A., L. S. JACOBS, C. T. HAMMER, AND W. H. DAUGHADAY. 1970. Catecholamine inhibition of prolactin secretion by isolated rat adenohypophyses. Endocrinology, 86:120-130.

BOLTON, N. J., A. CHADWICK, AND C. G. SCANES. 1974. The effect of thyrotrophin-releasing factor on the secretion of thyroid stimulating hormone and prolactin from the chicken anterior pituitary gland. J. Physiol. (London), 238:78-79P.

BOLTON, N. J., A. CHADWICK, T. R. HALL, AND C. G. SCANES. 1976. Effect of chicken and rat hypothalamic extracts on prolactin secretion in the chicken. IRCS Med. Sci., 4:495.

BORDER, G., AND A. CHADWICK. 1977. Effect of serotonin on prolactin release from chicken pituitary glands incubated *in vitro*. IRCS Med. Sci., 5:343.

BOWERS, C. Y., H. G. FRIESEN, P. HWANG, H. J. GUYDA, AND K. FOLKERS. 1971. Prolactin and thyrotrophin release in man by synthetic pyroglutamyl-histidyl-proline amide. Biochem. Biophys. Res. Commun., 45:1033-1041.

BOYD, A. E., E. SPENCER, I. M. D. JACKSON, AND S. REICHLIN. 1976. Prolactin releasing factor (PRF) in porcine hypothalamic extract distinct from TRH. Endocrinology, 99:861-871.

BROWNSTEIN, M. J., J. M. SAAVEDRA, M. PALKOVITS, AND J AXELROD. 1974. Histamine content of hypothalamic nuclei of the rat. Brain Res., 77:151-156.

BROWNSTEIN, M. J., M. PALKOVITS, J. M. SAAVEDRA, AND J. S. KIZER. 1976. Distribution of hypothalamic hormones and neurotransmitters within the diencephalon. Pp.1-41, *in* Frontiers in neuroendocrinology (L. Martini and W. F. Ganong, eds.), Raven Press, New York, x + 399 pp.

CALAS, A. 1977. Radioautographic studies of aminergic neurones terminating in the median eminence. Pp. 79-88, *in* Advances in biochemical psychopharmacology (E. Costa and G. L. Gessa, eds.), Raven Press, New York, xx + 708 pp.

CARLSON, H. E., AND A. F. IPPOLITI. 1977. Cimetidine, an H_2-antihistamine, stimulates prolactin secretion in man. J. Clin. Endocrinol. Metab., 45:367-370.

CARR, L. A., P. M. CONWAY, AND J. L. VOOGT. 1977. Role of norepinephrine in the release of prolactin induced by suckling and estrogen. Brain Res., 133:305-314.

CHADWICK, A. 1980. Comparative aspects of the action of the hormone prolactin with particular reference to the domestic fowl. Gen. Comp. Endocrinol., 40:317-318.

CHADWICK, A., AND T. R. HALL. 1975. Thyrotrophin-releasing factor (TRF) and prolactin release in the pigeon *in vivo*. J. Endocrinol., 64:26-27.

CHADWICK, A., C. G. SCANES, S. HARVEY, G. BORDER, AND N. J. BOLTON. 1977. Hypothalamic control of prolactin secretion in the domestic fowl. Proc. First Internat. Symp. Prolactine, Nice, France, Abstract 41.

CHADWICK, A., N. J. BOLTON, AND T. R. HALL. 1978a. The effect of hypothalamic extract on prolactin secretion by chicken pituitary glands *in vivo* and *in vitro*. Gen. Comp. Endocrinol., 34:70.

———. 1978b. The effect of hypothalamic extract and K on prolactin secretion by the pituitary gland of the domestic fowl. IRCS Med. Sci., 6:238.

CHEN, C. L., E. J. BIXLER, A. I. WEBER, AND J. MEITES. 1968. Hypothalamic stimulation of prolactin release from the pituitary of turkey hens and poults. Gen. Comp. Endocrinol., 11:489-494.

CLEMENS, J. A., AND J. MEITES. 1977. Control of prolactin secretion. Pp. 139-174, *in* Hormonal proteins and peptides (C. H. Li, ed.), Academic Press, New York, xi + 270 pp.

DAVIES, D. T., AND R. J. BICKNELL. 1976. The effect of testosterone on the responsiveness of the quail's pituitary to luteinizing hormone-releasing hormone (LH-RH) during photoperiodically induced testicular growth. Gen. Comp. Endocrinol., 30:487-499.

DAVIES, D. T., AND B. K. FOLLETT. 1980. Neuroendocrine regulation of gonadotrophin-releasing hormone secretion in the Japanese quail. Gen. Comp. Endocrinol., 40:220-225.

DAVIES, D. T., L. P. GOULDEN, B. K. FOLLETT, AND N. L. BROWN. 1976. Testosterone feedback on luteinizing hormone (LH) secretion during a photoperiodically induced breeding cycle in Japanese quail. Gen. Comp. Endocrinol., 30:477-486.

DAVIS, S. L., M. S. ANFINSON, J. KLINDT, AND D. L. OHLSON. 1977. Influence of prostaglandins and thyrotropin releasing hormone (TRH) on hormone secretion and growth in wether lambs. Prostaglandins, 13:1209-1219.

DONOSO, A. O., AND A. M. BANNZA. 1976. Acute effects of histamine on plasma prolactin and luteinizing hormone levels in male rats. J. Neur. Transm., 39:95-101.

DONOSO, A. O., W. BISHOP, C. P. FAWCETT, L. KRULICH, AND S. M. McCANN. 1971. Effects of drugs that modify brain monoamines concentrations on plasma gonadotropin and prolactin levels in the rat. Endocrinology, 89:774-784.

DULAR, R., F. LABELLA, S. VIVIAN, AND L. EDDIE. 1974. Purification of prolactin-releasing and inhibiting factors from beef. Endocrinology, 94:563-567.

ENJALBERT, A., M. PRIAM, AND C. KORDON. 1977. Evidence in favour of the existence of a dopamine-free prolactin-inhibiting factor (PIF) in rat hypothalamic extracts. Eur. J. Pharm., 41:243-244.

ENSOR, D.M. 1978. Comparative endocrinology of prolactin. Chapman and Hall, London, vi + 309 pp.

FENSKE, M., AND W. WUTTKE. 1976. Effects of intraventricular 6-hydroxidopamine injections on serum prolactin and LH levels: absence of stress induced pituitary prolactin release. Brain Res., 104:63-70.

FOLLETT, B. K. 1970. Gonadotrophin-releasing activity in the quail hypothalamus. Gen. Comp. Endocrinol., 15:165-179.

GALA, R. R., AND R. P. REECE. 1965. Influence of hypothalamic fragments and extracts on lactogen production *in vitro*. Proc. Soc. Exp. Biol. Med., 117:833-836.

GANONG, W. F. 1980. Prolactin: a general overview. Pp. 1-10, *in* Central and peripheral regulation of prolactin function (R. M. MacLeod and U. Scapagnini, eds.), Raven Press, New York, xxiv + 394 pp.

GOURDJI, D. 1970. Prolactine et relations photosexuelles chez les oiseaux. Pp. 233-258, *in* La photoregulation de la reproduction chez les oiseaux et les mammifres (J. Benoit and I. Assenmacher, eds.), C.N.R.S., Paris, 588 pp.

GOURDJI, D., AND A. TIXIER-VIDAL. 1966. Mise en évidence d'un contrôle hypothalamique stimulant de la prolactine hypophysaire chez le canard. C. R. Acad. Sci. Ser. D (Paris), 263:162-165.

GREIBROKK, T., B. L. CURRIE, K. JOHANSSON, J. J. HANSEN, K. FOLKERS, AND C. BOWERS. 1974. Purification of a prolactin inhibiting hormone and the revealing of a hormone D-GHIH which inhibits the release of growth hormone. Biochem. Biophys. Res. Commun., 59:704-709.

HALL, T. R. 1975. A comparative study of the control of secretion of prolactin, growth hormone and other pituitary gland hormones in fishes, amphibians, reptiles and birds. Ph.D. thesis, University of Leeds, iv + 467 pp.

———. 1980. Neurotransmitter regulation of release of prolactin and growth hormone in the pigeon. Am. Zool., 161:602.

HALL, T. R., AND A. CHADWICK. 1974. The effect on prolactin secretion of thyrotrophin-releasing factor (TRF). J. Endocrinol., 63:45.

——. 1975. Effect of methallibure on prolactin release from pigeon pituitaries incubated *in vitro*. J. Endocrinol., 64:72P.

——. 1976. Effects of growth hormone inhibiting factor (somatostatin) on the release of growth hormone and prolactin from pituitaries of the domestic fowl *in vitro*. J. Endocrinol., 68:163-164.

——. 1979a. Effect of methallibure on the pigeon hypothalamus. Proc. Leeds Phil. Lit. Soc., 10:243-260.

——. 1979b. Hypothalamic control of prolactin and growth hormone secretion in different vertebrate species. Gen. Comp. Endocrinol., 37:333-342.

——. 1981. Hypothalamic control of prolactin and growth hormone secretion in pituitary gland of the pigeon and the chicken: *in vitro* studies. Gen Comp. Endocrinol., in press.

HALL, T. R., A. CHADWICK, N. J. BOLTON, AND C. G. SCANES. 1975. Prolactin release *in vitro* and *in vivo* in the pigeon and the domestic fowl following administration of synthetic thyrotrophin-releasing factor (TRF). Gen. Comp. Endocrinol., 25:298-306.

HALL, T. R., A. CHADWICK, AND N. J. BOLTON. 1976. Assay of prolactin releasing and inhibiting activities in the hypothalami of vertebrates. Gen. Comp. Endocrinol., 28:242-243.

HARMS, P. G., S. R. OJEDA, AND S. M. MCCANN. 1973. Prostaglandin involvement in hypothalamic control of gonadotropin and prolactin release. Science, 181:760-761.

HARRIS, G. W. 1955. Neural control of pituitary gland. Arnold Press, London, ix + 298 pp.

HARVEY, S., AND C. G. SCANES. 1981. Effects of prostaglandins E₁ and E₂ and F₂α on circulating concentrations of growth hormone and prolactin in the domestic fowl (*Gallus domesticus*). J. Endocrinol., in press.

HARVEY, S., C. G. SCANES, N. J. BOLTON, AND A. CHADWICK. 1978a. Effect of thyrotropin-releasing hormone (TRH) and somatostatin (GHRIH) on growth hormone and prolactin secretion *in vitro* and *in vivo* in the domestic fowl (*Gallus domesticus*). Neuroendocrinol., 26:249-260.

HARVEY, S., C. G. SCANES, A. CHADWICK, AND N. J. BOLTON. 1978b. Effect of reserpine on plasma concentrations of growth hormone and prolactin in the domestic fowl. J. Endocrinol., 79:153-154.

HARVEY, S., C. G. SCANES, A. CHADWICK, G. BORDER, AND N. J. BOLTON. 1979a. Effect of chicken hypothalamus on prolactin and growth hormone secretion in male chickens. J. Endocrinol., 82:193-197.

HARVEY, S., C. G. SCANES, A. CHADWICK, AND N. J. BOLTON. 1979b. *In vitro* stimulation of chicken pituitary growth hormone and prolactin secretion by chicken hypothalamic extract. Experientia, 35:694-695.

HÖFELT, T., O. JOHANSSON, A. LJUNGDAHL, J. M. LUNDBERG, AND M. SCHULTZBERG. 1980. Peptidergic neurones. Nature (London), 284:515-521.

HORROBIN, D. F. 1975. Prolactin 1975. Eden Press, Montreal, v + 198 pp.

——. 1976. Prolactin 1976. Eden Press, Montreal, vi + 208 pp.

JACKSON, I. M. D., AND S. REICHLIN. 1974. Thyrotropin-releasing hormone (TRH): distribution in hypothalamic and extra-hypothalamic brain tissues of mammalian and submammalian chordates. Endocrinology, 95:854-862.

KAMBERI, I. A., R. S. MICAL, AND J. C. PORTER. 1971. Effect of melatonin and serotonin on the release of FSH and prolactin. Endocrinology, 88:1288-1293.

KANEMATSU, S., J. HILLIARD, AND C. H. SAWYER. 1963. Effect of reserpine on pituitary prolactin content and its hypothalamic site of action in the rabbit. Acta Endocrinol., 44:467-474.

KNIGHT, P. J., AND A. CHADWICK. 1975. The effect of *in vivo* administration of chicken hypothalamic extract on the crop sac and the pituitary gland of the pigeon. Gen. Comp. Endocrinol., 27:488-494.

KOCH, Y., K. H. LU, AND J. MEITES. 1970. Biphasic effects of catecholamines on pituitary prolactin release *in vitro*. Endocrinology, 87:673-675.

KORDON, C., AND A. ENJALBERT. 1980. Prolactin inhibiting and stimulating factors. Pp. 67-77, *in* Central and peripheral regulation of prolactin function (R. M. MacLeod and U. Scapagnini, eds.), Raven Press, New York, xxiv + 394 pp.

KRAGT, C. L. 1966. Stimulation of pigeon pituitary prolactin release by pigeon hypothalamic extract *in vitro*. Endocrinology, 76:1169-1176.

KRAGT, G. L., AND J. MEITES. 1965. Stimulation of pigeon pituitary prolactin release by pigeon hypothalamic extract *in vitro*. Endocrinology, 76:1169-1176.

LABRIE, F., G. PELLETIER, P. BORGEAT, J. DROUIN, L. FERLAND, AND A. BELANGER. 1976. Mode of action of hypothalamic regulatory hormones in the adenohypophysis. Pp. 63-93, *in* Frontiers in neuroendocrinology (L. Martini and W. F. Ganong, eds.), Raven Press, New York, ix + 399 pp.

LANGLELIER, P., AND S. M. McCANN. 1977. The effects of interruption of the ventral noradrenergic pathway on the proestrous discharge of prolactin in the rat (39717). Proc. Soc. Exp. Biol. Med., 154:553-557.

LIBERTUN, C., AND S. M. McCANN. 1976. The possible role of histamine in the control of prolactin and gonadotropin release. Neuroendocrinol., 20:110-120.

MA, R. C. S., AND A. V. NALBANDOV. 1963. Hormonal activity of the autotransplanted adenohypophysis. Pp. 306-312, in Advances in neuroendocrinology (A. V. Nalbandov, ed.), Univ. Illinois Press, Urbana, xii + 525 pp.

MACHLIN, L. J., L. S. JACOBS, M. CIVULIS, R. KINEES, AND R. MILLER. 1974. An assay for growth hormone and prolactin-releasing activities using a bovine cell culture system. Endocrinology, 95:1350-1358.

MACLEOD, R. M. 1969. Influence of norepinephrine and catecholamine-depleting agents on the synthesis and release of prolactin and growth hormone. Endocrinology, 85:916-923.

———. 1976. Regulation of prolactin secretion. Pp. 169-194, *in* Frontiers in neuroendocrinology (L. Martini and W. F. Ganong, eds.), Raven Press, New York, ix + 399 pp.

MACLEOD, R. M., AND J. E. LEHMEYER. 1974. Studies on the mechanisms of the dopamine-mediated inhibition of prolactin secretion. Endocrinology, 94:1077-1085.

MEITES, J. 1972. Hypothalamic control of prolactin secretion. Pp. 325-388, *in* Lactogenic hormones (G. E. W. Wolstenholme and J. Knight, eds.), Churchill Livingstone, London, xi + 416 pp.

MEITES, J., AND C. S. NICOLL. 1966. Adenohypophysis: prolactin. Ann. Rev. Physiol., 28:57-88.

MEITES, J., AND J. A. CLEMENS. 1972. Hypothalamic control of prolactin secretion. Vitamins Horm. (New York), 30:165-221.

MIODUSZEWSKI, I. R., L. GRANDISON, AND J. MEITES. 1976. Stimulation of prolactin release in rats by GABA. Proc. Soc. Exp. Biol. Med., 151:44-46.

MÜLLER, E. E., V. LOCATELLI, D. COCCHI, S. SPAMPINATO, J. APUD, S. FERRI, AND G. RACAGNI. 1980. Neural factors involved in the central regulation of prolactin. Pp. 79-96, *in* Central and peripheral regulation of prolactin function (R. M. MacLeod and U. Scapagnini, eds.), Raven Press, New York, xxiv + 394 pp.

NICOLL, C. S. 1965. Neural regulation of adenohypophysial prolactin secretion in tetrapods: indications from *in vitro* studies. J. Exp. Zool., 158:203-210.

NICOLL, C. S., AND R. P. FIORINDO. 1969. Hypothalamic control of prolactin secretion. Gen. Comp. Endocrinol., Suppl., 2:26-31.

NICOLL, C. S., R. P. FIORDINO, C. T. McKENEE, AND J. A. PARSONS. 1970. Assay of hypothalamic factors which regulate prolactin secretion. Pp. 115-144, *in* Hypophysiotropic hormones of the hypothalamus: assay and chemistry (J. Meites, ed.), Williams and Wilkins, Baltimore, ix + 338 pp.

NISTICO, G., G. GERMANA, E. CIRIACO, B. BRONZETTI, D. ROTIROTI, AND U. SCAPAGNINI. 1979. Crop sac response after systemic and intraventricular administration of neuroleptic drugs. Neuroendocrinol., 29:418-425.

NISTICO, G., G. GERMANA, E. CIRIACO, D. ROTIROTI, F. FARAONE, AND U. SCAPAGNINI. 1980. Ultrastructural changes of pituitary lactotrophs and crop-sac after intraventricular injections of drugs enhancing serotoninergic transmission in pigeons. Pp. 339-346, *in* Central and peripheral regulation of prolactin function (R. M. MacLeod and U. Scapagnini, eds.), Raven Press, New York, xxiv + 394 pp.

OJEDA, S. R., P. G. HARMS, AND S. M. McCANN. 1974. Possible role of cyclic AMP and prostaglandin E_1 in the dopaminergic control of prolactin release. Endocrinology, 95:1694-1703.

PASS, K. A., AND J. G. ONDO. 1977. The effects of α-aminobutyric acid on prolactin and gonadotropin secretion in the unanesthetized rat. Endocrinology, 100:1437-1442.

RIVIER, C., AND W. VALE. 1974. *In vitro* stimulation of prolactin secretion in the rat by thyrotropin releasing factor, related peptides and hypothalamic extracts. Endocrinology, 95:978-983.

———. 1977. Effect of γ-aminobutyric acid on histamine on prolactin secretion in the rat. Endocrinology, 101:506-511.

SATO, T., T. JYUJO, T. IESAKA, J. ISHIKAWA, AND M. IGARASHI. 1974. Follicle stimulating hormone and prolactin release induced by prostaglandins in rat. Prostaglandins, 5:483-490.

SCANES, C. G., A. CHADWICK, AND N. J. BOLTON. 1976. Radioimmunoassay of prolactin in the plasma of the domestic fowl. Gen. Comp. Endocrinol., 30:12-21.

SCANES, C. G., S. HARVEY, A. CHADWICK, AND N. J. BOLTON. 1977. Studies on the hypothalamic control of growth hormone and prolactin secretion in the chicken. J. Endocrinol., 73:10P.

SCANES, C. G., S. HARVEY, N. J. BOLTON, AND A. CHADWICK. 1978. Effect of prostaglandin E_2 on prolactin and growth hormone secretion in the domestic fowl. IRCS Med. Sci., 6:58.

SCHALLY, A. V. AND A. J. KASTIN. 1972. Hypothalamic releasing and inhibiting hormones. Gen. Comp. Endocrinol., Suppl., 3:76-85.

SCHALLY, A. V., A. ARIMURA, AND A. J. KASTIN. 1974a. Hypothalamic regulatory hormones. Science, 179:341-350.

SCHALLY, A. V., A. ARIMURA, J. TAKAHARA, T. REDDING, AND A. DUPONT. 1974b. Inhibition of prolactin release *in vitro* and *in vivo* by catecholamines. Fed. Proc., 33:237.

SCHALLY, A. V., A. DUPONT, A. ARIMURA, J. TAKAHARA, T. REDDING, J. CLEMENS, AND C. J. SHAAR. 1976. Purification of a catecholamine-rich fraction with prolactin release-inhibiting factor (PIF) activity from porcine hypothalami. Acta Endocrinol., 82:1-14.

SCHALLY, A. V., T. W. REDDING, A. ARIMURA, A. DUPONT, AND G. L. LINTHICUM. 1977. Isolation of gamma-aminobutyric acid from pig hypothalami and demonstration of its prolactin release-inhibiting (PIF) activity *in vivo* and *in vitro*. Endocrinology, 100:681-691.

SHAAR, C. J., AND J. CLEMENS. 1974. The role of catecholamines in the release of anterior pituitary prolactin *in vitro*. Endocrinology, 95:1202-1212.

SMYTHE, G. A., AND L. LAZARUS. 1973. Growth hormone regulation by melatonin and serotonin. Nature, 244:230-231.

SUBRAMANIAN, M. G., AND R. R. GALA. 1976. Further studies on the effects of adrenergic, serotoninergic and cholinergic drugs on the afternoon surge of plasma prolactin in ovariectomized, estrogen-treated rats. Neuroendocrinol., 22:240-249.

SUNBERG, D. F., C. P. FAWCETT, P. ILLNER, AND S. M. McCANN. 1975. The effect of various prostaglandins and a prostaglandin synthetase inhibitor on rat anterior pituitary cyclic AMP levels and hormone release *in vitro*. Proc. Soc. Exp. Biol. Med., 148:54-59.

SZABO, M., AND L. A. FROHMAN. 1976. Dissociation of prolactin-releasing activity from thyrotropin releasing hormone in procine stalk median eminence. Endocrinology, 98:1451-1459.

TIXIER-VIDAL, A. 1967. Modifications cytologiques ultrastructuals de la prehypophyse après divers procédés de deconnexion hypothalamo-hypophysaires chez les oiseaux. Biol. Med., 56:318-331.

TIXIER-VIDAL., AND D. GOURDJI. 1965. Evolution cytologique de l'hypophyse du canard en culture organotypique. Elaboration autonome de prolactine par les explants. C. R. Acad. Sci. Ser. D (Paris), 261:805-808.

———. 1972. Cellular aspects of the control of prolactin secretion in birds. Gen. Comp. Endocrinol., Supp., 3:51-64.

TIXIER-VIDAL, A., AND R. PICART. 1967. Etude quantitative par radioautographie au microscope electronique de l'utilisation de la DL-leucine-³H par les cellules de l'hypophyse du canard en culture organotypique. J. Cell. Biol., 35:501-509.

TIXIER-VIDAL, A., B. K. FOLLETT, AND D. S. FARNER. 1968. The anterior pituitary of the Japanese quail, *Coturnix coturnix japonica*. The cytological effects of photoperiodic stimulation. Z. Zellforsch. Mikrosk. Anat., 92:610-635.

VALVERDE, R. C., V. CHIEFFO, AND S. REICHLIN. 1972. Prolactin releasing factor in porcine and rat hypothalamic tissue. Endocrinology, 91:982-993.

VAN MAANEN, J. H., AND P. G. SMELIK. 1968. Induction of pseudopregnancy in rats following local depletion of monoamines in the median eminence of the hypothalamus. Neuroendocrinol., 3:177-186.

WEINER, R. I. 1973. Hypothalamic monoamine levels and gonadotrophin secretion following deafferentation of the medial basal hypothalamus. Prog. Brain Res., 39:165-170.

WEINER, R. I., AND W. F. GANONG. 1978. Role of brain monoamines and histamine in regulation of anterior pituitary secretion. Physiol. Rev., 58:905-976.

ZIMMERMAN, E. A. 1976. Localization of hypothalamic hormones by immunocytochemical techniques. Pp. 25-62, *in* Frontiers in neuroendocrinology (L. Martini and W. F. Ganong, eds.), Raven Press, New York, ix + 399 pp.

Reprinted from
Aspects of Avian Endocrinology:
Practical and Theoretical Implications (C. G. Scanes *et al.*, eds.)
Grad. Studies, Texas Tech Univ., 1982, 26:1-411.

THE ANTIADRENERGIC ACTIVITY OF METHALLIBURE

J. E. Taylor, R. A. Long, R. L. Tolman, R. M. Weppelman, and G. Olson

Merck, Sharp, and Dohme Research Laboratories, Rahway, New Jersey 07065 USA

Methallibure, whose structure is shown at the top of Fig. 1, is a nonsteroidal compound which has antifertility activity in a variety of species (Walpole, 1965). In domestic chickens it stops ovulation within four days (Ishigaki *et al.*, 1972) and causes a marked ovarian regression within 12 days (Imai, 1972). Since its antigonadal effect can be reversed by injections of partially purified avian gonadotropins, it appears to act at the hypothalamic or hypophyseal level rather than at the ovarian level (Imai, 1972).

In rats, methallibure reduces circulating luteinizing hormone (LH) levels (Labhsetwar and Walpole, 1972) but does not affect the LH surge induced by luteinizing hormone releasing factor (Mueh *et al.*, 1975). This suggests that its action is within the hypothalamus rather than within the pituitary. The observation that methallibure is more effective when implanted into the hypothalamus than when implanted into the pituitary is consistent with this proposal (Malven, 1971). To date, however, the biochemical basis of methallibure's antifertility action has not been determined.

It is noteworthy that methallibure is a thiourea derivative. Several thiourea derivatives are effective inhibitors of dopamine hydroxylase, the enzyme which converts dopamine to norepinephrine (Johnson *et al.*, 1969). Because central adrenergic activity appears to be required for fertility in both avian (Ferrando and Nalbandov, 1969; Kao and Nalbandov, 1972; Sharp, 1975; El Halawani and Burke, 1975) and mammalian species (Drouva and Gallo, 1976; Kalra *et al.*, 1978), it is tempting to speculate that the antifertility action of methallibure results from its inhibition of dopamine hydroxylase and its consequent lowering of central norepinephrine levels. This proposal is especially attractive in view of a recent publication from this laboratory which has shown that several presumed dopamine hydroxylase inhibitors have marked antifertility activity in laying hens (Weppelman *et al.*, 1980). To test this proposal we have determined the effects of methallibure on brain norepinephrine levels *in vivo* and its effects on purified avian dopamine hydroxylase *in vitro*.

When methallibure was fed to laying hens for one week at levels between 60 ppm and 750 ppm in the diet, both ovarian weight and brain norepinephrine levels were reduced (Table 1). The decrease in ovarian weight was due to the nearly complete loss of mature yellow follicles as noted previously (Imai, 1972). Egg production was dramatically reduced and no hens treated with 125 or more ppm laid an egg after the fourth day of treatment. In contrast to norepinephrine, brain dopamine was not consistently affected by

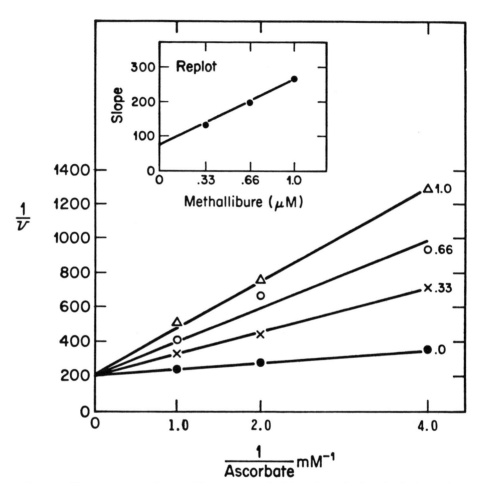

Fig. 1.—Above, structure of methallibure; below, double reciprocal plot of velocity against ascorbate concentration at different methallibure concentrations. The reaction mixtures contained 400 units of catalase, 1.25 μmoles fumarate, 2.5 μg pargyline, 5 nM dopamine, 0.1 μCi dopamine tritiated in the benzylic position (5-10 Ci per mm), and ascorbate and methallibure at the indicated concentrations. Enzyme activity was determined by measuring release of tritiated water (R. A. Long, *manuscript in preparation*).

methallibure. This suggests that the conversion of dopamine to norepinephrine, which is catalyzed by dopamine hydroxylase, is blocked. Thus, these data indicate that methallibure functions as a centrally active dopamine hydroxylase inhibitor *in vivo*.

TABLE 1.—*Effect of methallibure on ovarian weight and egg production and on brain norepinephrine and dopamine. Hens received diet containing methallibure at the levels indicated for one week. They were then sacrificed, their ovaries weighed, and norepinephrine and dopamine were extracted from the brains by the procedure of Anton and Sayre (1962) and assayed fluorometrically by the procedure of Laverty and Taylor (1968). All other procedures were as described previously (Weppelman et al., 1980).*

Methallibure in diet (ppm)	No. hens	Ovarian weight ± sd (g)	Brain norepinephrine ± sd (μg/g)	Brain dopamine ± sd (μg/g)	% egg production on test day						
					1	2	3	4	5	6	7
0	10	52.5±14.8	0.953±0.115	0.557±0.103	60	100	70	100	90	80	80
60	4	26.5±14.0*	0.920±0.086	0.580±0.050	100	50	0	0	25	25	0
125	4	34.3± 8.3*	0.796±0.245	0.296±0.130*	100	75	25	0	0	0	0
250	10	26.3±16.8*	0.756±0.173*	0.601±0.107	60	0	20	30	0	0	0
500	10	18.0± 7.6*	0.616±0.115*	0.653±0.165	70	60	30	0	0	0	0
750	4	32.5±18.9	0.559±0.194*	0.535±0.049	50	75	25	0	0	0	0

*Significantly different from controls by *t*-test ($P < 0.05$).

To investigate the effects of methallibure on avian hydroxylase *in vitro*, the enzyme was purified from avian adrenals and characterized (R. A. Long, manuscript in preparation). Like mammalian enzyme (Goldstein *et al.*, 1968; Craine *et al.*, 1973), the avian enzyme is a tetramer of about 300,000 daltons, utilizes ascorbate as reductant, and is activated by fumarate. Fig. 1 shows Lineweaver-Burke plots of reciprocal velocity versus reciprocal ascorbate at different concentrations of methallibure. The lines converge which indicates that methallibure is competitive with ascorbate. Since ascorbate is the first substrate to bind to the free enzyme, this implies that methallibure also binds to free enzyme. The insert in Fig. 2, which is a replot of slopes versus methallibure concentrations, indicates that the k_i of methallibure is about 0.33 μM. For comparison, the k_i of the most inhibitory thiourea derivative previously described is 1.5 μM (Johnson *et al.*, 1969).

Taken together, Table 1 and Fig. 2 indicate that methallibure is an effective inhibitor of avian dopamine hydroxylase *in vivo* and *in vitro*. This activity might account for its antifertility action, especially in view of prior reports that other dopamine hydroxylase inhibitors can have rather potent antifertility activity (El Halawani and Burke, 1975; Weppelman *et al.*, 1980).

SUMMARY

Methallibure is an effective inhibitor of avian dopamine hydroxylase and decreases the level of norepinephrine in the brains of laying hens. The antiadrenergic action of methallibure could account for its well-known antifertility activity.

LITERATURE CITED

ANTON, A. H., AND D. F. SAYRE. 1962. A study of the factors affecting the aluminum oxide-trihydroxyindole procedure for the analysis of catecholamines. J. Pharma. Expt. Therap., 138:360-375.

CRAINE, J. E., G. H. DANIELS, AND S. KAUFMAN. 1973. Dopamine-β-hydroxylase: the subunit structure and anion activation of the bovine adrenal enzyme. J. Biol. Chem., 248:7838-7844.

DROUVA, S. V., AND R. V. GALLO. 1976. Catecholamine involvement in episodic luteinizing hormone release in adult ovariectomized rats. Endocrinology, 99:651-658.

EL HALWANI, M. E., AND W. H. BURKE. 1975. Role of catecholamines in photoperiodically-induced gonadal development in *Coturnix* quail. Biol. Reprod., 13:603-609.

FERRANDO, G., AND A. NALBANDOV. 1969. Direct effect on the ovary of the adrenergic blocking drug dibenzyline. Endocrinology, 85:38-42.

GOLDSTEIN, M., T. H. JOH, AND T. G. GARVEY. 1968. Kinetic studies of the enzymatic dopamine β-hydroxylation reaction. Biochemistry, 7:2724-2730.

ISHIGAKI, R., Y. OHORI, S. EBISAWA, K. KINBARA, AND Y. YAMADA. 1972. Forced molting by methallibure (II). Japanese Poult. Sci., 9:79-83.

IMAI, K. 1972. Effects of avian and mammalian pituitary preparations of follicular growth in hens treated with methallibure or fasting. J. Reprod. Fert., 31:387-397.

JOHNSON, G. A., S. J. BOUKMA, AND E. G. KIM. 1969. Inhibition of dopamine β-hydroxylase by aromatic and akyl thioureas. J. Pharm. Exp. Therap., 168:229-234.

KALRA, S. P., P. S. KALRA, C. L. CHEN, AND J. A. CLEMENS. 1978. Effect of norepinephrine synthesis inhibitors and a dopamine agonist on hypothalamic LH-RH, serum gonadotropin and prolactin levels in gonadal steroid treated rats. Acta Endocrinol., 89:1-9.

KAO, L. W. L., AND A. NALBANDOV. 1972. The effect of antiadrenergic drugs on ovulation in hens. Endocrinology, 90:1343-1349.

LABHSETWAR, A. P., AND A. L. WALPOLE. 1972. Effects of methallibure on pituitary and serum levels of immunoreactive LH in spayed rats. J. Reprod. Fert., 31:147-149.

LAVERTY, R., AND K. M. TAYLOR. 1968. The fluorometric assay of catecholamines and related compounds: improvements and extensions of the hydroxyindole technique. Anal. Biochem., 22:269-279.

MALVERN, P. V. 1971. Hypothalamic sites of action for methallibure (ICI 33838): inhibition of gonadotropin secretion. J. Anim. Sci., 32:912-918.

MUEH, J., J. M. SHANE, AND F. NAFTOLIN. 1975. The effects of methallibure (ICI 33828) on luteinizing hormone release in castrate male rats challenged with luteinizing hormone releasing hormone. Endocrinology, 97:493-495.

SHARP, P. J. 1975. The effect of reserpine on plasma LH concentrations in intact and gonadectomised domestic fowl. Brit. Poult. Sci., 16:79-82.

WALPOLE, A. L. 1965. Non-steroidal agents inhibiting pituitary gonadotropic function. Pp. 159-179, *in* Agents affecting fertility (C. R. Austin and J. S. Perry, eds.), Churchill, London, viii + 319 pp.

WEPPELMAN, R. M., R. A. LONG, A. VAN IDERSTINE, J. E. TAYLOR, R. L. TOLMAN, L. PETERSON, AND G. OLSON. 1980. Antifertility effects on dithiocarbamates in laying hens. Biol. Reprod., 23:40-45.

Reprinted from
ASPECTS OF AVIAN ENDOCRINOLOGY:
PRACTICAL AND THEORETICAL IMPLICATIONS (C. G. Scanes *et al.*, eds.)
Grad. Studies, Texas Tech Univ., 1982, 26:1-411.

THE AVIAN HYPERCALCEMIC ASSAY FOR PARATHYROID HORMONE AND SELECTED APPLICATIONS

ALEXANDER D. KENNY

Department of Pharmacology and Therapeutics, Texas Tech University Health Sciences Center, Lubbock, Texas 79430 USA

The development in 1973 of simple, economical, and reasonably sensitive and precise assays of parathyroid hormone (PTH) in birds represented a significant improvement in the available *in vivo* assays. This advance was reported independently from two laboratories (Dacke and Kenney, 1973; Parsons *et al.*, 1973). Prior to that time the only *in vivo* assays available used mammalian species and were less sensitive (Table 1). In the following discussion, the development, description, pharmacological basis, and selected applications of one of these avian bioassays for PTH, that of Dacke and Kenny (1973), will be described.

Development of the Avian Hypercalcemic Assay for PTH

The United States Pharmacopeia—The National Formulary (1980) bioassay method for PTH, first described by Collip (1925-26), involves the use of dogs. These animals are too insensitive for most research purposes, requiring at least 100 U.S.P. units per dog. The inconvenience and expense of maintaining a dog colony is another limiting factor. Prior to the development of the avian bioassays the most popular *in vivo* bioassay for experimental purposes was that of Munson and his associates (Munson, 1955, 1960, 1961; Munson *et al.*, 1953) which used the parathyroidectomized rat. The development of this assay played a major role during the early 1960's in the final purification of PTH, which had remained essentially unpurified for 40 years since the first crude extracts were prepared in the early 1920's.

The avian hypercalcemic assay represents a significant advance over the parathyroidectomized rat assay with respect to sensitivity, simplicity, and economy. Polin *et al.* (1957) were the first to emphasize the rapidity (3-4 hours) of the hypercalcemic response to PTH in 5- to 7-week-old chickens and to suggest this response as a basis for a bioassay. However, their assay lacked sufficient sensitivity and precision to become widely used. In 1971, Dacke and Kenny reported that both 2- to 3-week-old Japanese quail and chickens responded to PTH with a hypercalcemic response which was both rapid (30-60 min) and sensitive (2 U.S.P. units/bird). This led to the development of a successful assay method using either 2- to 3-week-old Japanese quail or 5- to 6-day-old chickens (Dacke and Kenny, 1972, 1973). It was found that incorporation of $CaCl_2$ (51 mM) into all injection media was an important element in the development of the assay (Fig. 1). The Japanese

TABLE 1.—*Summary of representative assays for parathyroid hormone. The index of precision was calculated as the standard deviation/slope.*

Type of assay	Test system	Route of administration	Dose range U.S.P. units	Index of precision (λ)	Reference
		In Vivo Assays			
Hypercalcemia	Intact dog	sc	100-300	0.20	Collip, 1925-1926
	PTX rat	sc	10-100	0.23	Munson, 1961
	Japanese quail	iv	0.5-4.0	0.20	Dacke and Kenny, 1973
	Chicken	iv	1-10	0.14	Parsons *et al.*, 1973
Phosphaturia	Parathyroidectomized rat	sc	5-50	0.30	Kenny and Munson, 1959
		In Vitro Assays			
^{40}Ca release from bone	Mouse calvarium		0.01-1.0	0.15	Zanelli *et al.*, 1969
Formation of 3', 5'-cyclic AMP	Rat renal adenylate cyclase		0.2-1.0	0.08	Marcus and Aurbach, 1969
Radioimmunoassay	^{131}I-PTH/antibody		0.001-0.006		Berson *et al.*, 1963

quail assay is rapid (60 min), sensitive (0.4 U.S.P. units/bird), and reasonably precise (λ = 0.20).[1] The standard concentrations of 1, 3, and 9 U.S.P. units/ml, given at a dose of 0.4 ml/bird, yield a logarithmic dose-arithmetic response relationship which gives a significant regression with no significant departure from linearity (Fig. 2). Details of the method as run currently in Kenny's laboratory are presented in Table 2. Cumulative experience with the quail assay is summarized in Table 3.

Pharmacological Basis for the Hypercalcemic Response to PTH in Avian Species

In mammalian species the pharmacological basis for the hypercalcemic response to PTH is considered to be due mainly to bone resorption with minor contributions coming from renal tubular reabsorption and gut calcium transport (Kenny and Dacke, 1975). Parsons *et al.* (1973) claimed that bone resorption is also the major factor underlying the hypercalcemic response in avian species.

The hypercalcemic response to intravenously administered PTH in birds is rapid in onset and short in duration. Little is known about the source of the calcium contributing to the response. There are at least two major mechanisms that must be considered: 1) mobilization of calcium from bone, and 2) inhibition of the removal of calcium from extracellular fluid. It was anticipated that, by injecting ^{45}Ca simultaneously with PTH and following both the plasma total calcium and the plasma ^{45}Ca levels, some understanding of the mechanism would emerge. If bone resorption was the only source of the calcium, then only plasma total calcium would be expected to rise; bone would be essentially unlabelled with respect to ^{45}Ca. In contrast, if inhibition of removal of calcium from extracellular fluid was the only

[1]Lambda (λ), the index of precision, is obtained by dividing the standard deviation of the assay by the slope. The slope is the difference between the mean responses to the high and low doses divided by the logarithm of the dose interval. For two, three, and 4-fold intervals, the logarithms are 0.301, 0.477, and 0.602, respectively. The lower the value of λ, the more precise is the assay.

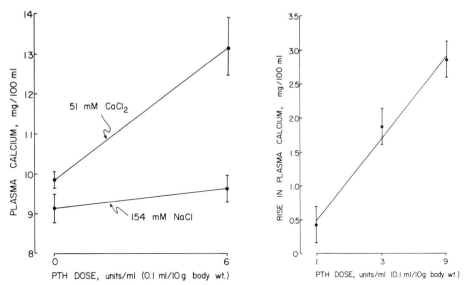

FIG. 1 (left).—Response of plasma calcium concentration to PTH 6 U.S.P. units/ml (3 U.S.P. units/bird) in fasted Japanese quail with or without injected calcium. The calcium-free injection medium consisted of 154 mM NaCl adjusted to pH 3.0. The high calcium medium consisted of 51 mM CaCl₂ (pH 3.0). Only the high calcium PTH solution gave a significant hypercalcemic response ($P < 0.002$) at 60 minutes after injection. There were 6 birds per dosage group. Data taken from Dacke and Kenny (1973).

FIG. 2 (right).—PTH dose-response slope in 2- to 3-week-old quail. The regression was highly significant ($P < 0.001$) and the slope showed no significant departure from linearity over a dose range of 1 to 9 U.S.P. units/ml. Values are expressed as the mean ± SE. Data taken from Dacke and Kenny (1973).

mechanism, then both plasma total calcium and [45]Ca levels would be expected to rise.

Unlike Parsons *et al.* (1973), Kenny and Dacke (1974) are less certain about the origin of the avian response. Using immature Japanese quail and chickens acutely labelled with [45]Ca, they followed the plasma [45]Ca and total calcium levels in the 2-hour period immediately following intravenous injection of synthetic bovine PTH (1-34). The [45]Ca was injected simultaneously with the bPTH (1-34). Some of their data are presented in Figs. 3 (Japanese quail) and 4 (chickens). In both species there was a rapid, marked rise in plasma [45]Ca levels, which peaked at 15 (Japanese quail) to 30 (chicken) minutes followed by a fall to control levels at 60 minutes. The plasma total calcium, on the other hand, rose more slowly and declined more slowly, remaining significantly elevated at 60 minutes.

The results indicated that the hypercalcemic response to PTH in birds appears to be complex, involving two or more underlying mechanisms. The initial phase (30 minutes or less) is characterized by a rise in both plasma total calcium and [45]Ca levels. This type of response may be interpreted as resulting from an inhibition of the exit of calcium from the extracellular

TABLE 2.—*Details of avian bioassay method for parathyroid hormone as modified from Dacke and Kenny (1973).*

1. Japanese quail, 2 to 3 weeks old (average weight 40 g) are fed Purina Turkey Startena *ad libitum* until the time of injection.
2. The birds, 5 or 6 per group, are injected intravenously in wing vein, 0.4 ml/bird, with either standard PTH (1, 3, and 9 U.S.P. units/ml) or unknown preparation dissolved in a solution containing 51 mM $CaCl_2$, 0.01% bovine serum albumin, and 0.01% cysteine HCl.
3. The birds are placed in individual cages without access to food or water.
4. The birds are bled by cardiac puncture under halothane anesthesia 60 minutes after injection.
5. The heparinized plasmas are analyzed for calcium by atomic absorption following precipitation of protein with 5% trichloroacetic acid.
6. The bioassay statistics are computed by means of a computer, using a BASIC program devised by Kenny (unpublished).
7. An 80-bird assay, allowing 1 control, 3 standard, and 12 unknown groups, can be completed in one working day.

compartment. The later phase (30 minutes or more) is characterized by elevated plasma total calcium levels only. This type of response may be interpreted as being largely due to bone resorption. Parson *et al.* (1973), using a different experimental design in which chickens were injected with radiocalcium (^{47}Ca) 60 minutes before the PTH, found no rise in plasma ^{47}Ca at 30 and 60 minutes, time intervals at which the plasma total calcium levels were elevated above those of control birds. This type of response was interpreted as being due to bone resorption.

If the above broad interpretation of the initial phase, namely inhibition of exit from extracellular fluid, is accepted, a discussion of the specific mechanism or mechanisms is appropriate. Major sites of exit of calcium from extracellular fluid may include bone, kidney, or gut, or any combination of these sites. Available information does not permit elimination of any of these sites from serious consideration. The findings of Clark and Wideman (1977) support a renal mechanism in starlings (*Sturnus vulgaris*). In this species parathyroidectomy results in a hypercalciuric response in the face of hypocalcemia; injection of PTH decreased urinary calcium clearance and raised the plasma calcium. These authors conclude that in the starling, "the kidney is the primary and immediate site of parathyroid hormone action." In fact, calculations led them to further conclude that the changes in plasma calcium concentration following manipulation of PTH status "can be accounted for by shifts in urinary excretion."

One further point deserves comment. Parsons and his associates have consistently observed an immediate hypocalcemic response to PTH in both dogs (Parson *et al.*, 1971) and rats (Parsons and Robinson, 1971). Such a response has never been observed in our laboratory in Japanese quail following PTH (Kenny and Dacke, 1974; Table 4). It has been observed in chickens under special conditions, that is, in laying hens on a very high (5%) calcium diet (Mueller *et al.*, 1973). No hypocalcemic response to PTH was seen in laying hens on a lower (2.26%) calcium diet.

TABLE 3.—*Cumulative experience with the Dacke and Kenny (1973) avian bioassay method for parathyroid hormone. All assays were done using details described in Table 2, except where noted otherwise. Standard: assays 1-20, Lilly Parathyroid Injection (U.S.P.); assays 21-24, Beckman synthetic bovine PTH (1-34). Data taken from Kenny and Dacke (1975).*

| Assay number code | Age[1] days | Body weight[2], g mean (range) | PTH dose[3], U.S.P. units/ml | | | | SD | Index of precision (SD/slope) |
| | | | 0 | 1 | 3 | 9 | | |
			Plasma Ca, mg/dl					
1. B90	23	44(40-50)	9.2	*9.5*	*12.0*	11.9	0.51	0.10
2. B108	20	41(35-49)	9.9	*10.3*	*11.8*	12.7	0.60	0.20
3. B112	21	38(33-46)	10.0	11.1	*12.2*	*13.5*	0.93	0.34
4. B117	22	43(36-50)	10.4	*10.9*	*12.5*	13.6	0.60	0.20
5. B120	23	43(37-52)	10.8	11.2	*11.9*	*14.0*	0.70	0.16
6. B184	18	49(46-55)	10.3	*10.5*	*12.2*	15.3	0.70	0.19
7. B188	23	62(50-70)	9.9	*10.5*	*12.2*		0.78	0.22
8. B192	17	51(45-60)	10.7	*11.1*	*12.9*	*14.0*	0.69	0.23
9. B216	16	40(35-48)	10.5	10.8	*11.8*	*13.4*	0.68	0.20
10. B220	15	40(32-49)	10.8	10.8	*11.5*	*13.1*	0.69	0.20
11. B224	16	39(30-48)	11.2	12.2	*12.5*	*15.3*	1.36	0.23
12. B228	17	44(36-53)	10.4	10.6	*11.3*	*13.1*	0.74	0.19
13. B256	16		9.7	10.2	*10.2*	*11.8*	1.00	0.30
14. B260	17		10.2	10.3	*10.6*	*12.5*	0.67	0.18
15. B268	15		9.6	9.5	*9.5*	*11.3*	0.67	0.19
16. B276	16		9.0	9.1	*9.9*	*11.6*	0.96	0.28
17. B280	17		9.5	10.1	*10.6*	*11.6*	1.00	Rejected
18. G2	17		9.0	9.7	*9.6*	*11.2*	0.95	0.28
19. G6	18		9.4	9.9	*10.6*	*12.0*	0.65	0.24
20. G36	16		10.5	*10.6*	*12.3*	12.9	0.96	0.27
21. G40	17		9.9	10.5	*11.0*	*12.5*	0.68	0.21
22. G44	17		9.9	10.7	*11.2*	*12.6*	0.76	0.26
23. G48	18		10.3	10.3	*10.8*	*12.1*	0.74	0.27
24. G56	17		9.9	9.7	*10.3*	*11.8*	0.66	0.21
Means			10.0	10.4	11.3	12.8	0.79	0.22

[1]Assays 1-5, birds were fed Purina Turkey Startena. Assays 6-24, birds were fed Purina Game Bird Startena.

[2]Assays 13-24, birds were not weighed.

[3]Dose basis: Assays 1-9, 0.1 ml/10 g body weight; assays 10-24, 0.4 ml/bird; only values in Italic type were used for calculation of slope and index of precision.

In conclusion, present evidence suggests that the initial hypercalcemic response to PTH in avian species is likely to be renal in origin rather than due to bone resorption, which may play a larger role during the later phase of the response.

Selected Applications of the Avian Hypercalcemic Assay for PTH

Since its inception we have applied the *in vivo* Japanese quail hypercalcemic assay for PTH to several problems. These have included the coupling of chromatographic separation of PTH and its biological active fragments with bioassay of the fractions so generated. PTH fragments so examined have included: 1) intact PTH (1-84) and the synthetic PTH (1-34) fragment; 2) those contained in a crude commercial parathyroid extract (Parathyroid

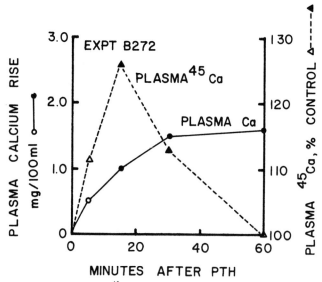

Fig. 3.—Plasma total calcium and ^{45}Ca responses in 3-week-old Japanese quail 5, 15, 30, and 60 minutes after an i.v. injection of synthetic bovine PTH (1-34) containing ^{45}Ca such that each bird received approximately 2 μCi ^{45}Ca simultaneously with 10 units of PTH. Plasma total calcium values are expressed in terms of the increase of the mean PTH-treated levels over the control levels at each time interval. Plasma ^{45}Ca data are presented as the mean of the PTH-treated group expressed as a percentage of that of the control group for each interval. Those points indicated by solid circles or triangles were significant responses; circles represent plasma total calcium; triangles, plasma ^{45}Ca. A significant elevation in plasma ^{45}Ca levels was seen at 15 and 30 minutes. Data taken from Kenny and Dacke (1974).

Injection, U.S.P., Lilly); and 3) those contained in urines obtained from normal and uremic human subjects. More recently we have applied the assay in a study of the structure-activity relationships associated with the PTH polypeptide. A brief description of these applications follow.

Chromatographic Separation of PTH (1-84) and its Fragments

The combination of Sephadex G-50 gel filtration with the *in vivo* Japanese quail hypercalcemic assay for PTH represents a powerful and relatively simple approach to the analysis of various substances and fluids for PTH-like activity. This technique has been applied to the following problems.

Separation of PTH (1-84) and PTH (1-34).—A combination of intact bPTH (1-84) and the synthetic fragment, bPTH (1-34), may be cleanly separated using Sephadex G-50 gel filtration. This has been demonstrated by Kenny *et al.* (1976) who subjected a mixture of a partially purified preparation (200 U.S.P. units/mg) of bovine PTH (1-84) obtained from Wilson Laboratories (Chicago, Illinois; IV-155.1A, lot no. 146873) and a synthetic preparation of bovine PTH (1-34) obtained from Beckman (Palo Alto, California) to gel filtration followed by bioassay of the fractions using the Japanese quail hypercalcemic assay. The results are presented in Fig.. 5. The

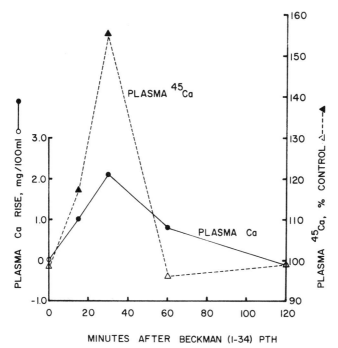

FIG. 4.—Plasma total calcium and ^{45}Ca responses in 9-day-old chickens 15, 30, 60, and 120 minutes after an i.v. injection of synthetic bovine PTH (1-34) containing ^{45}Ca such that each bird received approximately 2 μCi ^{45}Ca simultaneously with 20 units of PTH. Plasma total calcium values are expressed in terms of the increase of the mean PTH-treated levels over the control levels at each interval. Plasma ^{45}Ca data are presented as the mean of the PTH-treated group expressed as a percentage of that of the control group for each interval. Those points indicated by solid circles or triangles were significant responses; circles represent plasma total calcium; triangles, plasma ^{45}Ca. A significant rise in plasma ^{45}Ca levels was seen at 15 minutes (1-tailed test) and at 30 minutes. Data taken from Kenny and Dacke (1974).

hypercalcemic activity associated with PTH (1-84) peaked in fraction 28 and was cleanly separated from that linked with PTH (1-34) which peaked in fraction 44.

  Chromatographic Analysis of a Commercial Parathyroid Extract.—A crude pharmaceutical preparation of parathyroid hormone, Parathyroid Injection, U.S.P. (Lilly), the manufacture of which was discontinued in June, 1980, has been available for decades for clinical use. In spite of its crude nature (it was less than 0.5% pure), its stability and apparent constancy of its unitage, as determined by the U.S.P. dog assay, made the preparation extremely popular in experimental research. It essentially served as an unofficial international PTH standard as few research laboratories could afford to conduct the tedious and relatively insensitive U.S.P. dog assay. When a batch of Parathyroid Injection, U.S.P. (Lilly, lot no. 4RG76, 200 U.S.P. units/ml) was subjected to Sephadex G-50 gel filtration and the fractions assayed for hypercalcemic activity, two major peaks of biological activity were found (Fig. 6).

TABLE 4.—*Plasma calcium responses in Japanese quail at short intervals after parathyroid hormone injection. Data taken from Kenny and Dacke (1974).*

Expt. no.	PTH dose (units/bird, i.v.)	Mean plasma calcium response† mg/dl			
		1 min	5 min	15 min	30 min
B 272	10	+ 0.8	+ 0.5	+ 1.0*	+ 1.5*
G 126	20		+ 0.6*	+ 1.0*	+ 1.5*
G 100	25			+ 2.0*	+ 2.2*

*$P < 0.05$, significance of difference from control values.
†Increase of PTH-treated levels over control levels.

The position of the first peak indicated that the biologically active fragment or fragments it represented were slightly smaller than the intact PTH (1-84) molecule but larger than the PTH (1-34) fragment. The second peak represented a biologically active fragment or fragments which were smaller than those in the first peak but were still slightly larger than the PTH (1-34) fragment.

Chromatographic Analysis of Urines from Uremic Human Subjects.—A series of urines obtained from normal human subjects and from patients with chronic renal failure was subjected to Sephadex G-50 gel filtration followed by determination of the PTH-like activity in the fractions using the Japanese quail hypercalcemic assay (Kenny *et al.*, 1976). No PTH-like activity was detected in the five normal urines even when urinary aliquots of between 80 and 90 milliliters were processed. In contrast, the urines of four of the six uremic patients had detectable activity when aliquots of between 15 and 80 milliliters were examined (Table 5). Of interest are the elution profiles of those urines from the four patients with detectable PTH (Fig. 7). In two urines, those of JR and DW, PTH activity was associated with fractions which would include materials close in size to the intact PTH (1-84) molecule. On the other hand, in the remaining two urines, those of MW and TM, PTH activity was associated with fractions which would include fragments of a size which is considerably less than that of PTH (1-84) and is closer to that of PTH (1-34). These findings are not inconsistent with those of Reiss and Canterbury (1974) who identified two active fragments (9500 and 4500-5000, molecular weight) in pooled sera obtained from patients with primary hyperparathyroidism.

In the study of Kenny *et al.* (1976), an observation, peripheral to the intent of the study, was made that is worthy of mention. A biologically active fraction of low molecular weight was uncovered in the course of an experiment designed to indicate qualitative recovery of PTH (1-84) added to urine obtained from a uremic patient. This fraction represented material whose size was considerably smaller than that of PTH (1-34). The results are presented in Fig. 8. The urine in question had not been stored under ideal conditions and for this reason had not been included in the regular series. Nevertheless, the presence of this activity associated with a fragment or fragments of low molecular weight is interesting from several points of view.

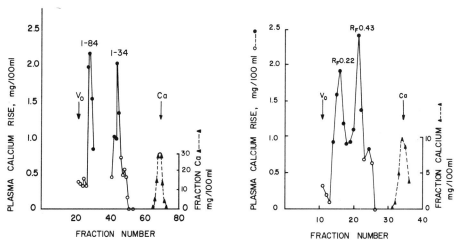

FIG. 5 (left).—Sephadex G-50 gel filtration of a mixture of 1 mg of crude bovine PTH (1-84) (Wilson TCA powder) and 120 μg of synthetic bovine PTH (1-34) (Beckman) on a 2.5 × 89-cm column eluted with 0.1 M formic acid containing 0.01% bovine serum albumin and 0.01% cysteine HCl; 6-ml fractions were collected. Protein (added bovine serum albumin) peaked at fraction 22, indicated by V_o (void volume); added Ca peaked at fraction 69. The 1-84 and 1-34 PTH peptides were adequately separated (R_F values of 0.13 and 0.47, respectively), as indicated by the peak hypercalcemic activities found in fractions 28 and 44 using the avian bioassay method for PTH. Significant ($P < 0.05$) hypercalcemic activity is represented by solid circles. For a definition of R_F see footnote to Table 5. Data taken from Kenny et al. (1976).

FIG. 6 (right).—Sephadex G-50 gel filtration of 500 USP units of Parathyroid Injection, U.S.P. (Lilly, lot no. 4RG76, 200 U.S.P. units/ml) applied to a 2.5 × 87-cm column; 12-ml fractions were collected; V_o, fraction 11; Ca peak, fraction 34. Five-milliliter aliquots of fractions 11 to 26 were lyophilized separately and bioassayed as individual fractions. Two major peaks of biological activity were revealed with R_F values of 0.22 and 0.43 respectively, both of which were outside the range (0.11-0.17) found in four runs with bovine PTH (1-84). Closed circles represent significant ($P < 0.05$) hypercalcemic activity. For a definition of R_F see footnote to Table 5. Data taken from Kenny et al. (1976).

Firstly, its existence is not an isolated observation in our total experience. Secondly, Reiss and Canterbury (1974) reported a low molecular weight (2500) immunoreactive fragment present in pooled sera obtained from chronically uremic patients. Thirdly, if it is of PTH origin, it might represent a fragment of human PTH smaller than the 2-34 or 1-31 peptides considered minimally necessary for hypercalcemic activity in bovine PTH (Tregear et al., 1973). Although the structural requirements for biological activity of bovine PTH have been extensively studied, no similar systematic study of human PTH has been done. Human and bovine PTH (1-34) peptides differ in their amino acid sequences at three positions: 1, 7, and 16. Position 1 is unnecessary for biological activity (Tregear et al., 1973) and may be eliminated from consideration. Although the other two substitutions have been considered to be of little biological importance, experimental data in support of this concept is still needed.

TABLE 5.—*Bioassayable PTH-like activity in human normal and uremic urines. Data taken from Kenny et al. (1976).*

Sample number	Subject		Urine volume applied [a] (ml)	PTH activity, U.S.P. units		
	Initials	Sex		units/mg of creatinine	units/24 hr	R_F [b]
Normal Human Urines						
1	TL	M	86 (17)	ND[c,d]		
2	DH	M	88 (18)	ND[d]		
3	AK	M	80 (16)	ND[d]		
4	GT	M	86 (17)	ND[d]		
5	JF	M	85 (17)	ND[d]		
Uremic Patient Urines						
1	JR	M	16 (4)	1.7	107	0.10
2	MW	F	76 (15)	0.4[d]	100[d]	0.43
3	DW	M	38 (30)	0.5	92	0.21
4	TM	M	50 (10)	1.4	286	0.42
5	JF	F	50 (67)	ND		
6	JS	M	43 (13)	ND[d]		

[a]Urine volume equivalent applied to Sephadex G-50 column; actual volume of formic/urea solution applied is given in parentheses. Normally the urine sample was lyophilized and dissolved in a smaller volume of 1 M formic/urea; all, or an aliquot, of this solution was applied.

[b]$R_F = (PTH - V_o)/(Ca \cdot V_o)$ where PTH, V_o and Ca represent the peak fractions of biological activity, protein (albumin), and Ca, respectively.

[c]ND: not detectable ($P < 0.05$).

[d]Parathyroid Injection, U.S.P. (Lilly, lot no. 4RG76d) used as the standard; in all other instances Beckman synthetic bovine PTH (1-34) (lot no. 21013, 1745 units/mg) was used as the standard.

Structure Activity Studies of PTH

Recently, in collaboration with Dr. Peter K. T. Pang and Ms. May Yang in our department, we have contributed to the further delineation of the structure-activity relationships of the PTH polypeptide. Pang and his associates (Pang *et al.*, 1980*a*, 1980*b*, 1980*d*; Crass and Pang, 1980) have extended considerably the original observation of Charbon (1968) that bovine PTH has a hypotensive action in the dog. It was important to determine in the early phases of this more recent study whether or not the hypercalcemic and hypotensive activities had the same structural requirements. Although several avenues were pursued to settle this point, only the following approach, emanating from our laboratory, will be described. It is well established that mild treatment of PTH (1-84) with H_2O_2 leads to inactivation of the hypercalcemic activity (Tashjian *et al.*, 1964). We have made the important observation that when synthetic bovine PTH (1-34) obtained from Peninsula Laboratories is treated with H_2O_2, in contrast to bovine PTH (1-84), it does not lose its hypercalcemic activity (Kenny and Pang, 1980; Pang *et al.*, 1981). It does, however, lose its hypotensive activity (Pang *et al.*, 1980*c*). The hypercalcemic data are presented in Table 6. Not only is there no destruction of hypercalcemic activity in the synthetic bovine PTH (1-34), there is a tendency, albeit not statistically significant, for the hypercalcemic activity to be

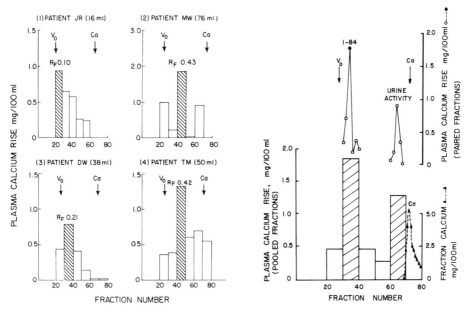

FIG. 7 (left).—Sephadex G-50 gel-filtration elution patterns obtained from four uremic urines exhibiting detectable PTH-like activity. Shaded bars indicate significant ($P < 0.05$) hypercalcemic activity. Numbers and initials refer to urine sample numbers and patients listed in Table 5; equivalent urine volumes applied to column are given in parentheses. Mean R_F values calculated using the middle fraction of the pooled samples. Two samples (1 and 3) had activity (mean R_F values of 0.10 and 0.21) which would include material close in size to the PTH (1-84) peptide. Two samples (2 and 4) had active fragments (mean R_F values of 0.43 and 0.42) closer in size to the PTH (1-34) peptide. For a definition of R_F see footnote to Table 5. Data taken from Kenny *et al.* (1976).

FIG. 8 (right).—Sephadex G-50 gel filtration of a mixture of 1 mg of crude bovine PTH (1-84) (Wilson TCA powder) and 10 ml of a uremic urine sample known to exhibit only low molecular weight biological activity. The mixture was lyophilized and dissolved in 5 ml of 1 M formic/ urea. Four milliliters of this solution were applied to a 2.5 × 87-cm column and eluted with 0.1 M formic acid; 6-ml fractions were collected; V_o, fraction 28; Ca peak, fraction 72. Fractions were pooled for bioassay either as 10 consecutive fractions (1.5-ml aliquots) or as consecutive pairs (4.0-ml aliquots). The histogram represents the former; the curves represent the latter. Significant ($P < 0.05$) hypercalcemic activity is represented by shaded bars or by solid circles. The added PTH (1-84) and urinary activities peaked at fractions 33-34 and 63-64 and had R_F values of 0.12 and 0.80, respectively. The former value is similar to that (R_F 0.13) obtained in the absence of added urine (Fig. 5). For a definition of R_F see footnote to Table 5. Data taken from Kenny *et al.* (1976).

enhanced. A pentapeptide sequence, PTH (24-28), which still possesses hypotensive activity, is devoid of hypercalcemic activity (Table 6).

SUMMARY

The development of a simple, economical, and reasonably sensitive and precise assay for PTH is described. This assay, based on the rapid hypercal-

TABLE 6.—*Assay of H₂O₂-treated bPTH (1-34) and of bPTH (24-28) in the Japanese quail hypercalcemic assay. Data taken from Pang et al. (1981).*

Treatment groups	Plasma Ca[a] mg/dl
Experiment A	
Control	8.0 ± 0.39 (6)
bPTH(1-34) 1.3 μg[b]	8.9 ± 0.31 (6)
bPTH(1-34) 4.0 μ	10.7 ± 0.31 (6)*
H₂O₂-treated bPTH(1-34) 1.3 μg	9.0 ± 0.33 (6)
H₂O₂-treated bPTH(1-34) 4.0 μ	10.2 ± 0.29 (6)*
Experiment B	
Control	11.2 ± 0.22 (6)
Crude bPTH(1-84)[c] 6.7 μg (1.2 U)[b]	11.2 ± 0.42 (6)
Crude bPTH(1-84) 20.1 μg (3.6 U)	13.2 ± 0.49 (6)**
Crude bPTH(1-84) 60.3 μg (10.8 U)	15.0 ± 0.25 (6)*
H₂0₂-treated bPTH(1-84) 35.7 μg	11.6 ± 0.44 (6)

[a]Data are given as mean ± SE (no. of animals).
[b]Dose of peptide given per bird; potency assigned by manufacturer was 10 U/μg.
[c]Trichloroacetic acid preparation (179 U/mg) obtained from Wilson Laboratories (IV-155.1A; lot no. 156712).
*Significantly different from control: $P < 0.05$ (Student's t-test); **$P < 0.01$.

cemic response to PTH in the Japanese quail (or chicken), represents a significant improvement over *in vivo* hypercalcemic assays heretofore available.

A modest study of the mechanism of the rapid avian hypercalcemic response to PTH is discussed. This study, using simultaneous administration of radiocalcium and PTH in Japanese quail, led to the conclusion that the initial rapid rise in plasma calcium following intravenous injection of PTH can best be described as inhibition of the exit of calcium from extracellular fluid rather than as originating from bone. Studies in the starling by others point to the kidney as the site of this action. Bone resorption may play a significant role in the later phases of the hypercalcemic response.

Several applications of the *in vivo* avian hypercalcemic assay are mentioned. The coupling of Sephadex G-50 gel filtration with the avian assay for the screening of the fractions for PTH-like activity yielded the following findings: 1) PTH (1-84) and PTH (1-34) can be cleanly separated; 2) a crude pharmaceutical preparation of parathyroid hormone (Parathyroid Injection, U.S.P., Lilly) was analyzed and found to contain two major peaks of biological activity representing fragments of the hormone which were both smaller than PTH (1-84) and larger than PTH (1-34); and 3) analysis of urines from uremic patients with chronic renal failure yielded detectable PTH-like activity in the majority of the urines examined. The PTH-like activity was associated with two peaks which represented materials of molecular size close to PTH (1-84) and PTH (1-34), respectively. Worthy of note is the existence in a urine, which had been inappropriately stored, of a third peak of biological activity of molecular size considerably smaller than the PTH (1-34) molecule. This latter finding suggests that avian hypercalcemic activity might be associated with a fragment of human PTH which is much smaller than the 2-34 or 1-31 peptides considered necessary for hypercalcemic activity in bovine PTH. Finally, the avian hypercalcemic assay has revealed the fact that mild treatment with H_2O_2 inactivates the hypercalcemic activity of PTH (1-84) but not that of PTH (1-34).

ACKNOWLEDGMENTS

The author wishes to acknowledge the collaboration of his various colleagues whose contributions to the work described is recorded in the publications cited. The investigations described were supported in part by the National Institutes of Health, Eli Lilly and Company, and the Dalton Research Center, University of Missouri at Columbia.

LITERATURE CITED

BERSON, S. A., R. S. YALOW, G. D. AURBACH, AND J. T. POTTS, JR. 1963. Immunoassay of bovine and human parathyroid hormone. Proc. Nat. Acad. Sci., U.S.A., 49:613-617.

CHARBON, G. A. 1968. A rapid and selective vasodilator effect of parathyroid hormone. Eur. J. Pharm., 3:275-278.

CLARK, N. B., AND R. F. WIDEMAN, JR. 1977. Renal excretion of phosphate and calcium in parathyroidectomized starlings. Amer. J. Physiol., 233:F138-F144.

COLLIP, J. B. 1925-26. The parathyroid glands. Harvey Lectures, 21:113-172.

CRASS, M. F., III, AND P. K. T. PANG. 1980. Parathyroid hormone: a coronary artery vasodilator. Science, 207:1087-1089.

DACKE, C. G., AND A. D. KENNY. 1971. Marked rapidity and sensitivity of the hypercalcemic response to parathyroid hormone in birds. Fed. Proc., 30:417.

———. 1972. An avian bioassay for parathyroid hormone. Fed. Proc., 31:225.

———. 1973. Avian bioassay method for parathyroid hormone. Endocrinology, 92:463-470.

KENNY, A. D., AND C. G. DACKE. 1974. The hypercalcemic response to parathyroid hormone in Japanese quail. J. Endocrinol., 62:15-23.

———. 1975. Parathyroid hormone and calcium metabolism. World Rev. Nutrition Dietetics, 20:231-298.

KENNY, A. D., AND P. L. MUNSON. 1959. A method for the biological assay of phosphaturic activity in parathyroid extracts. Endocrinology, 64:513-521.

KENNY, A. D., AND P. K. T. PANG. 1980. Failure of bPTH (1-34) to be inactivated by oxidation with hydrogen peroxide. Calc. Tiss. Int., 31:73.

KENNY, A. D., D. J. AHEARN, AND J. F. MAHER. 1976. Improved method for determining parathyroid hormone in biological material. Biochem. Med., 16:201-210.

MARCUS, R., AND G. D. AURBACH. 1969. Bioassay of parathyroid hormone in vitro with a stable preparation of adenyl cyclase from rat kidney. Endocrinology, 85:801-810.

MUELLER, W. J., K. L. HALL, C. A. MAURER, JR., AND I. G. JOSHUA. 1973. Plasma calcium and inorganic phosphate response of laying hens to parathyroid hormone. Endocrinology, 92:853-856.

MUNSON, P. L. 1955. Studies on the role of the parathyroids in calcium and phosphorus metabolism. Ann. New York Acad. Sci., 60:776-795.

———. 1960. Recent advances in parathyroid hormone research. Fed. Proc., 19:593-601.

———. 1961. Biological assay of parathyroid hormone. Pp. 94-113, in The parathyroids (R. O. Greep and R. V. Talmage, eds.), Charles C. Thomas, Springfield, xvii+473 pp.

MUNSON, P. L., A. D. KENNY, AND O. ISERI. 1953. Biological assay of calcium-mobilizing hormone (CMH) based on maintenance of serum calcium in parathyroidectomized rats. Fed. Proc., 12:249.

PANG, P. K. T., H. F. JANSSEN, AND J. A. YEE. 1980a. Effects of synthetic parathyroid hormone on vascular beds of dogs. Pharmacology, 21:213-222.

PANG, P. K. T., T. E. TENNER, JR., J. A. YEE, M. YANG, AND H. F. JANSSEN. 1980*b*. Hypotensive action of parathyroid hormone preparations on rats and dogs. Proc. Nat. Acad. Sci., U.S.A., 77:675-678.

PANG, P. K. T., C. M. YANG, H. T. KEUTMANN, AND A. D. KENNY. 1980*c*. The distinction between hypotensive and hypercalcemic actions of parathyroid hormone. Calc. Tiss. Int., 31:74.

PANG, P. K. T., M. YANG, C. OGURO, J. G. PHILLIPS, AND J. A. YEE. 1980*d*. Hypotensive actions of parathyroid hormone preparations in vertebrates. Gen. Comp. Endocrinol., 41:135-138.

PANG, P. K. T., C. M. YANG, H. T. KEUTMANN, AND A. D. KENNY. 1981. Structure activity relationship of parathyroid hormone: separation of the hypotensive and the hypercalcemic properties. Endocrinology, Submitted.

PARSONS, J. A., AND C. J. ROBINSON. 1971. Calcium shift into bone causing transient hypercalcemia after injection of parathyroid hormone. Nature (London), 230:581-582.

PARSONS, J. A., R. M. NEER, AND J. T. POTTS, JR. 1971. Initial fall of plasma calcium after intravenous injection of parathyroid hormone. Endocrinology, 89:735-740.

PARSONS, J. A., B. REIT, AND C. J. ROBINSON. 1973. Bioassay for parathyroid hormone using chicks. Endocrinology, 92:454-462.

POLIN, D., P. D. STURKIE, AND W. HUNSAKER. 1957. The blood calcium response of the chicken to parathyroid extracts. Endocrinology, 60:1-5.

REISS, E., AND J. M. CANTERBURY. 1974. Emerging concepts of the nature of circulating parathyroid hormones: implications for clinical research. Recent Prog. Horm. Res., 30:391-421.

TASHJIAN, A. H., D. A. ONTJES, AND P. L. MUNSON. 1964. Alkylation and oxidation of methionine in bovine parathyroid hormone: effects of hormonal activity and antigenicity. Biochemistry, 3:1175.

TREGEAR, G. W., J. VAN RIETSCHOTEN, E. GREENE, H. T. KEUTMANN, H. D. NIALL, B. REIT, J. A. PARSONS, AND J. T. POTTS, JR. 1973. Bovine parathyroid hormone: minimum chain length of synthetic peptide required for biological activity. Endocrinology, 93:1349-1353.

UNITED STATES PHARMACOPEIA (Twentieth Revision)—The National Formulary (Fifteenth Edition). 1980. United States Pharmacopeial Convention, Rockville, 1ii+1453 pp.

ZANELLI, J. M., D. J. LEA, AND J. A. NISBET. 1969. A bioassay method *in vitro* for parathyroid hormone. J. Endocrinol. 43:33-46.

Reprinted from
ASPECTS OF AVIAN ENDOCRINOLOGY:
PRACTICAL AND THEORETICAL IMPLICATIONS (C. G. Scanes *et al.*, eds.)
Grad. Studies, Texas Tech Univ., 1982, 26:1-411.

STEROIDOGENESIS IN AVIAN GRANULOSA CELLS: EFFECTS OF AMMONIUM ON oLH AND DIBUTYRYL CYCLIC AMP PROMOTED PROGESTERONE PRODUCTION

F. HERTELENDY, T. ZAKÁR, V. W. FISCHER, AND B. RABB

Departments of Obstetrics/Gynecology and Anatomy, St. Louis University School of Medicine, St. Louis, Missouri 63104 USA

The central role attributed to cyclic AMP in gonadotropin promoted steroidogenesis has been widely accepted. Recently, however, we have observed some marked differences in the kinetics of ovine luteinizing hormone (oLH) and dibutyryl cyclic AMP (BU_2cAMP) induced progesterone production in enzymatically dispersed chicken granulosa cells (Zakár and Hertelendy, 1980*b*). Moreover, employing oLH or bovine LH (bLH), we have been unsuccessful in demonstrating increased cyclic AMP production even at hormone concentrations greatly exceeding those necessary to maximally stimulate steroidogenesis. Yet, substances known to affect the adenylate cyclase/cyclic AMP system (theophylline, isoproterenol, heparin, imidazole, and cholera toxin) provoked responses consistent with their recognized actions (Hammond *et al.*, 1980). Similarly, the steroidogenic effects of both oLH and BU_2cAMP depend on continuous protein synthesis (Zakár and Hertelendy, 1980*b*) and on adequate supply of glucose or its glycolytic products (Zakár and Hertelendy, 1980*a*).

In an attempt to shed more light on the mechanism of LH action, we investigated the influence of ammonium ion on LH and BU_2cAMP induced progesterone secretion in chicken granulosa cells. Previous studies have shown that ammonium salts affect growth hormone release in rat anterior pituitary explants (Hertelendy *et al.*, 1971), insulin secretion both *in vitro* (Feldman and Lebovitz, 1971; Sener *et al.*, 1978), and *in vivo* (Schlienger *et al.*, 1974, 1975), aldosterone secretion in rat adrenal slices (Müller, 1965), as well as inhibiting progesterone synthesis in rat ovarian cells (Strauss *et al.*, 1978).

MATERIALS AND METHODS

Granulosa cells were isolated from the two largest preovulatory follicles of White Leghorn laying hens (about 20 and 45 hours before ovulation) as described previously (Zakár and Hertelendy, 1980*a*). Isolated cells (10^5/ml) were incubated in open polystyrene tubes at 37°C in Medium 199 containing 10 mM Hepes and 0.1 per cent BSA in a shaking waterbath. Cell viability by the trypan blue exclusion method, before and after incubation, was 92-96 per cent. At the end of incubations, the tubes were placed in an ice-bath, then centrifuged in the cold (1000 × g for 10 minutes). Progesterone in the

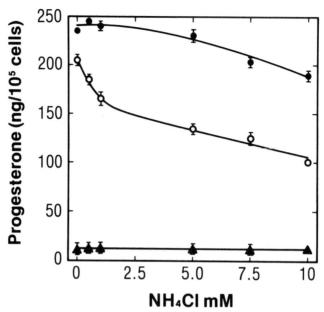

Fig. 1.—Effect of NH₄Cl on the oLH (50 ng/ml, closed circles) and BU₂cAMP (2.5 mM, open circles) stimulated progesterone production of granulosa cells. Cells (10⁵/ml) were incubated in Medium 199, 10 mM HEPES, 0.1% BSA at 37°C for 180 min in the presence of increasing concentrations of NH₄Cl. Progesterone in the medium was measured by radioimmunoassay. Steroidogenesis of unstimulated controls (closed triangles) is also presented. Experimental points are the means of five determinations ± sᴇ.

medium and in the cell pellet was determined by radioimmunoassay as previously described (Hammond *et al.*, 1980; Zakár and Hertelendy, 1980*a*).

Ovine LH (oLH-NIAMDD-LH 21), Actinomycin D (Sigma), N⁶, O²-dibutyryl adenosine 3':5'-cyclic monophosphoric acid sodium salt (BU₂cAMP) (Sigma), and NH₄Cl were dissolved in Medium 199 and added to incubation tubes before the addition of cell suspension.

Isolated granulosa cells were centrifuged and the resulting pellets were fixed at once with 3 per cent glutaraldehyde in Sorenson's phosphate buffer at pH 7.2. Following post-fixation in 1 per cent osmium tetroxide in Millonig's buffer, the pellets were dehydrated and embedded in Epon-Araldite. Ultra-thin sections, double-stained with uranyl acetate and lead citrate, were then examined in a Philips 300 electron microscope. For orientation, semi-thin sections (1 μm) were stained with methylene blue in sodium borate for viewing with a light microscope.

Results

Fig. 1 shows that NH₄Cl inhibited oLH- and BU₂cAMP-promoted steroidogenesis in a dose-dependent fashion. Despite the differences in the steroidogenic potency of the two agonists, and hence the somewhat higher base-

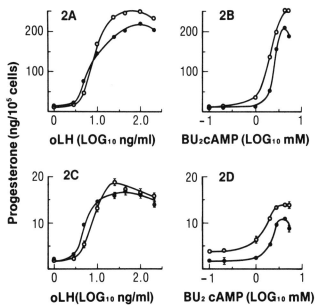

FIG. 2.—Effect of NH₄Cl on the dose-response relationship of progesterone synthesis to oLH (A) and BU₂cAMP (B). Incubation conditions were the same as described in the legend to Fig. 1. Total progesterone production (medium + granulosa cells) was determined in the presence (closed circles) and in the absence (open circles) of 5 mM NH₄Cl. Each point is the average of five determinations. The amount of progesterone associated with the cell pellet is also shown on C and D for oLH and BU₂cAMP stimulated cells, respectively (Mean + SE; $N = 5$).

line of the LH group, the parallelism between 2 mM and 10 mM NH₄Cl demonstrates a similarity of the dose-response relationships in this concentration range. However, at doses below 2 mM, NH₄Cl inhibited only BU₂cAMP-stimulated progesterone production and had no effect on gonadotropin-stimulated steroidogenesis.

A constant level of NH₄Cl (5 mM) produced markedly different effects on the dose-response curves generated by oLH and BU₂cAMP. In the presence of oLH above 10 ng/ml, NH₄Cl suppressed progesterone production, while at low oLH-doses (<10 ng/ml) progesterone synthesis was potentiated by ammonium (Fig. 2A). In BU₂cAMP-stimulated cells, NH₄Cl suppressed steroidogenesis throughout the entire effective concentration range (Fig. 2B). Figs. 2C and 2D show the cellular content of progesterone in the same experiment, and demonstrate similar characteristics of the NH₄Cl action. The potentiation by NH₄Cl of progesterone synthesis by low doses of oLH is particularly apparent.

To assess the temporal aspects of the action of NH₄Cl on oLH- and BU₂cAMP-promoted steroidogenesis, we incubated granulosa cells for 210 minutes with oLH or BU₂cAMP and, at various intervals during the first 150 minutes, NH₄Cl (10 mM) was added to the incubation tubes. It is clear that, depending on the time of addition, the effect of NH₄Cl on progesterone responses to oLH and BU₂cAMP was significantly different (Fig. 3). If added during the first hour of incubation, NH₄Cl reduced progesterone below that of the control in a similar fashion. Later addition of NH₄Cl to cells incubated with oLH failed to suppress progesterone secretion. On the other hand, NH₄Cl did decrease BU₂cAMP-promoted steroidogenesis. Furthermore, the stairlike shape of the curve, which was a reproducible phenomenon, suggests that the inhibitory effect of NH₄Cl on progesterone production of BU₂cAMP-stimulated cells has at least two components. One requires longer preincubation and appears to be analogous to that observed in LH-stimulated cells, and the other either has a much shorter or lacks the lag period, with a transient effect lasting only about 90 minutes under the present experimental conditions.

Recently we have described the kinetics of actinomycin and cycloheximide inhibition of progesterone production in oLH- and BU₂cAMP-stimulated cells (Zakár and Hertelendy, 1980*b*). Using the same approach, we compared in the present study the inhibitory effect of actinomycin and NH₄Cl (Fig. 4). While the maximal inhibition of both antagonists was similar when present throughout the incubation, the shape of the two curves is different. Actinomycin, when added during the first half hour or so, abolished the steroidogenic effect of oLH. This period coincides with the "induction-period" of steroidogenesis in these cells and suggests that a transcription-dependent step is involved (Zakár and Hertelendy, 1980*b*). The inhibitory effect of NH₄Cl, on the other hand, was directly related to the length of time the cells were exposed to this compound, suggesting a site of action different from that of actinomycin D. However, both antagonists were ineffective when the cells were preincubated with oLH for periods longer than one hour.

In red blood cells (Post and Jolly, 1957; Sachs, 1967), frog muscle (Fenn *et al.*, 1945), toad bladder (Guggenheim *et al.*, 1971), and several other systems, ammonium has been shown to interfere with sodium-potassium transport. The concentration of K⁺ in the medium is in the same range as the NH₄⁺ concentration which has effectively suppressed progesterone production in the preceding experiments. Indeed, if ammonium acts by interfering with potassium ions in intracellular processes, then it can be expected that variation of K⁺ concentration of the medium would modify the effect of a given NH₄Cl dose. As shown in Fig. 5, increasing concentrations of KCl partially reversed the inhibitory effect of NH₄Cl (5 mM) on progesterone production in oLH-stimulated cells. In BU₂cAMP-treated cells, on the other hand, variation of KCl content of the medium did not influence the effect of ammonium (Fig. 6).

Electron microscopic examination of control granulosa cells revealed a characteristic cellular profile with a normal complement of organelles (Fig. 7A). Exposure to 5 mM NH₄Cl caused the appearance of cytoplasmic vacuoles of varying size and shape (Fig. 7B). High magnification disclosed that these alterations were derived from dilatations of endoplasmic cisternae and were only partially membrane-delimited (Fig. 7C). Addition of 10 mM

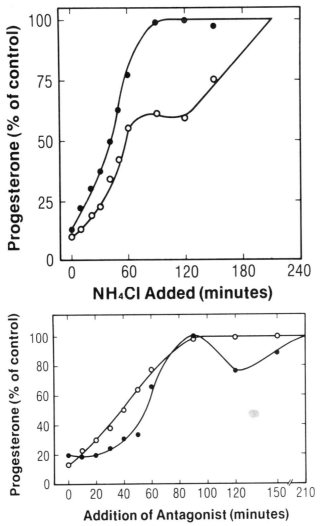

FIG. 3 (top).—Influence of preincubation with oLH or BU₂cAMP on the inhibitory effect of NH₄Cl on progesterone production. Granulosa cells were incubated for 210 min in the presence of either 50 ng/ml oLH (closed circles) or 2.5 mM BU₂cAMP (open circles). At the indicated points (open circles), NH₄Cl was added to five incubation mixtures of each treatment at 10 mM final concentration, and the incubation was continued up to 210 min. Progesterone was measured in the medium and expressed as per cent of control (that is, progesterone produced/210 min in the absence of NH₄Cl). Experimental points represent the average of five determinations. Standard errors were smaller than the diameter of the symbols.

FIG. 4 (bottom).—Comparison of the inhibitory effect of actinomycin and NH₄Cl on oLH-induced progesterone secretion. Granulosa cells (10^5/ml) were incubated with oLH (50 ng/ml). Between 0 and 150 min actinomycin D (closed circles; 8 μg/ml) or NH₄Cl (open circles; 10 mM) was added at times indicated by the symbols and incubation continued for an additional 60 min. LH-stimulated progesterone production at 210 min in the absence of antagonists equals 100%. Baseline progesterone output was 8.5% of hormone-stimulated value and has not been subtracted. Each point is the mean of five observations. Standard errors of the mean values were smaller than the diameter of the symbols.

TABLE 1.—*Effect of preincubation with NH₄Cl on subsequent progesterone production (ng/10⁵ cells) in response to oLH stimulation. Granulosa cells were preincubated for 1 h in M 199 alone or Medium + NH₄Cl. After incubation, the cells were centrifuged down, washed twice with fresh M 199 and incubated for 90 min in the presence or absence of the test compounds. A portion of the cells were incubated during this 90 min in a single batch in M 199. After this period, the cells were centrifuged, the medium decanted and the cells resuspended in fresh medium containing oLH (50 ng/ml) and incubated for a second 90 min. Note the increase in LH-stimulated progesterone production after the cells were allowed to recover for 90 min following preincubation with NH₄Cl (36% inhibition vs. 62% inhibition observed without the 90 min recovery period).*

60 min preincubation	Control	First 90 min NH₄Cl (10 mM)	oLH (50 ng/ml)	Second 90 min oLH (50 ng/ml)
M 199	.64 ± .08	.44 ± .03	5.79 ± .15	10.19 ± .48
NH₄Cl (10 mM)	.23 ± .04	.15 ± .05	2.21 ± .18	6.52 ± .23

NH_4Cl brought about a severely abnormal cytoplasmic architecture, with strikingly dilated spaces filling the cytoplasm (Fig. 7D). Incubation of cells with NH_4Cl for 60 minutes, followed by washing and subsequent incubation for 90 minutes in Medium 199 without ammonium, caused a reversal of vacuolation leading to an essentially normal appearance of the cytoplasm, with the exception of an increase in the number of lipid droplets (Fig. 7E). Parallel with these ultrastructural changes, the inhibitory effect of NH_4Cl on oLH-stimulated steroidogenesis was also largely reversible (Table 1).

DISCUSSION

The experimental results presented in this paper show that ammonium chloride influences progesterone production of chicken granulosa cells in a complex manner. The effect of the salt varies from inhibition to promotion of steroidogenesis, depending on dose, length of exposure, and the nature of the steroidogenic stimulant. NH_4Cl, considered to be a lysosomotropic agent (DeDuve et al., 1974), has been shown to raise the intralysosomal pH and accumulate in the lysosomes of mouse peritoneal macrophages (Ohkuma and Poole, 1978) and to inhibit phagosome-lysosome fusion in these cells (Gordon et al., 1980). In isolated rat hepatocytes, NH_4Cl causes lysosomal vacuolation as well as inhibiting the degradation of endogenous proteins (Seglen and Reith, 1976), which has also been observed in isolated lysosomes (Reijngoud et al., 1976). Moreover, there is evidence that the phagolysosomal system participates in hormonal responses (Szego, 1974), and several lysosomotropic agents, including NH_4Cl, have been shown to inhibit progesterone production by rat ovarian cells (Strauss et al., 1978). Collectively, these observations point to the possibility that the effects of ammonium on gonadotropin and cyclic nucleotide induced steroidogenesis is mediated by an action on the lysosomes.

In our system, however, a conspicuous feature of NH_4Cl application to granulosa cells was the cytoplasmic vacuolar derangement which was pro-

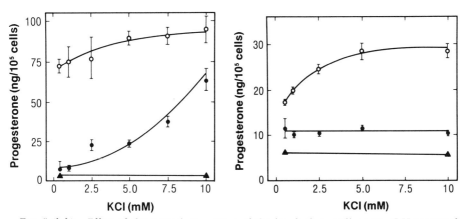

F<small>IG</small>. 5 (left).—Effect of the potassium content of the incubation medium on oLH-promoted steroidogenesis of granulosa cells in the presence and absence of NH₄Cl. Cells (10⁵/ml) were incubated for 180 min at 37° (in Hank's salt solution, 10 mM HEPES, 0.1% BSA:) containing 10 ng/ml oLH and various concentrations of KCl. Progesterone was measured in the medium. Each point is the mean of five determinations ± S<small>E</small>; open circles, no NH₄Cl; closed circles, 5 mM NH₄Cl; closed triangles, no oLH.

F<small>IG</small>. 6 (right).—Effect of potassium on BU₂cAMP-promoted steroidogenesis of granulosa cells in the presence and absence of NH₄Cl. Incubation conditions were the same as described in the legend to Fig. 5. Progesterone production was stimulated with 2 mM BU₂cAMP. Open circles: no NH₄Cl; closed circles: 5 mM NH₄Cl; closed triangles: no BU₂cAMP.

portional in its severity to the concentration of the salt. The widely dilated cisternae, combined with only a partial envelopment of the vacuoles by membranes, suggest endoplasmic reticular derivation of this abnormality. This observation coincides with the report by Palade (1956) who observed that NH₄⁺ exerted a specific effect on endoplasmic reticular membranes. In our study, the inhibition of granulosa cell function could be attributed to the severely deranged cytoplasmic architecture. However, involvement of the lysosomal system as another mechanism for cellular dysfunction cannot be ruled out entirely solely on morphological evidence.

Although the metabolic effects of NH₄Cl have not been investigated in the present study, comparison with the actions of actinomycin (Fig. 4) and cycloheximide (Zakár and Hertelendy, 1980b) suggest that neither transcriptional nor translational steps in protein synthesis are primarily involved.

Whatever the precise site and mode of action of ammonium in these cells is, there are significant differences in the way in which NH₄Cl influences oLH- and BU₂cAMP-promoted steroidogenesis. It is noteworthy that in cells stimulated with low concentration of LH (but not BU₂cAMP), NH₄Cl not only failed to inhibit progesterone production, but actually potentiated the action of the hormone. In this respect, the observation that NH₄Cl and other lysosomotropic agents inhibit the degradation of receptor-bound Human Chorionic Gonadotropin (HCG) in murine Leydig tumor cells (Ascoli and Puett, 1978) might be relevant. A similar effect in granulosa cells might then

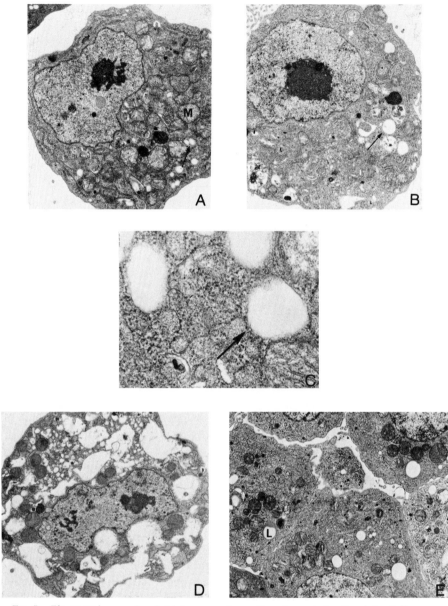

Fig. 7.—Electronmicrographs of chicken granulosa cells. A: control cell; minimal disarrangement of mitochondrial cristae (M) with a few small intracytoplasmic vacuoles (arrow); × 4,446. B: cell exposed to 5 mM NH₄Cl for 60 min; note enlarged vacuoles (arrow); × 5,720. C: close-up; dilated endoplasmic cisternae and vacuolated, partially membrane-bound spaces (arrow); × 20,800. D: 10 mM NH₄; cell with cytoplasm markedly disarranged by widely dilated spaces; × 5,265. E: cells preincubated with NH₄Cl, washed and then incubated with medium 199; normal appearance of cells; note lipid droplets (L); × 2,080.

explain the potentiating effect of NH_4Cl on submaximally stimulating doses of oLH and the lack of such an effect in cells treated with BU_2cAMP.

Differences in the sensitivity of BU_2cAMP- and oLH-stimulated cells to the inhibitory action of NH_4Cl have been consistently noted in the present study. It would appear that the physiochemical changes brought about by the interaction of the hormone with its receptors on the plasma membrane protects the cell to a certain degree, perhaps by rendering it less permeable to NH_4Cl. BU_2cAMP, on the other hand, which itself has to penetrate the cell membrane to act intracellularly, may in fact promote the influx of NH_4Cl. Consistent with this notion is the finding that increasing concentrations of potassium reversed the inhibitory effect of ammonium on LH-promoted progesterone production, but was completely ineffective when cells were stimulated with BU_2cAMP. Observations that NH_4^+ ions alter the K^+ handling of pancreatic cells (Sener et al., 1978) and compete with K^+ for the $(Na^+ + K^+)$-activated ATPase (Skou, 1965) indicate that these two ions may interfere with each other's uptake into the cells. Increasing K^+ concentration in the medium can consequently antagonize the effect of NH_4Cl by retarding its penetration across the otherwise intact plasma and/or lysosomal membranes. Furthermore, the kinetics of oLH- and BU_2cAMP-promoted steroidogenesis in the presence of NH_4Cl for varying periods was also markedly affected (Fig. 3). Kinetic studies of the effect of lysosomotropic weak bases on mouse peritoneal macrophages have shown that one of their effects (the elevation of intralysosomal pH) is very rapid, taking place within minutes, while the other (intralysosomal accumulation) is much slower and may proceed for hours (Ohkuma and Poole, 1978). Such an action may account for the complex time course effect of NH_4Cl on BU_2cAMP-stimulated progesterone secretion in our granulosa cell preparation.

In conclusion, NH_4Cl influences progesterone synthesis and release in response to LH and BU_2cAMP by acting at multiple sites. The observed differences in the response of granulosa cells to oLH and BU_2cAMP in the presence of ammonium suggests that the action of the two agonists is analogous rather than identical.

Acknowledgments

The experimental birds used in this study were generously provided by Dr. H. V. Biellier, University of Missouri at Columbia. This work was supported in part by NIH grant HD-09763.

Literature Cited

Ascoli, M., and D. Puett. 1978. Degradation of receptor-bound human choriogonadotropin by murine Leydig tumor cells. J. Biol. Chem., 253:4892-4899.

DeDuve, C., T. DeBarsy, B. Poole, A. Trouet, P. Tulkens, and F. Van Hoof. 1974. Lysosomotropic agents. Biochem. Pharm., 23:2495-2531.

Feldman, J. M., and E. Lebovitz. 1971. Ammonium ion, a modulator of insulin secretion. Amer. J. Physiol., 221:1027-1032.

Fenn, W. O., L. F. Haege, E. Sheridan, and J. B. Flick. 1945. The penetration of ammonia into frog muscle. J. Gen. Physiol., 28:53-77.

Gordon, A. H., P. D'Arcy Hart, and M. R. Young. 1980. Ammonia inhibits phagosome-lysosome fusion in macrophages. Nature, 286:79-80.

Guggenheim, S. J., J. Bourgoignie, and S. Klahr. 1971. Inhibition by ammonium of sodium transport across isolated toad bladder. Amer. J. Physiol., 220:1651-1659.

Hammond, R. W., H. Todd, and F. Hertelendy. 1980. Effects of mammalian gonadotropins on progesterone release and cyclic nucleotide production by isolated avian granulosa cells. Gen. Comp. Endocrinol., 41:467-476.

Hertelendy, F., H. Todd, G. T. Peake, L. J. Machlin, G. Johnston, and G. Pounds. 1971. Studies on growth hormone secretion: I. Effects of dibutyryl cyclic AMP, theophylline, epinephrine, ammonium ion and hypothalamic extracts on the release of growth hormone from rat anterior pituitaries in vitro. Endocrinology, 89:1256-1262.

Müller, J. 1965. Aldosterone stimulation in vitro. II. Stimulation of aldosterone production by monovalent cations. Acta Endocrinol., 50:301-309.

Ohkuma, S., and B. Poole. 1978. Fluorescence probe measurement of the intralysosomal pH in living cells and the perturbation of pH by various agents. Proc. Nat. Acad. Sci., U.S.A., 75:3327-3331.

Palade, G. E. 1956. The fixation of tissues for electron microscopy. Pp. 129-142, in Proceedings of international conference of electron microscopy, London, 1954 (R. Ross, ed.), Royal Microsc. Soc., London, 705 pp.

Post, R. L., and P. C. Jolly. 1957. The linkage of sodium, potassium, and ammonium active transport across the human erythrocyte membrane. Biochim. Biophys. Acta, 25:118-128.

Reijngoud, D. J., P. S. Oud, Y. Kas, and J. M. Tager. 1976. Relationship between medium pH and that of the lysosomal matrix as studied by two independent methods. Biochim. Biophys. Acta, 448:290-302.

Sachs, J. R. 1967. Competitive effects of some cations on active potassium transport in human red blood cell. J. Clin. Invest., 46:1433-1441.

Schlienger, J. L., M. Imler, and J. Stahl. 1974. Diminution de la tolérance glucosée et de l'insulino-sécrétion après perfusion de sels d'ammonium chez l'homme normal et chez des malades atteints de stéatose hépatique ou de cirrhose. Biol. Gastro-Enterol., 7:101-110.

———. 1975. Diabetogenic effect and inhibition of insulin secretion induced in normal rats by ammonium infusions. Diabetologia, 11:430-443.

Seglen, P. O., and A. Reith. 1976. Ammonia inhibition of protein degradation in isolated rat hepatocytes. Exp. Cell. Res., 100:276-280.

Sener, A., J. C. Hutton, S. Kawasu, A. C. Boschero, G. Somers, G. Devis, A. P. Herchuelz, and W. J. Malaisse. 1978. Metabolic and functional effects of NH₄Cl in rat islets. J. Clin. Invest., 62:868-878.

Skou, J. C. 1965. Enzymatic basis for active transport of Na⁺ and K⁺ across cell membrane. Physiol. Rev., 45:496-617.

Strauss, J. F., T. Kirsch, and G. L. Flickinger. 1978. Effects of lysosomotropic agents on progestin secretion by rat ovarian cells. Steroid Biochem., 8:731-738.

Szego, C. M. 1974. The lysosome as a mediator of hormone action. Rec. Progr. Hormone Res., 30:171-233.

Zakár, T., and F. Hertelendy. 1980a. Effects of mammalian LH, cyclic AMP and phosphodiesterase inhibitors on steroidogenesis, lactate production, glucose uptake and utilization by avian granulosa cells. Biol. Reprod., 22:810-816.

———. 1980b. Steroidogenesis in avian granulosa cells: early and late kinetics of oLH and dibutyryl cyclic AMP promoted progesterone production. Biol. Reprod., 23:974-980.

SECTION II

MECHANISMS IN THE HORMONAL CONTROL OF AVIAN BEHAVIOR

INTRODUCTION

J. Balthazart

Laboratoire de Biochimie Générale et Comparée, Université de Liège, 17, place Delcour, B-4020, Liège, Belgique

During the last decade, endocrinology, and more specifically avian endocrinology, has developed rapidly. Progress has been made possible largely by the appearance of new sensitive and specific techniques to measure hormones in biological fluids. The study of the physiological basis of behavior has also taken advantage of these developments, although advances in this field have been slower than in other types of endocrine research. This is explained to a large extent by the nature of the processes studied.

Behavior is indeed a very special dependent variable. It is affected by hormones and, in addition, the behavioral activity of one animal will in many instances influence the endocrine physiology of its congeners. The groups of Lehrman and Hinde have brilliantly illustrated, in the Ring dove and in the canary, the complexity of these networks of interactions which sometimes makes it difficult to perform a real causal analysis of the control mechanisms (see Lehrman, 1965; Hinde, 1965; and Silver *et al.*, 1979 for reviews). Furthermore, recent evidence suggests that the behavioral activity of one animal could also influence its own reproductive physiology (Cohen and Cheng, 1979), which complicates even more the analysis of the phenomenon. By its nature, behavior is also a dependent variable that appears to be more difficult to quantify than a morphological structure or a physiological characteristic. Every measure might be affected by problems of subjectivity. Finally, it must be stressed that hormones do not induce behavior but only increase the likelihood of its occurrence. Inadequate test conditions can thus always mask a potential hormonal effect and lead to erroneous conclusions.

Despite the difficulty of the task, the research on the hormonal controls of behavior is now entering an exciting new phase. New techniques are being used and the range of species studied is increasing rapidly. The papers in the following section examine new problems and the techniques currently used for study of these problems.

The use of radioimmunoassays of plasma hormones provides a valuable tool in the analysis of behavior (Pröve and Sossinka) and emphasizes the complexity of the mutual relationships between behavior and the hormonal status of an animal. Deviche then reviews parts of the literature on the role of nonsteroid hormones (gonadotropins in this case) in the control of avian behavior. The next paper (Guichard and Reyss-Brion) presents recent data showing a discrepancy between the peripheral levels of hormones and the behavioral output of birds, suggesting the presence of control mechanisms in the target cells for hormones. This idea is further illustrated by Ishii and

Tsutsui who show that in the quail, individual differences in aggressive behavior are not related to the circulating levels of testosterone. These authors also suggest that the hormonal status during infancy could be critical in determining the behavioral performances of the adult. The last papers are concerned with possible mechanisms which modulate the action of hormones in the brain cells. Massa describes the metabolism of androgens in the nervous system and attempts to correlate the enzymatic activities present in the brain with the behavior of the animals. Finally, the current literature on steroid receptors in the brain is reviewed (Balthazart). This author also shows how changes in concentration of steroid receptors regulate behavior.

Literature Cited

Cohen, J., and M. F. Cheng. 1979. Role of vocalizations in the reproductive cycle of ring doves (Streptopelia risoria): effects of hypoglossal nerve section on the reproductive behavior and physiology of the female. Horm. Behav., 13:113-127.

Hinde, R. A. 1965. Interaction of internal and external factors in integration of canary reproduction. Pp. 381-415, in Sex and behavior (F. Beach, ed.), John Wiley and Sons, New York, xvi + 592 pp.

Lehrman, D. S. 1965. Interaction between internal and external environments in the regulation of the reproductive cycle of the ring dove. Pp. 355-380, in Sex and behavior (F. Beach, ed.), John Wiley and Sons, New York, xvi + 592 pp.

Silver, R., M. O'Connell, and R. Saad. 1979. Effects of androgens on the behavior of birds. Pp. 223-278, in Endocrine control of sexual behavior (C. Beyer, ed), Raven Press, New York, x + 413 pp.

Reprinted from
ASPECTS OF AVIAN ENDOCRINOLOGY:
PRACTICAL AND THEORETICAL IMPLICATIONS (C. G. Scanes *et al.*, eds.)
Grad. Studies, Texas Tech Univ., 1982, 26:1-411.

RADIOIMMUNOASSAY OF PLASMA HORMONES AND ITS USE IN INVESTIGATIONS OF HORMONE AND BEHAVIOR CORRELATIONS IN BIRDS

EKKEHARD PRÖVE AND ROLAND SOSSINKA

Universität Bielefeld, Fakultät für Biologie, Verhaltensphysiologie, Postfach 8640, D-4800 Bielefeld 1, Federal Republic of Germany

Several behavioral patterns occur in the presence of specific releasers; thus, the central nervous system (CNS) and, to a certain degree, the endocrine system are concerned. On that topic the sexual behavior of male birds is well investigated. By castration and testosterone substitution experiments in several species of birds, it has been shown that sexual behavior in males is androgen dependent (for example, Berthold, 1849; Carpenter, 1933a, 1933b; Etienne and Fischer, 1964; Pröve, 1974; Arnold, 1975; Adkins, 1977; for a more detailed review see Silver *et al.*, 1979). But these kinds of investigations have only looked at one direction of the possible mutual relationships. For a successful breeding cycle to be completed, a complicated interrelation of internal and external factors must result by mutual interaction as has been shown in doves (Lehrman, 1965), budgerigars (Brockway, 1965), and canaries (Hinde and Steele, 1978). Thus, not only can specific hormones change the frequency or intensity of a specific behavior, but stimuli coming from certain environmental influences, from a conspecific, or by the acting bird itself are also able to change the endocrinological status of an individual. For a long time it was almost impossible to measure fluctuations of hormonal titers that had been caused by external stimuli because of the minute amounts of circulating hormones in the blood. However, the recent development of radioimmunological methods (which, by the use of highly specific antibodies and radioactive labelled hormones are capable of measuring hormone concentrations within the range of picograms) has now made it possible to detect such minute changes (For reviews, see Abraham, 1974; Jaffe and Behrman, 1974; Gupta, 1975). By the means of radioimmunoassay (RIA), correlations between the occurrence of sexual behavior patterns and the increase in the amount of gonadal hormones have been demonstrated in birds that are exposed to annual variations in their environment, and in turn show cyclic annual variations in reproduction (for example, Temple, 1974; Balthazart and Hendrick, 1976; Wingfield and Farner, 1978; Sossinka *et al.*, 1980).

We have been investigating the male Australian Zebra finch (*Taeniopygia guttata castanotis* Gould). This finch, found mainly in the semiarid areas of Australia, is especially adapted to irregular breeding seasons and exhibits

TABLE 1.—*Correlation between the number of courtship song motifs produced in a 30-minute courtship test and plasma androgen levels in male Zebra finches. Correlation coefficient* r = *0.674 (*P < *0.01). Data from Pröve (1978).*

Bird no.	Motifs	Testosterone (pg/ml)	Bird no.	Motifs	Testosterone (pg/ml)
1787	0	323	1784	39	1889
6	2	137	963	45	4865
1906	2	210	961	48	2741
3	3	220	1786	64	1596
2	5	543	772	67	2736
1905	7	480	1782	67	1712
4	9	501	1789	68	1463
1785	12	907	803	82	3103
1	16	1336	1788	101	2681
962	20	2168	14	111	5182
514	29	2289	804	117	6219
960	30	4572	1790	168	2972
966	32	1028			

opportunistic breeding periodicity (Immelmann, 1962; Sossinka, 1980). In the laboratory, males mature at a very early age and remain sexually active throughout the year (Sossinka, 1975). This species is well suited for experiments as it is easily bred in captivity, and its sexual behavior is highly stereotyped and thus is easily quantified (Immelmann, 1962).

Plasma androgens have been measured by RIA without chromatographical separation of testosterone and dihydrotestosterone. Blood samples were taken by wing vein puncture or in some investigations by decapitation. The blood was collected in heparinized capillary tubes, the plasma and blood cells separated by centrifugation, and the plasma stored frozen at −20°C until assayed. Androgens were extracted twice by ether, with a mean recovery rate of 92.6 ± 0.4%. The RIA was performed with a RIA pack (New England Nuclear), which contained testosterone antibody, (^{3}H)-labelled testosterone, and testosterone standard solution (Pröve, 1978). The intra-assay variation in our measurements was 10 per cent and the inter-assay variation was 12 per cent. As previously mentioned, this system does not discriminate between testosterone and dihydrotestosterone. For a more detailed analysis of endocrine events, a chromatography system has been used from which small plasma samples can be separated for various steroids (for description, see Wingfield and Farner, 1975).

By measuring plasma androgen titers, some correlations between the frequency of sexual behavior patterns and the amount of androgens in the plasma of individual male Zebra finches has been observed. Castration and testosterone substitution experiments have shown the performance of song is testosterone dependent (Pröve, 1974). By measuring "resting plasma androgen titers" (in other words, titers of males which are not influenced by ongoing sexual behavior), a significant positive correlation between the height of plasma androgen levels and the number of courtship song motifs produced

TABLE 2.—*Correlation between the number of undirected song motifs produced during the light phase of a L:D = 14:10 day and plasma androgen levels in male Zebra finches. Correlation coefficient* r = 0.853 (P < 0.01). *Data from Pröve (1978).*

Bird no.	Motifs	Testosterone (pg/ml)	Bird no.	Motifs	Testosterone (pg/ml)
810	41	44	806	153	1963
837	42	n.d.	1260	213	691
809	50	78	842	223	1255
1159	67	465	1262	281	1281
846	74	699	805	285	1019
806	75	184	841	353	2111
813	78	172	943	355	1991
838	91	n.d.	845	386	2145
811	96	347	840	392	1462
836	99	157	864	401	2018
808	103	513	844	413	1287
837	111	395	851	447	3074
990	119	239	1157	452	1717
847	132	703	863	793	2377
1290	132	724	850	812	6851
1213	147	1153			

n.d. = not detectable.

in a 30 minute courtship test was observed (Table 1). With the undirected song, a corresponding positive correlation between the height of plasma androgen levels and the number of song motifs could be shown (Table 2). These two song types differ in their hormonal thresholds (Pröve, 1974, 1978) and in some acoustical features (Sossinka and Böhner, 1980).

Correlation of actual plasma hormone titers with subsequent behavioral events, however, includes some irrelevant factors and suffers from some artifacts. Aside from circadian and circannual changes in the hormonal status of an individual, there are, first of all, short-term and midterm effects of external stimuli coming from conspecifics (or other environmental factors), which influence the endocrine system (for review, see Assenmacher and Farner, 1978). Therefore, we investigated the influence of different external factors on androgen titers. The following factors were tested: 1) the technique and time needed to get the blood samples, that is, wing vein puncture, which requires two to four minutes compared to decapitation which takes about 30 seconds; 2) the stress of the chasing situation; for example, blood samples were taken either immediately after a chase lasting 15 minutes or 2 hours later; 3) midterm changes in the housing conditions; for example, males were isolated in small cages from the aviary, and blood samples were taken 12 hours after isolation; or 4) extreme housing conditions: males were housed in small cages and examined over a period of several weeks.

The results indicate that the method of taking blood samples has no significant influence on androgen titers in the males (Fig. 1). Secondly, a slight decrease was seen 15 minutes after a chasing situation, but this difference is not significant in comparison with the large individual variations of the

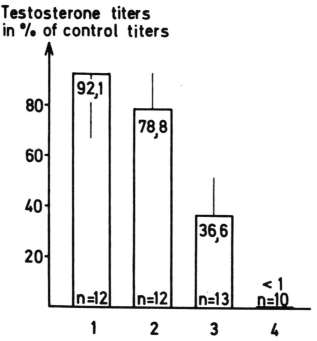

Fig. 1.—Influence of short and midterm stressors on plasma androgen titers in male Zebra finches. The columns show the deviation in per cent (mean ± SE) for the different experimental groups. The number at the top of each column indicates deviation (in per cent) from control titers; number at bottom of column indicates sample size. In the control group, designated 100%, blood samples were taken by wing venipuncture (absolute androgen titers 661 ± 106 pg/ml plasma, $N = 21$). Column 1, blood samples were taken by decapitation (609 ± 169 pg/ml plasma). Column 2, blood samples were taken immediately after a chasing situation which lasted 15 minutes (521 ± 93 pg/ml plasma). Column 3, blood samples were taken 2 hours after a chasing situation which lasted 15 minutes (242 ± 99 pg/ml plasma). Column 4, blood samples were taken 12 hours after isolation of males from an aviary to a small cage (absolute androgen titers near zero; not detectable in eight of 10 males).

control group. However, two hours after such a stressing situation, the androgen titers of the birds were significantly depressed ($P < 0.01$, t-test). Finally, the influence of midterm changes in the housing conditions was much more pronounced. In eight of 10 males the strange environment and isolation caused a depression of androgen titers to below detectable levels. Overcrowding over a long period has been observed to have a strong negative influence on androgen titers of the males. This was seen when males housed in groups of eight individuals in small cages were compared with males housed with a single female (Pröve, 1978). In addition to the artificial influence of such external factors, the sexual behavior patterns performed by an animal itself also have an influence on the sexual status of the individual. For example, in the budgerigar, the performance of song causes an increase in the volume of the testes (Brockway, 1974). These changes also should be detectable by means of RIA. In male Zebra finches, we found that the activity during courtship tests caused differing changes in androgen levels of sexually experienced and unexperienced birds (Pröve, 1978).

In summary, measurement of peripheral plasma steroid hormone levels can provide valuable data concerning factors which can cause interference. The endocrine system of an individual is very sensitive and reacts very quickly (within periods of about half an hour in the case of androgens) to changes in the environment. It is only by considering all the above mentioned factors which affect steroid hormone measurements by the RIA method that one can cautiously interpret the findings of some mutual relationships between the behavioral output and the hormonal status of an individual.

Furthermore, the use of RIA for measuring circulating plasma hormone levels has two disadvantages with respect to conclusions concerning correlations with the behavioral output of an animal: 1) peripheral hormone levels are only able to give indications, but not hard evidence, on the amount of hormones which are present in the target tissues (in the case of behavior neuron nuclei of specific centers in the CNS); and 2) we are not able to tell whether the hormone itself or its metabolites are responsible for the relevant behavioral phenomenon we are looking for (for example, the conversion of testosterone into dihydrotestosterone or the aromatization into estrogens can take place within the target tissues). To answer such questions other methods are needed, for example, a quantitative autoradiography (Stumpf and Grant, 1975) or the analysis of steroids and their metabolites in the target tissues directly (Balthazart and Hirschberg, 1979; Balthazart et al., 1979). Hence, because the costs of time and work for such methods are much higher, RIA will be an often used method in the field of investigations on hormones and behavior. But the results of RIA in interpreting correlations in hormones and behavioral events should be done only with great caution.

Acknowledgments

We thank Richard Zann for correcting the English manuscript. The investigations were funded in part by the Deutsche Forschungsgemeinschaft.

Literature Cited

Abraham, G. E. 1974. Radioimmunoassay of steroids in biological materials. Acta Endocrinol., Suppl. 183:1-41.

Adkins, E. K. 1977. Effects of diverse androgens on the sexual behavior and morphology of castrated male quail. Horm. Behav., 8:201-207.

Arnold, A. P. 1975. The effects of castration and androgen replacement on song, courtship, and aggression in Zebra Finches (Poephila guttata). J. Exp. Zool., 191:309-326.

Assenmacher, I., and D. S. Farner, eds. 1978. Environmental endocrinology. Berlin, Heidelberg, New York, xv + 334 pp.

Balthazart, J., and J. Hendrick. 1976. Annual variation in reproductive behavior, testosterone, and plasma FSH levels in the Rouen duck, Anas platyrhynchos. Gen. Comp. Endocrinol., 28:171-183.

Balthazart, J., and D. Hirschberg. 1979. Testosterone metabolism and sexual behavior in the chick. Horm. Behav., 12:253-263.

BALTHAZART, J., R. MASSA, AND P. NEGRI-CESI. 1979. Photoperiodic control of testosterone metabolism, plasma gonadotrophins, cloacal gland growth, and reproductive behavior in the Japanese quail. Gen. Comp. Endocrinol., 39:222-235.

BERTHOLD, A. A. 1849. Transplantation der Hoden. Arch. Anat. Physiol. U. Wiss. Med., 2:42-46.

BROCKWAY, B. F. 1965. Stimulation of ovarian development and egg laying by male courtship vocalisation in budgerigars (Melopsittacus undulatus). Anim. Behav., 13:575-578.

BROCKWAY, B. F. 1974. The influence of some experimental and genetic factors, including hormones, on the visible courtship behavior of budgerigars (Melopsittacus). Behaviour, 51:1-18.

CARPENTER, C. R. 1933a. Psychobiological studies of social behavior in Aves. I. The effect of complete and incomplete gonadectomy on the primary sexual activity of the male pigeon. J. Comp. Psychol., 16:25-57.

———. 1933b. Psychobiological studies of social behavior in Aves. II. The effect of complete and incomplete gonadectomy on secondary sexual activity with histological studies. J. Comp. Psychol., 16:59-90.

ETIENNE, A., AND H. FISCHER. 1964. Untersuchung über das verhalten kastrierter stockenten (Anas platyrhynchos L.) und dessen beeinflussung durch testosteron. Tierpsychol., 21:348-358.

GUPTA, D., ED. 1975. Radioimmunoassay of steroid hormones. Verlag Chemie, Weinheim, xiv + 224 pp.

HINDE, R. A., AND E. STEEL. 1978. The influence of day-length and male vocalizations on the estrogen-dependent behavior of female canaries and budgerigars, with discussion of data from other species. Pp. 40-73, in Advances in the study of behavior, Vol. 7 (J. S. Rosenblatt, R. A. Hinde, E. Shaw, and C. Beer, eds.), Academic Press, New York, xiv + 261 pp.

IMMELMANN, K. 1962. Beiträge zu einer vergleichenden biologie australischer prachtfinken (Spermestidae). Zool. Jb. Syst., 90:1-196.

JAFFE, B. M., AND H. R. BEHRMAN. 1974. Methods in hormone radioimmunoassay. Academic Press, New York, London, xxi + 520 pp.

LEHRMAN, D. S. 1965. Interaction between internal and external environments in the regulation of the reproductive cycle of the ring dove. Pp. 355-380, in Sex and behavior (F. A. Beach, ed.), John Wiley and Sons, New York, xvi + 592 pp.

PRÖVE, E. 1974. Der einflub von kastration und testosteronsubstituation auf das sexualverhalten männlicher Zebrafinken (Taeniopygia guttata castanotis Gould). J. Ornithol., 115:338-347.

———. 1978. Quantitative untersuchungen zu wechselbeziehungen zwischen balzaktivität und testosterontitern bei männlichen Zebrafinken (Taeniopygia guttata castanotis Gould). Z. Tierpsychol., 48:47-67.

SILVER, R., M. O'CONNELL, AND R. SAAD. 1979. Effect of androgens on the behavior of birds. Pp. 223-278, in Endocrine control of sexual behavior (C. Beyer, ed.), Raven Press, New York, x + 413 pp.

SOSSINKA, R. 1975. Quantitative untersuchungen zur sexuellen reifung des Zebrafinken, Taeniopygia castanotis Gould. Verh. Dtsch. Zool. Ges., 1974:344-347.

SOSSINKA, R. 1980. Ovarian development in an opportunistic breeder, the Zebra Finch Poephila guttata castanotis. J. Exp. Zool., 211:225-230.

SOSSINKA, R., AND J. BÖHNER. 1980. Song types in the Zebra Finch Poephila guttata castanotis. Z. Tierpsychol., 53:721-730.

SOSSINKA, R., E. PRÖVE, AND K. IMMELMANN. 1980. Hormonal mechanisms in avian behavior. Pp. 533-548, in Second international symposium on avian endocrinology (A. Epple and M. H. Stetson, eds.), Academic Press, New York, xv + 577 pp.

STUMPF, W. E., AND L. D. GRANT, EDS. 1975. Anatomical Neuroendocrinology. S. Karger, München, Paris, London, New York, Sydney, xiii + 472 pp.

TEMPLE, S. A. 1974. Plasma testosterone titers during the annual reproductive cycle of starlings (*Sturnus vulgaris*). Gen. Comp. Endocrinol., 22:470-479.

WINGFIELD, J. C., AND D. S. FARNER. 1975. The determination of five steroids in avian plasma by radioimmunoassay and competitive protein-binding. Steroids, 26:311-327.

———. 1978. The endocrinology of a natural breeding population of the white-crowned sparrow (*Zonotrichia leucophrys pugetensis*). Physiol. Zool., 51:188-205.

Reprinted from
ASPECTS OF AVIAN ENDOCRINOLOGY:
PRACTICAL AND THEORETICAL IMPLICATIONS (C. G. Scanes *et al.*, eds.)
Grad. Studies, Texas Tech Univ., 1982, 26:1-411.

ARE GONADOTROPINS DIRECTLY INVOLVED
IN THE CONTROL OF AVIAN ACTIVITIES?

PIERRE DEVICHE

*Laboratoire de Biochimie Générale et Comparée, Université de
Liège, 17, place Delcour, B-4020, Liège, Belgique*

During the last several decades, birds have been used increasingly as models for the study of the endocrine regulation of behavior. The majority of information collected in this field so far has dealt with the influence of steroidal hormones, (especially androgens, estrogens, and progestagens) on reproduction associated activities such as agonistic, sexual, and parental behavior. In addition, evidence has also been collected suggesting that several peptide hormones, and especially the adenohypophyseal secretions, might directly affect (independent of their influence on peripheral target endocrine glands) some aspects of the behavior of birds. This has been the case for the adrenocorticotrophic hormone (ACTH), which has been shown to facilitate imprinting behavior of ducklings (Martin, 1975, 1978a, 1978b; Landsberg and Weiss, 1977). Administration of ACTH also appeared to affect displacement activities of pigeons (Delius *et al.*, 1976; Deviche and Delius, 1981). Finally, prolactin plays an important role in the maintenance of incubation behavior of birds in the family Columbidae (Lehrman and Brody, 1961, 1964; Cheng, 1979). Prolactin also has been suggested as an influence in the migratory behavior of some species of birds (Meier *et al.*, 1965, 1969). The present paper will deal with the effects of the gonadotropins, luteinizing hormone (LH) and follicle-stimulating hormone (FSH), on avian activities.

At present, gonadotropins, to our knowledge, have never been shown to exert a direct influence on the behavior of any mammalian species. It has been suggested that they take some part in the regulation of behavior of an Anuran (Rey, 1948) and fish (Hoar, 1962; Blum and Fiedler, 1965; Baggerman, 1968; Kramer, 1971). From an evolutionary point of view, the confirmation of such a difference between mammals and other classes of vertebrates is of interest.

Up to now, a behavioral influence of the gonadotropins in birds has been proposed for social activities only (agonistic behavior, precopulatory courtship of Columbidae, social displays of male ducks). In this paper, I shall first present the results obtained for each of these categories of activities; then I shall discuss the possible mode of action of the gonadotropins on avian behavior.

LH and Agonistic Behavior

Considerable evidence has been accumulated to show that in several species of birds, androgens induce an important facilitating influence on agonistic activities (Ring dove, *Streptopelia risoria*: Vowles and Harwood, 1966; Hutchison, 1974a; Japanese quail, *Coturnix coturnix japonica*, Selinger and Bermant, 1967; chicken, *Gallus domesticus*, Young and Rogers, 1978). In male Zebra finches (*Poephila guttata*) placed in a standard test situation, for example, castration decreased the frequency and delayed aggressive behavior, whereas testosterone propionate (TP) injections into castrated birds induced changes in the opposite direction (Arnold, 1975).

Some observations, however, are hardly compatible with the hypothesis of a total control by testosterone of the agonistic activities. In pigeons, for example, castration did not result in a rapid disappearance of aggressive behavior (Carpenter, 1933). When starlings (*Sturnus vulgaris*) were castrated, they continued to sing, and social ranking was not altered for up to one month after gonadectomy (Davis, 1957). These results might be explained in several ways. Previous experience and learning factors could have induced the persistence of agonistic behavior even in castrated birds. On the other hand, avian adrenals are known to synthesize small amounts of sex steroids (chicken, Tanabe *et al.*, 1979). It might therefore be that, under some circumstances, these steroids are produced in sufficient quantities to maintain some behavioral patterns, although detailed information on this point is not available for birds at the present time. Finally, it is possible that non steroidal hormones, particularly gonadotropins, take some part in the regulation of avian agonistic behavior.

Experimental study of behavioral regulation by gonadotropins was first performed on the starling (Mathewson, 1961). In this work, single injection of mammalian LH into the subordinate bird of a pair of either intact or castrated starlings resulted in a reversal of dominance. These effects appeared within 15 minutes following the hormone administration, suggesting that in intact birds, LH did not alter social ranking through increased testosterone levels.

A stimulatory action of LH treatment on aggression was subsequently reported for intact males of another species, the red-billed weaverbird (*Quelea quelea*, Crook and Butterfield, 1968, 1970). In this species, important changes in the frequency of agonistic and nest building activities, as well as of plumage coloration, occur in the course of the year (Butterfield and Crook, 1968). Periodic variations in plumage coloration, which are considered to reflect modifications in gonadotropic (LH) activity (Witschi, 1954), were shown to correlate with frequency of agonistic encounters, but to a lesser extent with nest building activity, which was itself correlated with testes size. From these observations, it was proposed that in the weaverbird, aggressive behavior and nest building activity were controlled by LH and by testosterone, respectively (Butterfield and Crook, 1968). This hypothesis has some experimental support (Crook and Butterfield, 1968). In the absence of nest material, LH but not testosterone propionate (TP) administration into male birds improved the social status of low ranking members of a hierarchy, while TP treatment was effective in this respect when the males were provided with nest material. It was concluded that in the weaverbird, LH controls aggression in nonreproductive situations, while testosterone stimulates aggression associated with reproduction.

This attractive idea should not be accepted without caution, because, on one hand, Crook and Butterfield's experiments were carried out on very few birds, and on the other hand, these birds were intact rather than gonadectomized. For this reason, some indirect behavioral effect of LH cannot be excluded (Arnold, 1975).

Despite these restrictions, it is noteworthy that Crook and Butterfield's hypothesis of a behavioral role of LH agrees with a later study performed on females of the same species (Lazarus and Crook, 1973). These authors showed that ovariectomy in the breeding season, but not outside it, enhanced the frequency of agonistic encounters. To explain the differential effect of an ovariectomy performed both during and outside the reproductive period, it was presumed that this operation enhanced the LH circulating levels to a larger extent when performed during the breeding season. Indeed, the gonadotropin titers of gonadectomized female quail maintained under long daylength reached much higher values than the corresponding levels in birds kept under short daylength (Gibson et al., 1975). In the same study, Lazarus and Crook (1973) also observed an increased frequency of agonistic behavior following the administration of LH to gonadectomized females. On the other hand, estradiol benzoate injections to such females depressed the frequency of agonistic encounters. It was concluded that in the female weaverbird, as in the male, LH exerts some control over aggressive behavior.

Up to the present time, such a role of LH has been suggested for only one other species of bird. Murton et al. (1969) observed that treatment of feral pigeons (Columba livia) with mammalian LH resulted in an increased probability of attack following a bowing display. As in Crook and Butterfield's work, however, the birds used for this study were intact rather than castrated. Despite the fact that, in the pigeon, testosterone administration did not induce the same behavioral alterations as LH injections (Murton et al., 1969), these data cannot serve as a demonstration of a direct involvement of LH in the control of agonistic behavior. It finally must be pointed out that the extent to which Crook and Butterfield's suggestion of a dichotomic control of aggression (by LH and by testosterone) can be generalized also remains unknown at the present time.

FSH and the Courtship of Doves and Pigeons (Columbidae)

The courtship activities of pigeons and doves have been described in detail by several authors (Fabricius and Jansson, 1963; Akerman, 1966; Murton et al., 1969; Hutchison, 1975; Erickson and Martinez-Vargas, 1975). A reproductively active male introduced to a receptive female often exhibits a rapid succession of several well-defined patterns which have been divided into three main groups. The first group consists of primarily aggressive behavior (bowing, attacking, chasing, or driving); the second group is sexual behavior, including displacement preening, soliciting, mounting, and copulation; and the third group includes activities which are associated with nesting behavior, such as nest demonstration or nest soliciting, and nest building. In the course of the breeding cycle, gradual transition from one group of activities to the next has been generally observed.

As demonstrated by numerous studies, androgens, especially testosterone, are important in the control of courtship behavior. Indeed, castration resulted in decreased display frequency, whereas peripheral administration of testosterone to castrated birds restored the behavior observed in intact males (Erpino, 1969; Cheng and Lehrman, 1975). Extensive studies performed by Hutchison (1967, 1969, 1970, 1971, 1974a, 1974b, 1975) have demonstrated the importance of hypothalamic areas in the mediation of the stimulating effects of testosterone on the displays in doves.

The endocrine regulation of male dove courtship represents a very complex phenomenon, as demonstrated by two experiments performed by Hutchison (1974b, 1975). In these experiments, groups of birds were maintained under either long (14 hours/day) or short (6 hours/day) daylengths for one month prior to castration and during the experimental period. Three months after castration, birds of both groups were implanted with TP into the hypothalamus and they then were tested behaviorally. In these conditions, the chasing display was more frequent in the group maintained in long than in short photoperiodic regimen. More recently, McDonald and Liley (1978) also observed that the nest building activity of castrated TP injected male doves is higher in birds maintained in long days (16 hours/day) than in short days (8 hours/day), after castration and during hormonal treatment (see also Hinde et al., 1974, and Gosney and Hinde, 1975 for similar observations in female birds). Results in the same direction, though not significant, were obtained for courtship behavior. As discussed by Hutchison (1976), the behavioral differences observed between the birds maintained in long and in short days might be due to several influences of the photoperiod. One of these influences could consist in a direct action of the gonadotropins, the levels of which were presumably higher in the doves exposed to long than to short days (for such results in other species, see Gibson et al., 1975; Wilson and Follett, 1977; Stokkan and Sharp, 1980).

This possibility has been investigated in the male pigeon (Murton et al., 1969; Murton and Westwood, 1975). In the first study, adult intact males were injected repeatedly with several hormones, namely mammalian LH, FSH, prolactin, or TP. All birds were observed in the presence of females during the period of hormonal treatments. In these conditions, FSH injections increased the frequency of the driving (aggressive) display which is often observed at the beginning of pair formation. This treatment also decreased the frequency of nest demonstration. The effects of FSH injections were to a certain extent specific, as they did not appear in groups of either LH- or prolactin-injected pigeons. In the same experiment, TP administration did not alter the frequency of driving, though they stimulated the bowing display and favored the transition from courtship towards sexual behavior.

These results were extended in a later study (Murton and Westwood, 1975). In this work, FSH injections into intact estradiol pretreated males were again shown to encourage the aggressive (attacking and chasing) component of the courtship display, while TP treatment to such birds instead favored the sexual-appeasement component (bowing) of this display. From these results (see also Murton and Westwood, 1977), it was proposed that, in the male pigeon, the gradual transition from one group of displays to another one during the normal breeding cycle is dependent on changes in the endocrine status of the birds. Accordingly, endogenous FSH titers would be high at the beginning of pair formation, and this hormone would stimulate the aggressive components of courtship display through some direct behavioral action. It must, however, be emphasized that this hypothesis is only partly supported by recent data obtained on the male dove (Cheng and Balthazart, unpubl. observ.). Indeed, in this species, no clearcut variations of the circulating levels of immunoreactive FSH were observed during the courtship phase of the breeding cycle, though a slight decrease seemed to occur just before the females laid the eggs. However, as pointed out already, androgens and especially testosterone are known to exert an important influence on the courtship behavior of pigeons and doves, including the precopulatory courtship patterns. Therefore, it seems that the influence of FSH on these activities should be considered not as determining, but rather as complementing the predominant action of testosterone. The way and the extent to which such a modulatory influence of FSH would occur and would depend on the presence of testosterone or of another androgen remains, however, entirely speculative at present.

Finally, it must also be stressed that, up to now, the study of the control of behavior of pigeons by FSH has been performed only with intact birds, rather than with hypophysectomized castrates or with males with controlled circulating levels of testosterone. Therefore, the possibility exists that some of the FSH-induced observed behavioral effects may have been mediated indirectly.

FSH and the Social Displays of Ducks (Anas platyrhynchos)

The social behavior of male ducks can be divided into three distinct categories: social displays or courtship, sexual activities which are associated with copulation, and aggressive behavior. Each category includes several well-defined patterns which have been submitted to detailed descriptive studies (Heinroth, 1910; Lorenz, 1951; MacKinney, 1969; Weidmann and Darley, 1971). Most of these patterns are not observed with the same frequency during the whole annual reproductive cycle, but, rather, their appearance is restricted to a limited period of the year (Weidmann, 1956; Johnsgard, 1960; Raitasuo, 1964).

Extensive studies have demonstrated that androgens, particularly testosterone, play an important part in the control of social displays and sexual behavior. For example, TP injections into immature (Balthazart, 1974, 1978) or very young birds (Balthazart and Stevens, 1975, 1976; Deviche, 1979a) stimulate sexual behavior and, when administered in high amounts, elicit social displays in ducklings (Deviche and Balthazart, 1976). Also, castration of adult birds leads to a sharp decrease of the frequency of social activities, while TP administration to castrated males restores behavior to that observed in sham-operated drakes (Deviche, 1979b).

It has also been suggested that, in male ducks, social displays might be partly regulated by nonsteroidal hormones, namely by FSH. This conclusion was drawn from the following observations: a) in very young birds, TP treatment often failed to induce social displays, though it stimulated sexual activities (Balthazart and Stevens, 1975, 1976); b) repeated injections of TP to adult birds in the springtime decreased, rather than increased, the frequency of social displays (Balthazart and Deviche, 1977) and at the same time decreased the plasma levels of FSH (Balthazart et al., 1977); c) in the course of the breeding cycle of intact males, sexual behavior is most frequent in spring, while plasma testosterone is high, and social displays are observed mainly in winter, when plasma testosterone levels are lower but plasma FSH levels are maximal. Sexual behavior thus correlates with testosterone levels, while social displays correlate with FSH levels (Balthazart and Hendrick, 1976); d) similar correlations appeared in the course of a study of the daily distribution of behavioral activities and hormone plasma levels of intact males (Balthazart and Hendrick, 1979a).

A hypothetical model has been proposed which accounts satisfactorily for most of these results (Balthazart, 1977). Accordingly, the social displays of ducks would be controlled both by FSH and testosterone. The sensitivity to testosterone of the neural mechanisms involved in the regulation of social displays would vary according to the period of the breeding cycle. Therefore, the relative importance of FSH and testosterone in the control of these activities would also depend on the reproductive state of the birds. This would, for example, explain why treatment of intact birds with TP does not elicit the same behavioral effects in all cases. This model also implies that FSH modulates social displays in conjunction with testosterone for ducks rather than by itself. This would provide an explanation to the observation that the frequency of social displays observed in castrated males is very low, although these birds possess enhanced levels of plasma FSH (Deviche, 1979b; Deviche et al., 1980). However, in the course of several experiments, repeated administration of mammalian gonadotropins to ducklings (Deviche and Balthazart, 1976; Balthazart and Hendrick, 1979b) or to adult birds (Balthazart and Deviche, 1977) have failed to induce significant behavioral modifications. Further experimental studies would therefore be necessary to confirm that these hormones play some direct behavioral role in combination with androgens.

Possible Mode of Action of Gonadotropins on Avian Behavior

Most studies in which a direct influence of the gonadotropins on avian activities has been proposed have been performed by peripheral injection of hormonal preparations of mammalian origin. If indeed mammalian gonadotropins directly affect behavior, two conditions must be fulfilled. The first condition is that the biological action of FSH and LH isolated from mammals is not entirely specific; that is, these hormones can induce some biological effects in birds too. Several types of evidence support this assumption. On one hand, FSH preparations isolated from different species of tetrapod vertebrates, including birds and mammals, present a high degree of immunochemical similarity (Licht and Bona Gallo, 1978). On another hand, mammalian FSH and LH bind specifically to the testes of several species of birds. The testicular binding of FSH is inhibited by mammalian FSH and also by avian adenohypophyseal extracts (Ishii and Farner, 1976; Ishii and Adachi, 1977; Tsutsui and Ishii, 1978; Bona Gallo and Licht, 1979; Bortolussi et al., 1979). Finally, mammalian LH and, to a lesser extent, FSH, possess steroidogenic properties in the male quail, as is generally the case in mammals (Jenkins et al., 1978; Dorrington and Armstrong, 1979; van der Molen et al., 1979).

The second condition is that gonadotropins are able to induce some direct effects at the level of the brain. Up to now, evidence supporting this point is completely lacking for birds. Examination of data obtained in mammals, however, provides three types of evidence in this direction. First, gonadotropins can exert a direct sensitive and specific negative feedback (short-loop feedback) on their own secretion (Szontagh and Uhlarik, 1964; Patritti-Laborde and Odell, 1978; Molitch et al., 1979; Patritti-Laborde et al., 1979). The site of this feedback mechanism has been suggested to be mainly hypothalamic, though recent work rather favored the hypothesis of a pituitary localization (Patritti-Laborde et al., 1979). However, up to now, such a mechanism has never been shown to exist in any species of bird. In addition, administration of mammalian LH to intact male ducklings did not alter their endogenous levels of either LH or FSH, although this treatment induced very high plasma levels of exogenous LH, as well as an increase of the testicular weight (Balthazart and Hendrick, 1979b). Secondly, gonadotropins administration to mammals can modify rapidly the electroencephalographic activity. Such changes have been observed in the hypothalamus (LH, Teresawa et al., 1969) and also in the hippocampus (LH, Gallo et al., 1972; see also Kawakami and Saito, 1965; Ramirez et al., 1967). If also present in birds, such alterations might explain the rapid behavioral modifications measured in some cases following LH treatment (starling, Mathewson, 1961). Thirdly, the testosterone uptake by the preoptic hypothalamic area of rats is decreased after hypophysectomy (Mc Ewen et al., 1970). In birds, this neural zone is directly involved in the regulation of many activities, including precopulatory (courtship) behavior (see Hutchison, 1976, for a review of this topic). Conceivably, therefore, the synergistic behavioral influence of gonadotropins and androgens which has been proposed by several authors might result from an alteration of the brain uptake of androgens by the former hormones. This hypothesis would explain why castrated birds generally display very little courtship activity although their plasma levels of gonadotropins are higher than those measured in sham-operated birds which, on another hand, are behaviorally active (Deviche, 1979b; Deviche et al., 1980). Future work should clarify this possibility.

Conclusions

Until now, the possibility that gonadotropins directly influence the behavior of birds has been proposed only for four species: the starling, the red-billed weaverbird, the domestic pigeon, and the male duck. All those studies except one (weaverbird, Lazarus and Crook, 1973) have been performed only on male birds. The data collected so far clearly suggest that FSH and LH might be directly implicated in the regulation of some aspects of social behavior. However, and as it has been discussed in this paper, alternative explanations often cannot be excluded from the interpretation of the results. This difficulty has arisen from the methodological procedure used for investigating the behavioral influence of the gonadotropins. Therefore, it does not seem that the hypothesis of a direct modulatory role of either FSH or of LH on avian activities should be definitely accepted at the present time. Future progress in this field may now depend on the research of the existence of brain receptors to the gonadotropins and on the examination of the possible influence within the nervous system of the gonadotropins on the uptake (and metabolism?) of behaviorally active steroids, such as the androgens. As pointed out previously, such an influence could provide an explanation for some of the experimental results.

Acknowledgments

We are indebted to Professor E. Schoffeniels for his continued interest in our work. This study was partly supported by grant 2.4544.76 from the "Fonds de la Recherche Collective" to Professor E. Schoffeniels. The author is Senior Research Assistant at the Belgian National Fund for Scientific Research.

Literature Cited

Akerman, B. 1966. Behavioral effects of electrical stimulation in the forebrain of the pigeon. I. Reproductive behavior. Behaviour, 26:323-338.

Arnold, A. P. 1975. The effects of castration and androgen replacement on song, courtship and aggression in Zebra Finches (Poephila guttata). J. Exper. Zool., 191:309-325.

Baggerman, B. 1968. Hormonal control of reproductive and parental behaviour in fishes. Pp. 351-404, in Perspectives in endocrinology (E. Barrington and C. Jorgensen, eds.), Academic Press, New York, xvi + 583 pp.

BALTHAZART, J. 1974. Short-term effects of testosterone propionate on the behaviour of young intact male domestic ducks (*Anas platyrhynchos*). Psychol. Belg., 14:1-10.

———. 1977. Le contrôle hormonal du comportement chez le canard domestique mâle (*Anas platyrhnchos* L.). Thèse de Doctorat, Université de Liège, 339 pp.

———. 1978. Behavioural and physiological effects of testosterone propionate and cyproterone acetate in immature male domestic ducks, *Anas platyrhynchos*. Z. Tierpsychol., 47:410-421.

BALTHAZART, J., AND P. DEVICHE. 1977. Effects of exogenous hormones on the reproductive behaviour of adult male domestic ducks. I. Behavioural effects of intramuscular injections. Behav. Processes, 2:129-146.

BALTHAZART, J., AND J. HENDRICK. 1976. Annual variations in reproductive behaviour, testosterone, and plasma FSH levels in the Rouen duck, *Anas platyrhynchos*. Gen. Comp. Endocrinol., 28:171-183.

———. 1979a. Effects of exogenous gonadotropic and steroid hormones on the social behaviour and gonadal maturation of male domestic ducklings *Anas platyrhynchos*. Arch. Internat. Physiol. Biochem., 87:742-761.

———. 1979b. Relationships between the daily variations of social behavior and of plasma FSH, LH and testosterone levels in the domestic duck *Anas platyrhynchos* L. Behav. Processes, 4:107-128.

BALTHAZART, J., AND M. STEVENS. 1975. Effects of testosterone on the social behaviour of groups of male domestic ducklings *Anas platyrhynchos* L. Anim. Behav., 23:926-931.

———. 1976. Social behaviour of testosterone injected ducklings in the presence of oestrogen-treated females. Behaviour, 57:288-306.

BALTHAZART, J., P. DEVICHE, AND J. HENDRICK. 1977. Effects of exogenous hormones on the reproductive behaviour of adult male domestic ducks. II. Correlation with morphology and hormone plasma levels. Behav. Processes, 2:147-161.

BLUM, V., AND K. FIEDLER. 1965. Hormonal control of reproductive behavior in some cichlid fishes. Gen. Comp. Endocrinol., 5:186-196.

BONA GALLO, A., AND P. LICHT. 1979. Differences in the properties of FSH and LH binding sites in the avian gonad revealed by homologous radioligands. Gen. Comp. Endocrinol., 37:521-532.

BORTOLUSSI, M., P. DEVICHE, L. COLOMBO, AND G. MARINI. 1979. In vitro binding of radioiodinated rat follicle-stimulating hormone to the testis of the mallard duck, *Anas platyrhynchos* L. Basic Appl. Histochem., 23:279-284.

BUTTERFIELD, P. A., AND J. H. CROOK. 1968. The annual cycle of nest building and agonistic behaviour in captive *Quelea quelea* with reference to endocrine factors. Anim. Behav., 16:308-317.

CARPENTER, C. 1933. Psychobiological studies of social behavior in Aves. J. Comp. Psychol., 16:25-90.

CHENG, M. F. 1979. Progress and prospects in ring dove research: a personal view. Pp. 97-130, in Advances in the study of behavior, Vol. 9 (J. S. Rosenblatt, R. A. Hinde, C. Beer, and M.-C. Busnel, eds.), Academic Press, New York, xii + 282 pp.

CHENG, M. F., AND D. LEHRMAN. 1975. Gonadal hormone specificity in the sexual behavior of ring doves. Psychoneuroendocrinol., 1:95-102.

CROOK, J. H., AND P. A. BUTTERFIELD. 1968. Effects of testosterone propionate and luteinizing hormone on agonistic and nest building behaviour of *Quelea quelea*. Animal Behav., 16:370-384.

———. 1970. Gender role in the social system of *Quelea*. Pp. 211-248, in Social behaviour in birds and mammals (J. H. Crook, ed.), Academic Press, London and New York, xi + 492 pp.

DAVIS, D. E. 1957. Aggressive behavior in castrated starlings. Science, 126:253.

DELIUS, J. D., B. CRAIG, AND C. CHAUDOIR. 1976. Adrenocorticotropic hormone, glucose and displacement activities in pigeons. Z. Tierpsychol., 40:183-193.

DEVICHE, P. 1979a. Effects of testosterone propionate and pituitary-adrenal hormones on the social behaviour of male ducklings (Anas platyrhynchos L.) in two test situations. Z. Tierpsychol., 49:77-86.

———. 1979b. Behavioral effects of castration and testosterone propionate replacement combined with ACTH in the male domestic duck (Anas platyrhynchos L.). J. Exp. Zool., 207:471-480.

DEVICHE, P., AND J. BALTHAZART. 1976 Behavioural and morphological effects of testosterone and gonadotropins in the young male domestic duck (Anas platyrhynchos L.). Behav. Processes, 1:217-232.

DEVICHE, P., AND J. D. DELIUS. 1981. Short-term modulation of domestic pigeon (Columba livia L.) behaviour induced by intraventricular administration of ACTH. In press.

DEVICHE, P., J. BALTHAZART, W. HEYNS, AND J.-C. HENDRICK. 1980. Endocrine effects of castration followed by androgen replacement and ACTH injections in the male domestic duck (Anas platyrhynchos L.). Gen. Comp. Endocrinol., 41:53-61.

DORRINGTON, J. H., AND D. T. ARMSTRONG. 1979. Effects of FSH on gonadal functions. Rec. Prog. Hormone Res., 35:301-342.

ERICKSON, C. J., AND M. C. MARTINEZ-VARGAS. 1975. The hormonal basis of cooperative nest building. Pp. 91-109, in Neural and endocrine aspects of behaviour in birds (P. Wright, P. G. Caryl, and D. Vowles, eds.), Elsevier, Amsterdam, x + 408 pp.

ERPINO, M. J. 1969. Hormonal control of courtship behaviour in the pigeon (Columba livia). Anim. Behav., 17:401-405.

FABRICIUS, E., AND A.-M. JANSSON. 1963. Laboratory observations on the reproductive behaviour of the pigeon (Columba livia) during the pre-incubation phase of the breeding cycle. Anim. Behav., 11:534-547.

GALLO, R. G., J. H. JOHNSON, D. I. KALRA, D. I. WHITMOYER, AND C. H. SAWYER. 1972. Effect of LH on multiple unit activity in the rat hippocampus. Neuroendocrinol., 9:149.

GIBSON, W. R., B. K. FOLLETT, AND B. GLEDHILL. 1975. Plasma levels of luteinizing hormone in gonadectomized Japanese quail exposed to short or to long daylengths. J. Endocrinol., 64:87-101.

GOSNEY, D., AND R. A. HINDE. 1975. An oestrogen-mediated effect of photoperiod on the reproductive behaviour of the budgerigar. J. Reprod. Fert., 45:547-548.

HEINROTH, O. 1910. Beitrage zur Biologie, namentlich Ethologie und Psychologie der Anatiden. Ver. 5 Int. Ornith. Kongr. Berlin, pp. 589-702.

HINDE, R. A., E. STEEL, AND B. K. FOLLETT. 1974. Effect of photoperiod on oestrogen-induced nest-building in ovariectomized or refractory female canary (Serinus canarius). J. Reprod. Fert., 40:383-399.

HOAR, W. 1962. Reproductive behaviour of fish. Gen. Comp. Endocrinol., Suppl., 1:206-216.

HUTCHISON, J. B. 1967. Initiation of courtship by hypothalamic implants of testosterone propionate in castrated doves (Streptopelia risoria). Nature, 216:591-592.

———. 1969. Changes in hypothalamic responsiveness to testosterone in male Barbary doves (Streptopelia risoria). Nature, 222:176-177.

———. 1970. Influence of gonadal hormones on the hypothalamic integration of courtship behaviour in the Barbary dove. J. Reprod. Fert., Suppl., 11:15-41.

———. 1971. Effects of hypothalamic implants of gonadal steroids on courtship behaviour in Barbary doves (Streptopelia risoria). J. Endocrinol., 50:97-113.

———. 1974a. Post-castration decline in behavioural responsiveness to intrahypothalamic androgen in doves. Brain Res., 81:169-181.

———. 1974b. Effect of photoperiod on the decline in behavioural responsiveness to intrahypothalamic androgen in doves (Streptopelia risoria). J. Endocrinol., 63:583-584.

———. 1975. Target cells for gonadal steroids in the brain: studies on steroid-sensitive mechanisms of behaviour. Pp. 123-137, in Neural and endocrine aspects of behaviour in birds (P. Wright, P. G. Caryl, and D. M. Vowles, eds), Elsevier, Amsterdam, x + 408 pp.

————. 1976. Hypothalamic mechanisms of sexual behavior, with special references to birds. Pp. 159-200, *in* Advances in the study of behavior (J. Rosenblatt, R. A. Hinde, E. Shaw, and C. Beer, eds.), Academic Press, New York, xii + 282 pp.

ISHII, S., AND T. ADACHI. 1977. Binding of avian testicular homogenate with rat follicle-stimulating hormone and inhibition of the binding by hypophyseal extracts of lower vertebrates. Gen. Comp. Endocrinol., 31:287-294.

ISHII, S., AND D. S. FARNER. 1976. Binding of follicle-stimulating hormone by homogenates of testes of photostimulated white-crowned sparrows, *Zonotrichia leucophrys gambelii*. Gen. Comp. Endocrinol., 30:443-450.

JENKINS, N., J. P. SUMPTER, AND B. K. FOLLETT. 1978. The effects of vertebrate gonadotropins on androgen release *in vitro* from testicular cells of Japanese quail and a comparison with their radioimmunoassay activities. Gen. Comp. Endocrinol., 35:309-321.

JOHNSGARD, P. 1960. A quantitative study of sexual behavior of the mallard and the black ducks. Wilson Bull., 72:133-153.

KAWAKAMI, M., AND H. SAITO. 1965. Unit activity in the hypothalamus of the cat: effect of genital stimuli, luteinizing hormone and oxytocin. Jap. J. Physiol., 17:466-486.

KRAMER, B. 1971. Zur hormonalen Steuerung von Verhaltensweisen der Fortpflanzung beim Sonnenbarsch. Z. Tierphyschol., 28:351-386.

LANDSBERG, J. W., AND J. WEISS. 1977. Stress and increase of the corticosterone level prevent imprinting in ducklings. Behaviour, 57:173-189.

LAZARUS, J., AND J. H. CROOK. 1973. The effects of luteinizing hormone, oestrogen and ovariectomy on the agonistic behaviour of female *Quelea quelea*. Anim. Behav., 21:49-60.

LEHRMAN, D. S., AND P. BRODY. 1961. Does prolactin induce incubation behaviour in the ring dove? J. Endocrinol., 22:269-275.

————. 1964. Effect of prolactin on established incubation behaviour in the ring dove. J. Comp. Physiol. Psychol., 57:161-165.

LICHT, P., AND A. BONA GALLO. 1978. Immunochemical relatedness among pituitary follicle-stimulating hormones of tetrapod vertebrates. Gen. Comp. Endocrinol., 36:575-584.

LORENZ, K. 1951. Comparative studies on the behaviour of the Anatidae. Avic. Mag., 57:157-182 (translated by Dr. C. H. D. Clarke).

MACKINNEY, F. 1969. The behaviour of ducks. Pp. 593-626, *in* The behaviour of domesticated animals (E. Hafez, ed.), 2nd edition, Bailliere, Tindall, and Cassel, London, xii + 617 pp.

MARTIN, J. T. 1975. Hormonal influences in the evolution and ontogeny of imprinting behavior in the duck. Pp. 357-366, *in* Progr. Brain Res., 42, xii + 425 pp.

MARTIN, J. T. 1978a. Embryonic pituitary adrenal axis, behavior development and domestication in birds. Amer. Zool., 18:489-499.

MARTIN, J. T. 1978b. Imprinting behavior: pituitary-adrenocortical modulation of the approach response. Science, 200:565-567.

MATHEWSON, S. F. 1961. Gonadotrophic hormones affect aggressive behavior in starlings. Science, 134:1522-1523.

MCDONALD, P. A., AND N. R. LILEY. 1978. The effects of photoperiod on androgen-induced reproductive behavior in male ring doves, *Streptopelia risoria*. Horm. Behav., 10:85-96.

MC EWEN, B. S., D. W. PFAFF, AND R. F. ZIGMOND. 1970. Factors influencing sex hormone uptake by rat brain regions. II. Effect of neonatal treatment and hypophysectomy on testosterone uptake. Brain Res., 21:17-28.

MEIER, A. H., D. S. FARNER, AND J. R. KING. 1965. A possible endocrine basis for migratory behaviour in the white-crowned sparrow, *Zonotrichia leucophrys gambelii*. Anim. Behav., 13:453-465.

MEIER, A. H., J. T. BURNS, AND J. W. DUSSEAU. 1969. Seasonal variations in the diurnal rhythm of pituitary prolactin content in the white-throated sparrow, *Zonotrichia albicollis*. Gen. Comp. Endocrinol., 12:282-289.

MOLITCH, M. E., M. EDMONDS, J. C. HENDRICK, P. FRANCHIMONT, AND W. D. ODELL. 1979. Specificity of short-loop feedback of luteinizing hormone in the rabbit. Neuroendocrinol., 29:49-53.

MURTON, R. K., AND N. J. WESTWOOD. 1975. Integration of gonadotropin and steroid secretion, spermatogenesis and behaviour in the reproductive cycle of male pigeons species. Pp. 51-89, *in* Neural and endocrine aspects of behaviour in birds (P. Wright, P. G. Caryl, and D. M. Vowles, eds.), Elsevier, Amsterdam, x + 408 pp.

———. 1977. Avian breeding cycles. Clarendon Press, Oxford, 594 pp.

MURTON, R. K., R. J. P. THEARLE, AND B. LOFTS. 1969. The endocrine basis of breeding behaviour in the feral pigeon (*Columba livia*): I. Effects of exogenous hormones on the pre-incubation behaviour of intact males. Anim. Behav., 17:286-306.

PATRITTI-LABORDE, N., AND W. D. ODELL. 1978. Short-loop feedback of luteinizing hormone: dose-response relationships and specificity. Fert. Steril., 30:456-460.

PATRITTI-LABORDE, N., A. R. WOLFSEN, D. HEBER, AND W. D. ODELL. 1979. Site of short-loop feedback for luteinizing hormone in the rabbit. J. Clin. Invest., 64:1066-1069.

RAITASUO, K. 1964. Sexual behaviour of the mallard duck *Anas platyrhynchos*, in the course of the annual cycle. Riistatieteellisia julkaisuga, 24:1-72.

RAMIREZ, D., B. KOMISARUK, D. WHITMOYER, AND C. SAWYER. 1967. Effects of hormones and vaginal stimulation on the EEG and hypothalamic units in rats. Amer. J. Physiol., 212:1376-1384.

REY, P. 1948. Sur le déterminisme hormonal du réflexe d'embrassement chez les males de batraciens anoures. J. Physiol. (Paris), 40:292-293.

SELINGER, H. E., AND G. BERMANT. 1967. Hormonal control of aggressive behavior in Japanese quail (*Coturnix coturnix japonica*). Behaviour., 28:255-268.

STOKKAN, J.-A., AND P. J. SHARP. 1980. The roles of day length and the testes in the regulation of plasma LH levels in photosenstitive and photorefractory willow ptarmigan (*Lagopus lagopus lagopus*). Gen. Comp. Endocrinol., 41:520-526.

SZONTAGH, F. E., AND S. UHLARIK. 1964. The possibility of a direct "internal" feedback in the control of pituitary gonadotropin secretion. J. Endocrinol., 29:203-204.

TANABE, Y., T. NAKAMURA, K. FUJIOKA, AND O. DOI. 1979. Production and secretion of sex steroid hormones by the testes, the ovary, and the adrenal glands of embryonic and young chickens (*Gallus domesticus*). Gen. Comp. Endocrinol., 39:26-33.

TERESAWA, E., D. I. WHITMOYER, AND C. H. SAWYER. 1969. Effects of luteinizing hormone (LH) on multiple activity in the rat hypothalamus. Amer. J. Physiol., 217:1119.

TSUTSUI, K., AND S. ISHII. 1978. Effects of follicle-stimulating hormone and testosterone on receptors of follicle-stimulating hormone in the testis of the immature Japanese quail. Gen. Comp. Endocrinol., 36:297-305.

VAN DER MOLEN, H. J., W. M. O. VAN BEURDEN, M. A. BLANKENSTEIN, W. DE BOER, B. A. COOKE, J. A. GROOTEGOED, F. H. A. JANSZEN, F. H. DE JONG, E. MULDER, AND F. F. G. ROMMERTS. 1979. The testis: biochemical actions of the trophic hormones and steroids on steroid production and spermatogenesis. J. Steroid Chem., 11:13-18.

VOWLES, D., AND D. HARWOOD. 1966. The effect of exogenous hormones on aggressive and defensive behaviour in the ring dove (*Streptopelia risoria*). J. Endocrinol., 36:35-51.

WEIDMANN, U. 1956. Verhaltenstudien an der Stockenten (*Anas platyrhynchos* L.). I. Das Aktionsystem. Z. Tierpsychol., 13:208-271.

WEIDMANN, U., AND J. DARLEY. 1971. The role of the female in social display of mallards. Anim. Behav., 19:287-299.

WILSON, F. E., AND B. K. FOLLETT. 1977. Testicular inhibition of gonadotropin secretion in photosensitive tree sparrows (*Spizella arborea*) exposed to a winter-like day length. Gen. Comp. Endocrinol., 32:440-445.

WITSCHI, E. 1954. Vertebrate gonadotropins. Mem. Soc. Endocrin., 4:149-163.

YOUNG, C. E., AND L. J. ROGERS. 1978. Effects of steroidal hormones on sexual, attack, and search behavior in the isolated male chick. Horm. Behav., 10:107-117.

Reprinted from
ASPECTS OF AVIAN ENDOCRINOLOGY:
PRACTICAL AND THEORETICAL IMPLICATIONS (C. G. Scanes *et al.*, eds.)
Grad. Studies, Texas Tech Univ., 1982, 26:1-411.

HORMONAL PATTERN AND BEHAVIOR
IN OVARIECTOMIZED FOWL

ARLETTE GUICHARD[1] AND MARYSE REYSS-BRION[2]

[1]*E. R. 123 CNRS—U. 166 Inserm, Maternité Port-Royal, 75014
Paris, France; and* [2]*Institut d'Embryologie du CNRS et du Collège
de France, 49 bis av. de la Belle Gabrielle, 94130 Nogent S/Marne,
France*

It has long been observed that the genital tract of the hen is characterized by asymmetric development; the left ovary and oviduct become functional and the right gonad remains undeveloped. A postnatal ovariectomy induces the development of the rudimentary right female gonad into a testislike gland (Benoit, 1923; Domm, 1927, 1929). The sex reversed females show secondary male sexual characters such as comb, wattles, spurs, and male plumage. The evolution of the steroidogenic cells during the masculinization have been studied by histological and histoenzymological techniques (Woods and Domm, 1966; Reyss-Brion and Scheib, 1980).

Recently, using radioimmunoassay, we have studied the level of steroids (progesterone, dehydroepiandrosterone, $\Delta_4$androstenedione, testosterone, dihydrotestosterone, estrone, and estradiol) in normal male and female chick gonads and in testislike right gonads from chicks ranging in age between nine days and five months (Reyss-Brion *et al.*, 1980).

In the present work we have investigated the sex hormone pattern of older animals (until 12 months) and the correlation between this steroidogenesis and the modifications of sexual characters and behavior occurring after ovariectomy.

MATERIAL AND METHODS

The investigations were performed on Hubbard chickens that were sinistrally ovariectomized at nine days post-hatching as previously described (Reyss-Brion and Scheib, 1980). Males, females, and ovariectomized fowls were sacrificed at varying ages from two to 12 months. The autopsies were always made during spring or summer to avoid seasonal changes; combs and wattles were weighed. Gonads were recovered, cut into small pieces, and incubated in synthetic medium (Parker 199 + 1 per cent glutamine) for two hours at 39°C, under an atmosphere of 90 per cent oxygen and 10 per cent CO_2. The gonads were then homogenized in their incubation medium. The homogenates were dispensed in aliquots to be measured by the radioimmunoassay procedures described previously (Guichard *et al.*, 1977).

Our paper reports results on testosterone, estrone, and estradiol content in ovary, testes, and transformed gonads. The mean of six assays was determined for each stage studied. The statistical significance of the data was

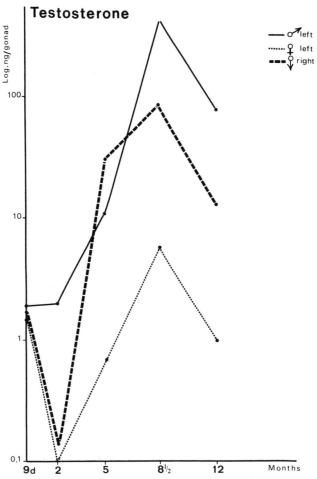

FIG. 1.—Changes of testosterone in left ovary (♀), left testis (♂), and right female gonad after ovariectomy (♀) from nine days to 12 months, expressed in Log ng/gonad.

tested by the nonparametric Mann-Whitney U-test (Mann and Whitney, 1947; Jacobson, 1963).

RESULTS AND DISCUSSION

Steroid Radioimmunoassay

The evolution of testosterone production was studied in the right female gonad in ovariectomized females and in the normal ovary and left testis in intact animals (Fig. 1; Table 1).

At 9 days, testosterone production was similar in ovary, testis, and right female gonad. This is in agreement with observations of the 18-day embryonic gonads in organotypic culture (Guichard *et al.*, 1977).

At 2 months, a significant difference appeared between male and female testosterone content; the level remained constant in the testis and dropped to

TABLE 1.—*Steroid production by incubated female, male, and transformed gonads (expressed in pg/gonad).*

Age	Female (left gonad)			Male (left gonad)			Transformed right gonad		
	Testosterone	E_1	E_2	Testosterone	E_1	E_2	Testosterone	E_1	E_2
9 days	1535	750	243	1883	36	0	1661[a]	83[a]	0[a]
2 months	0	5136	142	2047	129	3	141	884	1191
5 months	701	29921	12740	10670	318	0	31014	30	193
8½ months	5577	18080	9312	453399	4159	1808	88408	788	1179
12 months	1020	7896	19866	79311	1570	1590	12761	759	1585

[a] At 9 days, the results given for transformed gonad are those of the rudimentary female gonad at the time of surgery.

a nondetectable level in the ovary. A significant decrease of testosterone also was observed in the transforming right gonad which had female-like hormonal content in agreement with histological observations. At two months, we frequently observed remaining cortical components with atretic follicles, which later disappeared.

At 5 months, the difference between male and female gonads was maintained. Testosterone was chiefly present in the testis but also was found in the ovary. In the transformed gonad, the production of testosterone was significantly higher than that of the testis ($0.005 < P < 0.01$), in fact, twice as high as the sum of that produced by the two normal testes. At this stage, the masculinization of the right female gonad after ovariectomy seemed to be achieved.

This important production of testosterone was correlated with the growth of the comb. The mean weight of this erectile organ in all transformed animals (2500 mg) was significantly higher ($0.005 < P < 0.01$) than that of the normal female fowl (284 mg). It exceeded that in males (1890 mg), but the differences were not significant. This confirmed the fact that, in the chicken, the comb is a target organ for androgen (Balthazart *et al.*, 1979).

At 8½ months, testosterone reached maximum values in all gonads. This peak of testosterone might have occurred between five and eight and one-half months. In the testis-like gland, testosterone content stayed elevated, but became significantly lower ($0.01 < P < 0.05$) than that of the testis. These differences remained at 12 months.

Estrogens, specifically estrone (E_1) and estradiol (E_2) also were examined in animals between nine days and 12 months (Fig. 2; Table 1). The concentration of estrogens was always significantly higher ($0.01 < P < 0.05$) in ovaries than in other gonads. At the time of surgery, estrogens were mainly present in the functional ovary which was removed; testes and right female gonads had similar low levels. At two months, estrogens were predominant in the ovary but a notable amount appeared in the transforming gonad, which contained 2000 pg of estrogens ($E_1 + E_2$). The control right female gonad (in the unaltered female) of the same age only produced 96 pg. This is in agreement with the decrease of testosterone and with the feminized histology. At five months, an important increase of estrogens in the ovary was observed; this level remained constant after five months.

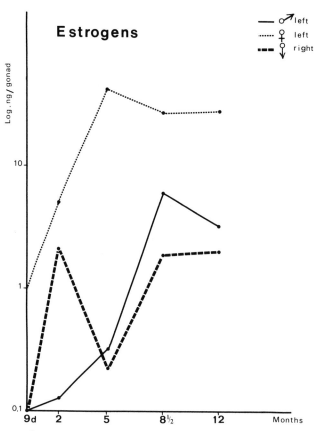

Fɪɢ. 2.—Changes in total estrogen in left ovary, left testis and right female gonad after ovariectomy from niné days to 12 months, expressed in Log ng/gonad.

In the testis, as in the transformed gonad, the estrogen concentration at nine days was very low, and the difference between the two gonads was not significant. From this period until 12 months, testis and the testis-like gland showed similar levels of estrogen. It must be noted that the male gonad generally produced more E_1 than E_2, unlike the transformed gonad in which E_2 was predominant (Table 1).

Behavioral Observations

Males and females of many species exhibit various patterns of behavior, especially related to courtship and reproduction (Tinbergen, 1951). Recent evidence suggests that hormones elicit these typical patterns of behavior. Variations in the hormonal system cause quantitative and qualitative changes in the behavioral output. Inversely, the level of circulating hormones can be modified by the perception of either self-performed behavior or partner behavior (Pröve and Sossinka, this vol.). The androgen dependence of male sexual characters was suggested by total or subtotal castration experiments. The capon lost the sexual behavior of the cock, whereas testosterone propionate-treated capons recovered male sexual behavior (Deviche, this vol.). Thus, our ovariectomized fowls with male secondary sexual characters and high levels of androgens were expected to show male-like behavior.

The behavior of the ovariectomized hen has been described previously by Benoit (1923) and Domm (1927). They found that those animals "showed none of the normal female sexual reactions and few or none of the male." This is in agreement with our observations. Our ovariectomized animals lost all female behavioral patterns; they did not stay on the nest or cackle like other hens; however, they did not acquire any male activity. They were not aggressive and they did not sing. Furthermore, they were not interested in females and did not fight cocks. Most were found to be quiet and inoffensive. When normal cocks were introduced into the hen house, they considered these poulards to be females. The mating behavior of these ovariectomized animals probably resulted from their passivity rather than from a real female comportment.

Our results suggest two hypotheses: 1) testosterone level, although very high at five months in transformed animals, was perhaps below the threshold of activity for male behavior; and 2) testosterone increased the comb size, but a higher dose might be necessary to induce the sexual behavior, and the metabolism of testosterone could be a necessary step in the expression of the physiological effects of this hormone (Balthazart and Hirschberg, 1979).

In the transformed animals, secretion of testosterone might occur too late to be able to modify female brain centers and corresponding sexual behavior patterns. Some evidence suggests that structural differences in male and female central nervous systems might contribute to these differences in behavior (Raisman and Field, 1971). It is well known that the brain is one of the most important organs to become sexually differentiated in fetal or neonatal life under the influence of testosterone (Short, 1979). Testosterone appears to be aromatized to estrogen in the brain, and there is now much evidence to suggest that estrogen may be the biologically active form in the central nervous system (Naftolin et al., 1975). Moreover it has been demonstrated that there are no sex differences in the distribution of steroid receptors in the brain but there are important physiological differences between male and female in the hypothalamic and pituitary response to injection of estrogen (Knobil, 1974; Van Look et al., 1977).

On the other hand, there is a striking sexual dimorphism in song control areas of the brain of the Zebra finch, which can be related to behavioral differences between the two sexes. Adult Zebra finch females do not sing, even after being administered testosterone, because some of their brain centers, which are responsible for singing, are in a less developed state than are those of the males (Nottebohm and Arnold, 1976). Females treated with steroids in early development will sing after testosterone treatment in adulthood (Gurney and Konishi, 1980). Thus, the maturation of the brain centers that regulate sexual behavior seems to be very important.

Conclusions

Despite a male steroidogenic profile induced by postnatal ovarietomy, we did not observe real male sexual behavior. As at puberty, the androgen level is higher in ovariectomized than in normal animals. It is concluded that hormonal changes in the transformed pullets occurred too late to modify completely the brain centers that were organized according to the genetic sex during ontogenesis.

Acknowledgment

This work was supported by grants from D.G.R.S.T. 79.7.1397.

Literature Cited

Balthazart, J., and D. Hirschberg. 1979. Testosterone metabolism and sexual behavior in the chick. Horm. Behav., 12:253-263.

Balthazart, J., D. Hirschberg, and C. Puts. 1979. Testosterone metabolism in the brain and the comb of chicks and its control of steroid hormones. In preparation.

Benoit, J. 1923. Transformation experimentale du sexe par ovariectomie précoce chez la poule domestique. C. R. Acad. Sci., 177:1074-1077.

Domm, L. V. 1927. New experiments on ovariectomy and the problem of sex inversion in the fowl. J. Exp. Zool., 48:31-173.

———. 1929. The effect of bilateral ovariectomy in the brown leghorn fowl. Biol. Bull., 56:459-497.

Guichard, A., L. Cedard, Th.-M. Mignot, D. Scheib, and K. Haffen. 1977. Radioimmunoassays of steroids produced by cultured chick embryo gonads: differences according to age, sex and side. Gen. Comp. Endocrinol., 32:255-265.

Gurney, M. E., and M. Konishi. 1980. Hormone-induced sexual differentiation of brain and behavior in Zebra finches. Science, 208:1380-1384.

Jacobson, J. E. 1963. The wilco on two samples statistic tables and bibliography. J. Amer. Stat. Assoc., 58:1086-1103.

Knobil, E. 1974. On the control of gonadotropin secretion in the rhesus monkey. Rec. Prog. Hormone Res., 30:1-36.

Mann, H. B., and D. R. Whitney. 1947. On a test of whether one of two random variables is stochastically larger than the other. Ann. Math. Stat., 18:50-60.

Naftolin, F., K. J. Ryan, I. J. Devis, V. V. Reddy, F. Flores, Z. Petro, R. J. White, Y. Takao, and L. Woolin. 1975. The formation of estrogens by central neuroendocrine tissues. Rec. Prog. Hormone Res., 31:295-315.

Nottebohn, F., and A. P. Arnold. 1976. Sexual dimorphism in vocal control areas of the songbird brain. Science, 194:211-213.

Raisman, G., and P. M. Field. 1971. Sexual dimorphism in the preoptic area of the rat. Science, 173:731-733.

Reyss-Brion, M., and D. Scheib. 1980. Development of the steroid producing cells during the transformation of the right gonad into a testis ovariectomized chicks. Gen. Comp. Endocrinol., 40:69-77.

Reyss-Brion, M., Th-M. Mignot, and A. Guichard. 1981. Evolution of steroidogenesis in the right gonad of the fowl masculinized by left ovariectomy. Gen. Comp. Endocrinol., in press.

Short, R. V. 1979. Sex determination and differentiation. Brit. Med. Bull., 35:121-127.

Tinbergen, N. 1951. The study of instinct. Oxford Univ. Press, Toronto, xii + 228 pp.

Van Look, P. F. A., W. M. Hunter, C. S. Carker, and D. T. Baird. 1977. Failure of positive feedback in normal men and subjects with testicular feminization. Clin. Endocrinol., 7:353-366.

WOODS, J. E., AND L. V. DOMM.   1966.   A histochemical identification of the androgen producing cells in the gonads of the domestic fowl and albino rat.   Gen. Comp. Endocrinol., 7:559-570.

Reprinted from
Aspects of Avian Endocrinology:
Practical and Theoretical Implications (C. G. Scanes *et al.*, eds.)
Grad. Studies, Texas Tech Univ., 1982, 26:1-411.

HORMONAL CONTROL OF AGGRESSIVE BEHAVIOR IN MALE JAPANESE QUAIL

Susumu Ishii and Kazuyoshi Tsutsui

*Department of Biology, Waseda University, Nishi-Waseda, Tokyo
160, Japan*

It has been established that circulating androgen must reach a certain level for the display of aggressive behavior to occur in male animals and birds. In his excellent monograph, Leshner (1978) introduced a number of studies suggesting that degrees of aggressiveness are correlated with levels of circulating androgen. However, he cautioned that we should not easily accept this relationship, in so far as there have been no experimental studies on the dose-response relationship between androgen dosages and levels of aggressiveness.

In contrast to most investigators, Eaton and Resko (1974) observed a lack of correlation between testosterone levels and levels of aggressiveness in male *Maccaca fuscata*. In ducks and Japanese quail, Balthazart *et al.* (1977, 1979) reported that frequencies of aggressiveness are not correlated with androgen levels. Tsutsui and Ishii (1981) also obtained a similar result in Japanese quail and further demonstrated that replacement therapy with either a constant dose or different doses of testosterone propionate (TP) to a group of castrated Japanese quail did not change their order of aggressiveness.

This communication summarizes our published and unpublished studies on the hormonal control of aggressive behaviors of the Japanese quail and discusses the hormonal mechanism that participates in the determination of the level of aggressiveness.

Method for Determining Levels of Aggressiveness

Five to eight Japanese quail were used in each experiment. Birds were paired in all possible combinations in each experimental group. Each pair was kept in a small observation cage (25 by 37 by 18 cm) for five or ten minutes and their behaviors were observed and recorded.

Aggressiveness of birds was scored according to the following criteria. If only one of the pair attacked (pecking, head-grabbing, mounting), and the other showed no attack, a score of $+2$ points was given to the aggressor and -2 to the passive bird. If both birds attacked, but the frequency of the attack was higher in one, $+1$ was given to the more aggressive bird and -1 was given to the less aggressive one. When both birds attacked at an equal frequency or both did not attack, a score of 0 was given to each. Points were then totaled for each bird. Accordingly, the points for an individual bird ranged between $-2(n-1)$ and $+2(n-1)$, in a group of n birds.

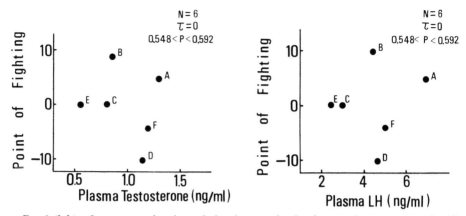

Fig. 1 (left).—Scattergram showing relation between levels of aggressiveness expressed with point of fighting and plasma testosterone levels of six adult male Japanese quail.

Fig. 2 (right).—Scattergram showing relation between levels of aggressiveness expressed with point of fighting and plasma luteininzing hormone levels of six adult male Japanese quail.

Levels of Aggressiveness and Hormone Levels

Orders of aggressiveness as well as plasma levels of testosterone and luteinizing hormone (LH) were observed and recorded for a group of adult male Japanese quail. Statistical analysis using Kendal's method for rank correlation revealed that there was no significant correlation between orders of aggressiveness as determined by points of fighting and levels of testosterone as determined by a radioimmunoassay (Fig. 1).

This result is in accordance with results reported by Balthazart et al. (1977, 1979) in ducks and Japanese quail; there was no correlation between the androgen level and the frequency of aggressive behaviors. In the sexual behavior of the mammal, Harding and Feder (1976) found that there were no differences in plasma testosterone levels between highly active and relatively inactive male guinea pigs in terms of sexual response. A similar result was obtained by Damassa et al. (1977) for sexual behavior of rats and by Balthazart et al. (1977, 1979) for sexual behavior of ducks and Japanese quail.

In our study as well as the study by Balthazart et al. (1979), it was found that plasma LH and FSH levels also showed no significant correlation with orders of aggressiveness in a group of adult male Japanese quail (Fig. 2).

Castration and Replacement Therapy

Aggressive behavior in a group of adult male Japanese quail disappeared completely several days after castration (Fig. 3). This behavior reappeared after subcutaneous injections of a constant dose (3 mg) of testosterone propionate. The order of aggressiveness re-established by the replacement therapy was identical to that observed before castration. Injections of different doses (0.4 to 2.0 mg) of testosterone propionate to the same birds also produced the same order of aggressiveness with those observed previously.

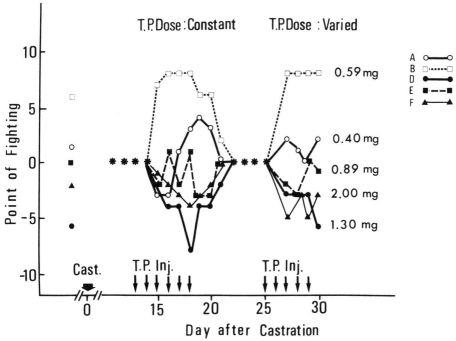

FIG. 3.—Effects of castration and replacement therapies with a constant (3 mg) and different doses of testosterone propionate (TP) on the order of aggressiveness of five adult male Japanese quail. The level of aggressiveness is shown in ordinate by points of fighting (see text). Average fighting points for seven days observed before castration are shown near the left end of the figure.

Another experiment was conducted to determine the effects of artificially increased levels of testosterone in lower-ordered male Japanese quail. Daily injections of two mg of testosterone propionate to fifth and sixth order individuals of a group of six birds did not change their orders (Fig. 4).

Thus, we demonstrated that 1) replacement therapies with either constant or different doses of testosterone propionate in castrated males could not change the order of aggressiveness, and 2) the increase in testosterone levels in lower-order males did not change their orders. These experiments strongly suggest that intensity of the behaviors is not related to the circulating androgen level, although a sufficient circulating androgen concentration is necessary for the display of the behavior.

In the sexual behavior of the rat, it was reported by Beach and Holz-Tucker (1949) that graded doses of testosterone propionate increased the frequency of copulations in castrated rats proportionally to the androgen dose up to the precastration level. However, the same treatment to intact rats neither increased nor decreased the frequency of copulations. Cunningham et al. (1977), using low and high mating lines of the Japanese quail, found that the mating frequency of castrated and androgen-substituted birds of the low mating line never reached the level of those of the high mating line at

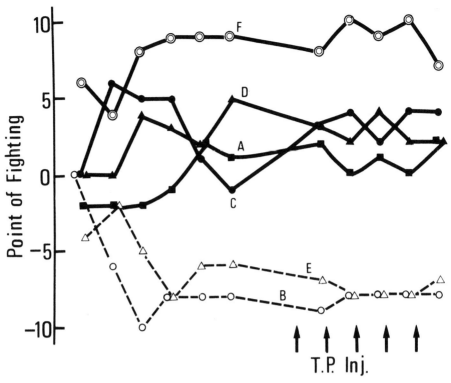

Fig. 4.—Effects of testosterone propionate (TP) injections on the order of aggressiveness in 5th and 6th order adult male Japanese quail ($n = 6$).

any androgen dosages employed. Although they failed to obtain consistent results on the dose-response relationship, they further showed that exogenous androgen had no influence on the mating frequency of intact birds from the low mating line but that it had a depressing effect on the mating frequency of those from the high mating line.

It may be concluded that testosterone secreted from the testis in adult male mammals and birds has a permissive effect on the aggressive behavior of the animal. It is also likely that the intensity of aggressiveness, as measured by the frequency of the behavior as an absolute value or by the order of aggressiveness as relative value, is independent from the circulating androgen level and may be intrinsic to each individual or predetermined by some unknown mechanism before sexual maturation. In terms of sexual behavior, a similar hormone mechanism can be considered, although some conflicting results still exist.

Steroid Species to Activate Aggressive Behaviors

Adkins has shown in a series of studies that estradiol-17β and aromatizable androgens activate male sexual behaviors in chickens and Japanese quail

TABLE 1.—*The order of aggressiveness of adult male Japanese quail receiving injections of 50 and 200 µg of testosterone propionate (TP) per day for three days or vehicle alone (from days 10 through 12 after hatch). The average order was used for birds having the same level of aggressiveness.*

Treatment	Order of aggressiveness of each bird				
TP, 200 µg	1.5	4	4	4	9
TP, 50 µg	1.5	7	8	10	13
Vehicle alone	6	11	12	14	15

(Adkins, 1975, 1977; Adkins and Adler, 1972), and antiestrogen, CI-628, inhibited testosterone-stimulated copulatory behavior in castrated male Japanese quail (Adkins and Nock, 1976).

We can show that aggressive behavior displayed between males was also activated in castrated Japanese quail by estradiol-17β and aromatizable androgens, i.e., testosterone and Δ^4-androstenedione, but not activated by non-aromatizable androgens, that is, 5α-dihydrotestosterone, and 5α-androstandione. These results support an idea that aggressive behavior of Japanese quail, as well as sexual behavior of mammals and birds, is activated by estradiol converted from testosterone in the brain, although the level of aggressiveness is believed to be determined by some other mechanism.

Effects of Androgens Administered at an Immature Stage on the Order of Aggressiveness at Adulthood

Although a number of investigators have studied effects of steroid hormone administered at early developmental stages on sexual behavior and aggressive behavior after maturation in mammals (see Leshner, 1978), only a few reports are available on comparable studies in avian species other than the chicken (Japanese quail, Adkins, 1975, 1977; Hutchison, 1978). We injected testosterone propionate or its vehicle into three groups of immature male Japanese quail. Injections were made on days 10, 11, and 12 after hatch. Aggressive behavior was observed soon after sexual maturation (approximately day 60). The first group received 50 µg, and the third group received only the vehicle (sesame oil). All the birds were isolated soon after the first injection and had had no chance to fight each other prior to the time observations of aggressiveness were made.

The order of aggressiveness throughout the three groups was determined from scores of paired fighting (Table 1). Statistical analysis using the Kruskal-Wallis test revealed that there was a significant difference ($P < 0.05$) in the level of aggressiveness among three groups.

In the second experiment, 5α-dihydrotestosterone (5α-DHT) and estradiol-17β were injected into male cockerels at two different dose levels of 125 µg and 500 µg on day 10. The order of aggressiveness differed significantly between two 5α-DHT injected groups and a control group (Table 2). Estradiol-17β induced no significant change in the order of aggressiveness.

TABLE 2.—*The order of aggressiveness of adult male Japanese quail receiving an injection of 500 or 125 µg of 5α-dihydrotestosterone (5α-DHT) or vehicle alone on day 10 after hatch. The average order was used for birds having the same level of aggressiveness.*

Treatment	Order of aggressiveness of each bird					
5α-DHT, 500 µg	4.5	4.5	4.5	8	8	8
5α-DHT, 125 µg	1	2	4.5	11	15	16
Vehicle alone	10	12.5	12.5	14	17	18

Thus, we found that treatment of cockerels of Japanese quail with testosterone and 5α-DHT, but not estradiol-17β (approximately day 10), increased aggressiveness to levels higher than those of noninjected birds, probably in a dose-dependent manner. These results suggest that the level of aggressiveness of adult male Japanese quail is predetermined by the circulating androgen level at immature stages.

Our results clearly differ from the finding by Adkins (1975) in that the embryonic steroid treatment demasculinized male Japanese quail in their sexual behaviors. The difference in the developmental stage at which the hormone was administered or in the behaviors observed might be the cause of the different results. Hutchison (1978) treated male and female chicks of Japanese quail with testosterone and estradiol on the day of hatch. She did not observe any change in differentiation of sexual behaviors. It might be considered that posthatch treatment with androgen induces only quantitative change in sexual behaviors as it did in aggressive behaviors.

Conclusions

Our results suggest that the androgen level at an early developmental stage determines the level of aggressiveness after maturation. Individual differences in the early androgen level may be the cause of individual differences in aggressiveness at adulthood. Circulating androgen at adulthood might enable a predetermined potential of aggressiveness to be expressed.

Acknowledgments

The authors are grateful to Dr. M. Wada for conducting radioimmunoassay of LH and to Miss N. Ohkubo for her technical assistance. This research is supported by Grant-in-Aid for Special Project Research from the Ministry of Education, Science and Culture, Japan.

Literature Cited

ADKINS, E. K. 1975. Hormonal basis of sex differentiation in the Japanese quail. J. Comp. Physiol. Psychol., 89:61-71.

———. 1977. Effects of diverse androgens on the sexual behavior and morphology of castrated male quail. Horm. Behav., 8:201-207.

ADKINS, E. K., AND N. T. ADLER. 1972. Hormonal control of behavior in the Japanese quail. J. Comp. Physiol. Psychol., 81:27-36.

ADKINS, E. K., AND B. K. NOCK. 1976. The effects of the antiestrogen CI-628 on sexual behavior activated by androgen or estrogen in quail. Horm. Behav., 7:417-429.

BALTHAZART, J., P. DEVICHE, AND J. C. HENDRICK. 1977. Effects of exogenous hormones on the reproductive behavior of adult male domestic ducks. II. Correlations with morphology and hormone plasma levels. Behav. Processes, 2:147-161.

BALTHAZART, J., R. MASSA, AND P. NEGRI-CESI. 1979. Photoperiodic control of testosterone metabolism, plasma gonadotropins, cloacal gland growth, and reproductive behavior in the Japanese quail. Gen. Comp. Endocrinol., 39:222-235.

BEACH, F. A., AND A. M. HOLZ-TUCKER. 1949. Effects of different concentrations of androgen upon sexual behavior in castrated male rats. J. Comp. Physiol. Psychol., 42:433-453.

CUNNINGHAM, D. L., P. B. SIEGEL, AND H. P. VAN KREY. 1977. Androgen influence on mating behavior in selected lines of Japanese quail. Horm. Behav., 8:166-174.

DAMASSA, D., E. SMITH, B. TENNENT, AND J. DAVIDSON. 1977. The relationship between circulating testosterone levels and male sexual behavior in rats. Horm. Behav., 8:275-286.

EATON, G. G., AND J. A. RESKO. 1974. Plasma testosterone and male dominance in a Japanese macaque (*Maccaca fuscata*) troop compared with repeated measure of testosterone in laboratory males. Horm. Behav., 5:251-259.

HARDING, D. F., AND H. H. FEDER. 1976. Relation between individual differences in sexual behavior and plasma testosterone levels in the guinea pig. Endocrinology, 98:1198-1205.

HUTCHISON, R. E. 1978. Hormonal differentiation of sexual behaviors in Japanese quail. Horm. Behav., 11:363-387.

LESHNER, A. I. 1978. An introduction to behavioral endocrinology. Oxford Univ. Press, New York, xi + 361 pp.

TSUTSUI, K., AND S. ISHII. 1981. Effects of sex steroids on aggressive behavior of male Japanese quail. Gen. Comp. Endocrinol., 44:480-486.

Reprinted from
Aspects of Avian Endocrinology:
Practical and Theoretical Implications (C. G. Scanes *et al.*, eds.)
Grad. Studies, Texas Tech Univ., 1982, 26:1-411.

BRAIN METABOLISM OF TESTOSTERONE AND BEHAVIOR

Renato Massa

*Department of Endocrinology, University of Milano, 21 Via A.
Del Sarto, 20129 Milano, Italy*

Male sexual behavior in birds and mammals has long been known to depend on the presence of testosterone; however, the mechanism by which testosterone activates sexual behavioral patterns has been better understood since the biochemical fate of androgens in the central nervous system of mammals and birds was established. In the central nervous system of the rat, testosterone undergoes a process of 5α-reduction resulting in the formation of 5α-androstane-17β-ol-3-one (5α-DHT) (Jaffe, 1969; Kniewald *et al.*, 1971; Denef *et al.*, 1973; Thieulant *et al.*, 1973); moreover, testosterone can also be oxidized into androstenedione (Kniewald *et al.*, 1971; Massa and Martini, 1974) or aromatized into estradiol and estrone (Naftolin *et al.*, 1971, 1972, 1976). All these steroids can be further converted into different metabolites covering somewhat different spectra of biological activities (5α-androstanedione, 5α-androstane-3α,17β-diol, and 5α-androstane-3β,17β-diol.)

There is evidence that, in the rat, 5α-DHT is more active than testosterone in the negative feedback control of luteinizing hormone (LH) secretion (Beyer *et al.*, 1971; Verjans and Eik-Nes, 1973; Zanisi *et al.*, 1973) and that, in other species of mammals, it can stimulate male sexual behavior when given alone or in combination with estradiol (Alsum and Goy, 1974; Phoenix, 1974; Agmo and Sodersten, 1975; Payne and Bennett, 1976). It is also well established that estradiol plays an important role in eliciting male sexual behavior in several species of mammals. For instance, estrogens can stimulate sexual behavior in rats (Celotti *et al.*, 1979). Moreover, among androgens, those molecules that can be converted into estrogens (such as testosterone and androstenedione) are behaviorally the most effective in rats, rabbits, and hamsters while those that cannot be aromatized (such as dihydrotestosterone or androsterone) are usually ineffective or weak (Cellotti *et al.*, 1979).

Testosterone Metabolism and Sexual Behavior in Birds.

In the central nervous system of birds, the conversion of testosterone into 5α-DHT occurs to a lesser extent than in mammals; however, in birds, testosterone is largely metabolized through pathways of 5β-reduction (Nakamura and Tanabe, 1974; Massa *et al.*, 1977; Davies *et al.*, 1980; Massa and Sharp, 1981) and 17β-oxidation (Massa *et al.*, 1977; Davies *et al.*, 1980). When tested by injection into castrated quails maintained in long photoperiods, only some of the metabolites of testosterone were found to exert feedback effects. Thus, androstenedione and the 5α-reduced metabolites of

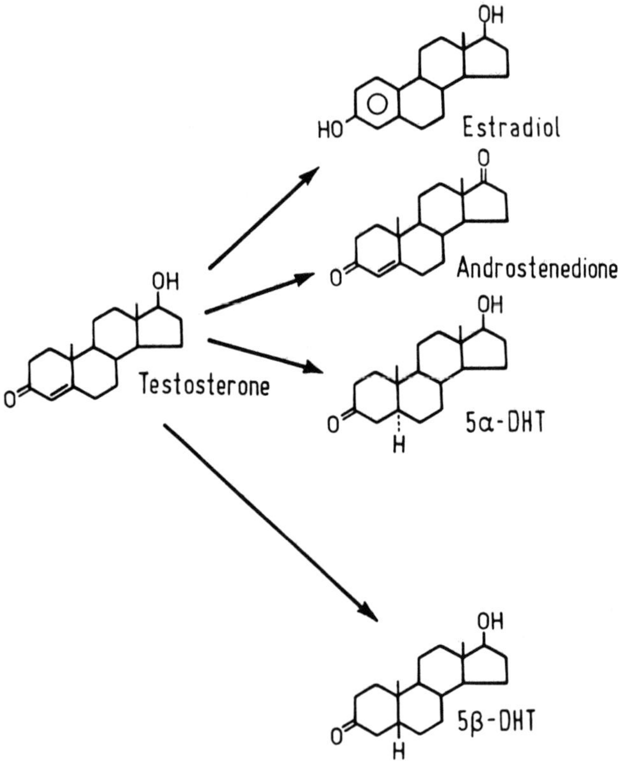

FIG. 1.—Pathways of testosterone metabolism in the central nervous system of birds.

testosterone (5α-androstane-3α,17β-diol, and 5α-androstane-3,17-dione) inhibit LH secretion and a significant decrease in plasma LH concentrations can be observed six hours after the injections. The 5β-reduced metabolites (5β-androstane-17β-ol-3-one, 5β-androstane-3α,17β-diol, and 5β-androstane-3,17-dione), however, have no negative feedback activity (Davies et al., 1980). In the central nervous system of birds, the conversion of androgens into estrogens has also been found to occur (Fig. 1; Callard et al., 1978a, 1978b).

Therefore, it can be hypothesized that estrogens might play a role in the activation of male sexual behavior in birds as they do in mammals. This hypothesis was first submitted to experimental verification by Adkins and Nock (1976) who found that the antiestrogen CI-628 was effective in suppressing copulation in castrated male quails treated with testosterone. Subsequently, the ability of various androgens in eliciting sexual behavior in castrated male quails was tested and it was found that copulation was activated to a significant extent only by testosterone, whereas strutting was also activated by androstenedione; androsterone, 5α-DHT, and 5β-DHT were completely ineffective (Adkins, 1977). Similar results were obtained by Balthazart and Hendrick (1978) who found that 5α-DHT was not able to stimulate sexual behavior of the male chick whereas testosterone was effective.

Adkins and Pniewski (1978) also studied the ability of different sex steroids to stimulate crowing, strutting, and copulation in castrated male quails: birds treated with 5α-DHT did not copulate but two of them strutted; estradiol treated males copulated but did not strut; males receiving a combined treatment with 5α-DHT and estradiol strutted and copulated. Crowing was activated to the highest extent by 5α-DHT, to a lesser extent by testosterone; no other steroid was effective. These results suggested that, in the quail, the different components of male reproductive behavior have different molecular requirements for their activation. A similar study on the sexual behavior of the barbary dove (*Streptopelia risoria*) was performed by Hutchison (1970) who observed the disappearance of different courtship displays following castration and injection with various steroids. Castrated doves injected with testosterone bowed, nest-solicited, and chased the female; birds injected with estradiol nest-solicited and chased but did not bow. In experiments with pigeons (*Columba livia*, Murton *et al.*, 1969) estradiol was shown to stimulate nest-demonstration behavior while testosterone, androsterone, and androstenedione activated bowing and chasing; 5α-DHT did not activate any type of behavior.

The pathway of 5β-reduction

In most experiments, either on the regulation of LH secretion or on the control of the sexual behavior, the 5β-reduced metabolites of testosterone did not appear to be biologically effective. However, because the pathway of 5β-reduction is irreversible, it is possible that the 5β-reductase activity might play an indirect role by acting as an inactivation shunt limiting the conversion of the circulating testosterone to such active androgens as 5α-DHT, 5α-androstane-3α,17β-diol (5α-3α-diol), and androstenedione, and thereby modulating the effects of testicular steroids on gonadotropin secretion or sexual behavior. However, some different possibilities should also be considered: 5β-DHT stimulated copulation in chicks (Balthazart and Hirschberg, 1979); and it decreases the following reaction of chicks during imprinting sessions. In view of these results, it is possible that 5β-DHT plays some role in eliciting sexual behavior in some species of birds and might also play some more general behavioral role even in a larger number of species. Besides testosterone, progesterone is also converted to its 5β-reduced metabolites in the brain of birds. Its conversion into 5β-pregnane-3,20-dione (5β-DHP) has been demonstrated in the brain of the laying hen by Sharp and Massa (1980). There is also evidence that the 5β-reduced derivatives of progesterone exert a marked depressant action on the central nervous system of mammals (Holzbauer, 1971; Glyermek *et al.*, 1967). Since the forebrain of birds is known to be involved in various aspects of reproductive and fearful behavior (Beach, 1951; Phillips, 1964; Salzen and Parker, 1975; Gentle *et al.*, 1978), it is possible that the 5β-reduced metabolities of progesterone also might be involved directly or indirectly in birds in the regulation of these behaviors.

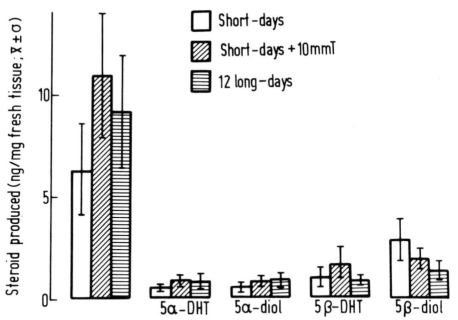

Fig. 2.—Changes in the conversion of testosterone into androstenedione (Δ_4A), 5α-dihydro-testosterone (5α-DHT) 5α-androstane-3α,17β-diol), 5β-dihydrotestosterone (5β-DHT), 5β-andro-stane-3α,17β-diol (5β-diol) in the cloacal gland of immature male Japanese quail following exposure to 12 "long days" (16L:8D) or implants of testosterone (10 mm silastic capsules) (from Balthazart *et al.*, 1979).

A different approach to the role of the various testosterone metabolites in the sexual behavior of birds consists of studying the possible correlations between sexual behavior and the enzymatic activities that convert testoste-rone or progesterone into other steroids in the brain. This approach was already adopted by Dessi-Fulgheri *et al.* (1976) in the study of the behavior of small laboratory rodents.

In a study of this type, the changes in reproductive behavior, plasma gonadotropins, and cloacal gland area occurring in male Japanese quail after transfer to long days or testosterone implantation were correlated with testosterone metabolism in the hyperstriatum, hypothalamus, pituitary, and cloacal gland (Balthazart *et al.*, 1979). In the cloacal gland, long days were found to enhance the conversion of testosterone into androstenedione (Fig. 2) and a positive correlation between this conversion and the cloacal gland area (but not with plasma testosterone) was observed. Similarly, the changes occurring in aggressive behavior and struts were positively correlated with hyperstriatal conversion of testosterone into androstenedione and negatively correlated to hypothalamic conversion of testosterone into 5β-DHT; how-ever, no correlation between behavioral variables and plasma testosterone levels was observed (Fig. 3; Balthazart *et al.*, 1979). The conversion of testos-terone into 5β-DHT, 5β,3α-diol, and 5β-androstane-5β-diol (5β-3β-diol) in the brain of ring dove was studied by Hutchison and Steimer (1980) who

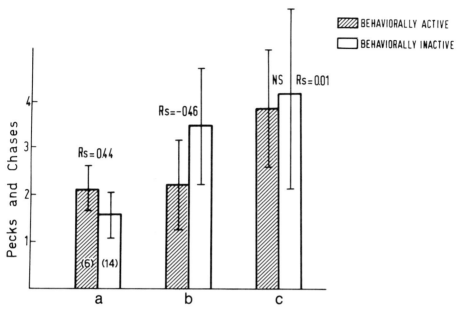

Fig. 3.—Correlation between aggressive behavior (pecks and chases) and   a) the conversion of testosterone into androstenedione in the hyperstriatum (HYS-Δ_4);   b) the conversion of testosterone into 5β-DHT in the hypothalamus (HYP-5β); and   c) plasma testosterone levels. (Adapted from Balthazart *et al.*, 1979).

found that these 5β-reduced metabolites are rapidly formed in and eliminated from the brain. Since the 5β-reduced derivatives of testosterone do not have any androgenic activity and are also the main product of testosterone metabolism in the liver, the authors conclude that the 5β-reduction pathway appears to be a major route for testosterone inactivation. A further support to this hypothesis is given by other data of the same authors showing that the 5β-reductase is highest in those areas of the dove brain not involved in the control of behavior (Steimer and Hutchison, 1980a, 1980b).

LITERATURE CITED

ADKINS, E. 1977.   Effects of diverse androgens on sexual behaviour and morphology of castrated male quail. Horm. Behav., 8:201-207.

ADKINS, E., AND L. NOCK. 1976.   The effects of the antiestrogen CI-628 on sexual behaviour activated by androgen or estrogen in quail. Horm. Behav., 7:417-429.

ADKINS, E. AND E. E. PNIEWSKI. 1978.   Control of the reproductive behaviour by sex steroids in male quail. J. Comp. Physiol. Psychol., 92:1169-1178.

AGMO, A., AND P. SODERSTEN. 1975.   Sexual behaviour in castrated rabbits treated with testosterone, oestradiol, DHT or oestradiol in combination with DHT. J. Endocrinol., 67:327-332.

ALSUM, P., AND R. W. GOY. 1974.   Action of esters of T, DHT and oestradiol on sexual behaviour in castrated male guinea pigs. Horm. Behav., 5:207-218.

BALTHAZART, J., AND J. C. HENDRICK. 1978.   Steroidal control of plasma luteinizing hormone, comb growth and sexual behaviour in male chicks. J. Endocrinol., 77:149-150.

BALTHAZART, J., AND D. HIRSCHBERG. 1979.   Testosterone metabolism and sexual behaviour in the chick. Horm. Behav., 12:253-263.

BALTHAZART, J., R. MASSA, AND P. NEGRI-CESI. 1979. Photoperiodic control of testosterone metabolism, plasma gonadotrophins, cloacal gland growth and reproductive behaviour in the Japanese quail. Gen. Comp. Endocrinol., 39:222-235.

BEACH, F. A. 1951. Effect of forebrain injury upon mating behaviour in male pigeons. Behaviour, 4:36-59.

BEYER, C., G. MORALI, AND H. L. CRUZ. 1971. Effects of 5α-dihydrotestosterone on gonadotrophin secretion and estrous behaviour in the female Wistar rat. Endocrinology, 89:1158-1161.

CALLARD, G. V., Z. PETRO, AND K. I. RYAN. 1978a. Conversion of androgen to estrogens and other steroids in the vertebrate brain. Amer. Zool., 18:511-523.

———. 1978b. Phylogenetic distribution of aromatase and other androgen-converting enzymes in the central nervous system. Endocrinology, 103:2283-2290.

CELOTTI, F., R. MASSA, AND L. MARTINI. 1979. Metabolism of sex steroids in the central nervous system. Pp. 41-53, in Endocrinology, Vol. I (L. J. De Groot, ed.), Grune and Stratton, New York, 000 + 000.

DAVIES, D. T., R. MASSA, AND R. JAMES. 1980. Role of testosterone and of its metabolites in regulating gonadotrophin secretion in the Japanese quail. J. Endocrinol., 84:211-222.

DENEF, C., C. MAGNUS, AND B. S. MC EWEN. 1973. Sex differences and hormonal control of testosterone metabolism in rat pituitary and brain. J. Endocrinol., 59:605-621.

DESSI-FULGHERI, F., N. LUCARINI, AND C. LUPO DI PRISCO. 1976. Relationships between testosterone metabolism in the brain, other endocrine variables and intermale aggression in mice. Aggressive Behav., 2:223-231.

GENTLE, M. J., D. G. M. WOOD-GUSH, AND J. GORDON. 1978. Behavioural effects of hyperstriatal ablations in Gallus domesticus. Behav. Processes, 3:137-148.

GLYERMEK, L., G. GENTHER, AND L. FLEMING. 1967. Some effects of progesterone and related steroids on the central nervous system. Internat. J. Neurochem., 6:191-198.

JAFFE, R. B. 1969. Testosterone metabolism in target tissues: hypothalamic and pituitary tissues of the adult rat and human fetus and the immature rat epiphysis. Steroids, 14:483-498.

HOLZBAUER, M. 1971. In vivo production of steroids with central depressant action by the ovary of the rat. Brit. J. Pharm., 43:560-569.

HUTCHISON, J. B. 1970. Influence of gonadal hormones on the hypothalamic integration of the courtship behaviour in the barbary doves. J. Reprod. Fert., Suppl., 11:15-41.

HUTCHISON, J. B., AND J. STEIMER. 1980. Metabolic inactivation of behaviourally effective androgens in the dove brain. Abstract. Physiol. Soc., July 1980, 69P.

KNIEWALD, Z., R. MASSA, AND L. MARTINI. 1971. Conversion of testosterone into 5α-androstane-17β-ol-3-one at the anterior pituitary and hypothalamic levels. Pp. 784-791, in Hormonal steroids (V. H. T. James and L. Martini, eds.), Excerpta Medica, Amsterdam, xvi + 1063 pp.

MASSA, R., L. CRESTI, AND L. MARTINI. 1977. Metabolism of testosterone in the pituitary gland and in the central nervous system of the European starling (Sturnus vulgaris). J. Endocrinol., 75:347-354.

MASSA, R. AND L. MARTINI. 1974. Testosterone metabolism: a necessary step for activity? J. Steroid Biochem., 5:941-947.

MASSA, R., AND P. J. SHARP. 1981. Conversion of testosterone to 5β-reduced metabolites in the neuroendocrine tissues of the maturing cockerel. J. Endocrinol., 88:263-269.

MURTON, R. K., R. J. P. THEARLE, AND B. LOFTS. 1969. The endocrine basis of the breeding behaviour in the feral pigeon (Columba livia). I. Effects of exogenous hormones on the pre-incubation behaviour of intact males. Anim. Behav., 17:286-306.

NAFTOLIN, F., K. J. RYAN, AND Z. PETRO. 1971. Aromatization of androstenedione by the diencephalon. J. Clin. Endocrinol. Metab., 33:368-370.

———. 1972. Aromatization of androstenedione by anterior hypothalamus of adult male and female rats. Endocrinology, 90:295-298.

NAFTOLIN, F., K. J. RYAN, AND I. J. DAVIES. 1976. Androgen aromatization by neuroendocrine tissues. Pp. 347-355, *in* Subcellular mechanisms in reproductive neuroendocrinology (F. Naftolin, K. J. Ryan, and I. J. Davies, eds.), Elsevier, Amsterdam, xii + 529 pp.

NAKAMURA, T., AND Y. TANABE. 1974. "In vitro" metabolism of steroid hormones by chicken brain. Acta Endocrinol., 75:410-416.

PAYNE, A. P., AND N. K. BENNETT. 1976. Effects of androgens on sexual behaviour and somatic variables in the male golden hamster. J. Reprod. Fert., 47:239-244.

PHILLIPS, R. E. 1964. "Wildness" in the mallard duck: effects of brain lesions and stimulation on escape behaviour and reproduction. J. Comp. Neurol., 127:89-100.

PHOENIX, C. 1974. Effects of DHT on sexual behaviour of castrated male Rhesus monkeys. Physiol. Behav., 12:1045-1056.

SALZEN, E. A., AND D. M. PARKER. 1975. Arousal and orientation functions of the avian telencephalon. Pp. 205-242, *in* Neural and endocrine aspects of behaviour in birds (P. Wright, P. G. Caryl, and D. N. Vowles, eds.), Elsevier, Amsterdam, x + 408 pp.

SHARP, P. J., AND R. MASSA. 1980. Conversion of progesterone to 5α and 5β reduced metabolites in the brain of the hen and its potential role in the induction of the pre-ovulatory release of luteinizing hormone. J. Endocrinol., 86:459-464.

STEIMER, T., AND J. B. HUTCHISON. 1980a. "In vitro" metabolism of testosterone in the dove brain and behaviour. Gen. Comp. Endocrinol., 40:336-337.

————. 1980b. Aromatization of testosterone within a discrete hypothalamic area associated with the behavioural action of androgen in the male dove. Brain Res., 192:580-591.

THIEULANT, M. L., S. SAMPEREZ, AND P. JOUAN. 1973. Binding and metabolism of H^3 testosterone in the nuclei of rat pituitary in vivo. J. Steroid Biochem., 4:677-685.

VERJANS, H. L., AND K. N. EIK-NES. 1976. Serum LH and FSH levels following an intravenous injection of a gonadotrophin-releasing principle in a normal and gonadectomized adult rats treated with estradiol 17β or 5α-dihydrotestosterone. Acta Endocrinol., 83:493-505.

ZANISI, M., M. MOTTA, AND L. MARTINI. 1973. Inhibitory effect of 5α-reduced metabolites of testosterone on gonadotropin secretion. J. Endocrinol., 56:315-316.

Reprinted from
ASPECTS OF AVIAN ENDOCRINOLOGY:
PRACTICAL AND THEORETICAL IMPLICATIONS (C. G. Scanes *et al.*, eds.)
Grad. Studies, Texas Tech Univ., 1982, 26:1-411.

STEROID RECEPTORS AND BEHAVIOR

JACQUES BALTHAZART

*Laboratoire de Biochimie Générale et Comparée, Université de
Liège, 17, place Delcour, B-4020 Liège, Belgique*

It has long been known that behavior is controlled at least partly by hormones, particularly from data generated from the laboratory of Frank Beach
in the late 1940s. At that time it was possible to list species in all vertebrate
orders in which reproductive behavior (aggressive or sexual) had been shown
to be hormone dependent. In the next two decades, the collection of experimental data continued, despite the fact that at that time endocrinologists had
mainly morphological techniques for an indication of circulating gonadal
steroids which might modulate a behavioral response. Indeed, most of the
data were derived from experiments using removal of the endocrine gland
and hormone replacement therapy as the single procedure. During the last
decade, research methodology has changed substantially, mainly as a result
of the appearance of radioimmunoassays which allows for the detailed analysis of relationships between plasma hormone levels and behavior (for testosterone, Furuyama *et al.*, 1970; Ismail *et al.*, 1972; for estradiol, Abraham,
1969; for progesterone, Niswender and Midgley, 1970; Abraham *et al.*, 1971).
These techniques, originally used in clinical endocrinology and in endocrine
studies in mammals, have been used with increasing frequency in experiments on avian behavior (for example, Korenbrot *et al.*, 1974; Silver *et al.*,
1974; Balthazart, 1976; Balthazart and Hendrick, 1976, 1979; Feder *et al.*,
1977; Ottinger and Brinkley, 1978; Pröve, 1978; Harding and Follett, 1979).
Refined techniques were also designed to analyze the metabolism of steroids
(Karavolas and Herf, 1971; Massa *et al.*, 1977; Callard *et al.*, 1978; see also
Massa, this vol.), and it was recognized that in some instances circulating
steroids (for example, testosterone) should be considered only as prehor-
mones because their metabolities are actually the effective compounds in the
target organs (see Feder, 1978; Luttge, 1979; for birds see Cheng and Lehrman, 1975; Adkins and Nock, 1976; Adkins *et al.*, 1980). Finally, extremely
sensitive assays (in the femtomole range) for both cytoplamic and nuclear
steroid receptors have become available and have been used to study the
interaction of steroids with the brain which is the main target organ in the
context of behavioral studies.

The Brain as a Target Organ for Steroids and Resulting Sexual Behavior

When considering the actions of steroid hormones relevant to sexual
behavior at the brain level, two classes of phenomena must be distinguished
according to the age of the animals. During the prenatal or perinatal period,

exposure to gonadal steroids affects the development of the nervous system in a permanent fashion so that behavior and endocrine physiology will be modified definitively (Wilson and Glick, 1970; Adkins, 1978, 1979; Harlan *et al.*, 1979; Mashaly and Glick, 1979). The mechanisms of these organizational effects of steroids have been reviewed recently by several authors (Adkins, 1978; Fox *et al.*, 1978; Lieberburg *et al.*, 1978; Beyer *et al.*, 1979; Arnold, 1980) and are beyond the scope of this paper.

Steroid hormones also activate sex behavior in mature animals in that they increase the probability of occurrence of specific behavior patterns in adult animals. Sterotaxic placement of small quantities of steroids in the brains of small animals was found to stimulate sexual behavior (Lisk, 1960, 1962). Thus, the brain was recognized as the main site of action of steroids as far as control of behavior is concerned. It is now generally accepted that sex steroids stimulate sexual behavior by increasing the excitability of specific neural circuits possibly related to the detection or integration of external stimuli. Numerous studies have defined with good precision the neural sites involved in the hormonal control of sexual behavior in mammals (Morali and Beyer, 1979; Larsson, 1979) and in birds (Komisaruk, 1967; Barfield, 1969, 1971; Hutchison, 1971, 1978; Haynes and Glick, 1974; Gibson and Cheng, 1979). Autoradiographic techniques, especially designed for diffusible substances (Stumpf and Sar, 1975) such as steroids, showed that these same areas of the brain retain radioactive steroids after an intravenous injection. As a result of the availability of various techniques, maps of brain areas, which accumulate testosterone, estradiol, or progesterone, are now available for a number of bird species (Meyer, 1973; Zigmond *et al.*, 1973; Arnold *et al.*, 1976; Martinez-Vargas *et al.*, 1976; Wood-Gush *et al.*, 1977).

Indirect evidence also suggests that the effects of steroids in the brain might be mediated by the same biochemical events proposed for the peripheral organs. Drugs which are known to block the binding of steroids with their receptor frequently act to suppress the behavioral effects of the hormone. For example, Adkins and Mason (1974) showed that the anti-androgen, cyproterone acetate, which competed with androgens for the binding with brain cytosol receptor (Attardi and Ohno, 1976; Lieberburg *et al.*, 1977), also suppressed the copulatory behavior and strutting in Japanese quail. Similarly, Silver (1977) demonstrated that cyproterone acetate markedly depressed the bow-cooing in male ring doves, a behavior pattern which is highly androgen-dependent (Erickson, 1965; Cheng and Lehrman, 1975). However, these effects could be attributed to other mechanisms as it is known that cyproterone acetate has antigonadotropic effects (such as depressed LH and FSH secretion by the pituitary) and thus could have decreased plasma testosterone rather than have prevented its binding to the receptor. Balthazart (1978) showed that in immature ducks, cyproterone acetate blocked the sexual behavior induced by exogenous testosterone without decreasing the plasma levels of the hormone. This demonstrated that the effect on behavior was not the result of antigonadotropic properties of cyproterone acetate (Balthazart, 1978). However, this did not prove that the behavioral inhibition could be explained by the blockage of the binding of testosterone with its receptor. Mc Ewen and co-workers (1979) suggested that cyproterone acetate could interfere with the conversion of testosterone to estrogens which could be the active hormone in the induction of copulation in male birds as demonstrated for the quail by Adkins *et al.* (1980). Similar data also exist for drugs which block the behavioral effects of estrogens. CI-628 is an anti-estrogen which has been shown to reduce nuclear binding of ^{3}H-estradiol in the preoptic area-hypothalamus-amygdala of ovariectomized female rats, although this action is less pronounced and of shorter duration than in a peripheral tissue such as the uterus (Luine and Mc Ewen, 1977). Adkins and Nock (1976) showed that in quail this compound suppressed behavioral events known to be estrogen-dependent (such as sexual receptivity in the female and the sexual behavior in the male, a process that is induced by testosterone injection, but acts through aromatization of testosterone into estrogen; Adkins *et al.*, 1980). Similar data demonstrating behavioral inhibitions induced by compounds which compete for the binding of steroids with their receptors are frequent in the mammalian literature (for review, see Larsson, 1979).

Steroid Receptors in the Bird Brain

Despite the fact that the chick oviduct is one of the best studied models for hormone-receptor interactions, very little is known about steroid receptors in the brain of birds. Receptors have been characterized in a number of peripheral organs, for example, androgen receptors in the comb and ear lobes of cocks (Dube and Tremblay, 1974). Progestin receptors have been so far characterized only in the brain and as such in only two kinds of birds: the hen (domestic fowl) and the Ring dove.

Using a high specific activity tritiated progesterone as a ligand (105 Ci/mmol), Kawashima *et al.* (1978) characterized a progestin binding component in the cytosol of the anterior lobe of the pituitary, in the hypothalamus, and in the oviduct magnum of laying White Leghorn hens. The dissociation constant was very similar in all three tissues (1-3 x 10^{-10}M), although it was significantly lower in the pituitary than in the hypothalamus, as demonstrated in a later study (Kawashima *et al.*, 1979*b*). This suggests that the nature of the receptors may differ slightly between the two tissues. As has been shown in mammals (MacLusky and Mc Ewen, 1980), the number of binding sites is much lower (about ten times) in the hypothalamus than in the pituitary. Kawashima *et al.* (1979*a*) later demonstrated that the concentrations of progestin receptors in the hypothalamus and pituitary change during an ovulatory cycle and showed peaks both 18 and eight hours before ovulation. These changes were probably related to the ovulatory process as they are not observed during a 24 hour period in non-laying hens. Recent evidence suggests that they are actually caused by changes in the amounts of sex steroids secreted during the cycle. The same authors (Kawashima *et al.*,

1979*b*) showed that testosterone, estradiol, or progesterone injections alter the progestin receptor concentrations in the hypothalamus and pituitary. These concentrations decreased one hour after a progesterone injection which probably reflects translocation of the receptor to the nucleus. Estradiol injections increased the hypothalamic and hypophyseal concentrations of progestin receptors; the effect appeared rapidly (within four to eight hours in the hypothalamus, one hour in the pituitary), and vanished quickly (four hours in the pituitary). Finally, testosterone injections markedly increased the receptor concentration in the pituitary after four hours, but this effect was not observed in the hypothalamus. Taken together these data suggest that the maximum number of progesterone receptors in the hypothalamus observed before ovulation results from the changes in the plasma concentrations of estradiol and progesterone (increasing amounts of estradiol secreted by the growing follicle followed by a surge of progesterone). The estradiol-induced increase in progestin receptors might act to sensitize the brain to the circulating progesterone and could thus play a critical role in the control of ovulation. The relevance for behavior of these hormonal interactions remains to be investigated.

Progestin Receptors in the Dove Brain: Possible Involvement in the Control of the Incubation

The Ring dove (*Streptopelia risoria*) is probably the best studied bird species in so far as interactions between hormones and behavior are concerned. Following the work of Lehrman, hundreds of papers have shown how the reproduction of males and females is synchronized by visual, auditory, and tactile stimuli, and how hormones are involved in this process. In collaboration with J. D. Blaustein, a receptor assay originally set up for the guinea pig was used to analyze a behavioral problem in the Ring dove, a problem which does not seem to find an explanation by classical techniques such as castration, hormone replacement therapy, and radioimmunoassay of plasma hormones.

In male and female Ring doves, injections of progesterone facilitated incubation behavior (Lehrman, 1958). This effect was probably centrally mediated since Komisaruk (1967) showed that progesterone implants in the preoptic area or in the lateral forebrain system were also behaviorally effective. In addition, Stern (1974) demonstrated that the effects of exogenous progesterone depend on the simultaneous presence of gonadal steroids. Incubation behavior was restored in gonadectomized animals only if the birds were primed with testosterone or estradiol (Stern and Lehrman, 1969; Stern, 1974). Moreover, in female but not in male doves, plasma progesterone increased markedly about the time when the birds engaged in incubation (Silver *et al.*, 1974). The same pattern has also been described in the Collared dove (*Streptopelia decaocto*, Peczely and Pethes, 1979).

Two questions thus emerged from this set of data: 1) What are the hormonal stimuli, if any, which induce male doves to sit on eggs? and 2) What is at the biochemical level the basis of the synergism between sex steroids (estradiol and testosterone) and progesterone in the induction of incubation behavior by exogenous hormones?

Data has shown that, in mammals, estradiol can increase the concentration of progestin receptors. This suggests a possible mechanism involving changes in progesterone binding which would then control the sensitivity of neural circuits to the circulating progesterone.

Using an *in vitro* assay based on a synthetic progestin as ligand (R 5020 = 17,21-dimethyl-19-nor-pregna-4,9-diene-3,20-dione = promegestone), we demonstrated the presence of a progestin binder in the brain of castrated male Ring doves. The affinity of the receptor was similar in the hyperstriatum and in the hypothalamus ($K_d = \pm 4 \times 10^{-10}$ mol/l), a value very close to the affinity measured with the same technique in the guinea pig hypothalamus (Blaustein and Feder, 1979) and to the affinity for ^{3}H-progesterone measured in the hen hypothalamus (Kawashima *et al.*, 1978). The binder was progestin specific; only progesterone and R 5020 were strong competitors. Corticosterone significantly decreased the binding of tritiated R 5020 to the receptor when present in excess in the incubation medium, but its relative binding affinity was less that 15% that of progesterone. Interestingly, corticosterone also competed for the binding of ^{3}H-progesterone with the receptor in the hen (Kawashima *et al.*, 1978) but not in the guinea pig (Blaustein and Feder, 1979) nor in the rat (MacLusky and Mc Ewen, 1980). Whether this represents a systematic difference between birds and mammals will have to be evaluated in future experiments.

The binding in the Ring dove brain was saturable, though its concentration was higher than values reported for mammals obtained using the same technique (up to 133 fmol/mg protein in the hypothalamus of testosterone primed Ring doves and 34 fmol/mg protein in the guinea pig brain). Progesterone injections given four hours before the sacrifice of the animals significantly decreased the binding observed in the hypothalamus (about 50%), probably as a result of the translocation of the receptors to the nucleus.

The binding for progestins, which possess all characteristics of a true receptor, was correlated with the biological function and more specifically correlated with the behavioral role of progesterone. The receptor concentration is higher in the preoptic area-hypothalamus, that is, a site of action of progesterone in the induction of incubation (Komisaruk, 1967), than in other areas studied. Furthermore, we found that testosterone-priming (200 μg testosterone propionate daily for seven days, a treatment which fully restored the facilitatory effects of progesterone on incubation; Stern and Lehrman, 1969) increased the concentration of progestin receptors in the hypothalamus but not in other brain areas. The same effect was observed and was even more pronounced in castrated male doves primed for seven days with a daily injection of 50 μg estradiol benzoate (63% increase instead of the 34% increase in the testosterone-treated birds). No effect was observed following a priming with 5α-dihydrotestosterone. This implies, although does not prove, that the effect of testosterone could be mediated by the aromatization of testosterone in the brain (Callard *et al.*, 1978). It was also interesting to note that priming either with testosterone propionate or estradiol benzoate for only two days did not significantly increase the hypothalamic concentrations of progestin receptors, at least when the assay was performed 24 hours after the last injection (short-term increases as demonstrated by Kawashima *et al.*, 1979*b*, could have been ignored). The increase, however, was started somehow, as revealed by the slightly higher mean concentrations of receptors compared to the control values (110 ± 23% and 121 ± 22% of controls in testosterone and estradiol treated birds, respectively). These data support the notion that changes in hypothalamic progestin receptor concentrations could play a role in the control of incubation in Ring doves.

It is likely that the synergism between exogenous testosterone and exogenous progesterone in the induction of incubation behavior in castrated doves (Stern and Lehrman, 1969) is related to the androgenic induction of hypothalamic progestin receptors. This effect could be mediated by way of aromatization, but more direct evidence would be needed to support this conclusion.

It is also possible that changes in hypothalamic progestin receptors play a role in the control of incubation behavior during the reproductive cycle of male doves (see Fig. 1). After a male dove was paired with a female, its plasma concentration of testosterone quickly increased within four hours and reached a peak three days later (Feder *et al.*, 1977). After seven to ten days, the female laid eggs and both male and female engaged in incubation. By that time, the female displayed increased progesterone plasma levels, but no change was observed in the male, whose progesterone plasma concentration remained at the baseline throughout the cycle (Silver *et al.*, 1974). It is possible that the increased androgen levels observed during courtship could have induced an increase in the hypothalamic concentration of progestin receptors in intact male birds. This could have resulted in an increased sensitivity to the baseline level of progesterone. During the cycle, endogenous progesterone thus affected the incubation behavior, not through a change in the plasma level of the hormone, but by an increase in the brain sensitivity to a constant signal. The time course of the increase in receptor concentration after a testosterone priming was also consistent with this idea; progestin receptors were increased after seven days, which is close to a normal egg laying latency in doves, while little or no effect was observed after two days. Direct evidence in support of this idea, however, is not yet available. It could be obtained rather easily by measuring the occupation of the progestin receptor at different stages of the reproductive cycle in intact doves. This would show whether more receptor is translocated to the nucleus around the time when birds start incubating.

CONCLUSIONS

The last five years have seen the rapid development of extremely sensitive techniques for the assay of steroid receptors, allowing for their measure in

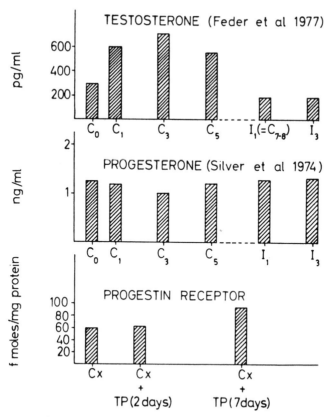

FIG. 1.—Summary of the changes in plasma testosterone and progesterone observed in male doves during the initial phase of a reproductive cycle. These changes have been compared with the variations in hypothalamic concentrations of progestin receptors induced by a priming with testosterone propionate (TP) for two or seven days (data from Balthazart et al., 1980). C indicates courtship; I, incubation; subscripts are for the different days within one period.

tissues with low receptor concentrations such as the brain. These techniques have been used in a number of behavioral studies on mammals; studies in birds are now in progress. Much circumstantial evidence now supports the notion that the interaction of steroids with cytosol receptors and their translocation to the nucleus are critical steps in the appearance of a behavioral response. There is, however, no definitive proof that behavioral responses are generated by the same mechanisms which are accepted for the steroid action in the peripheral target organs. Actually, a number of differences between brain and peripheral receptors are already evident and, furthermore, a number of steroid effects in the brain which are possibly not mediated by the interaction with intracellular receptors have been reported. We still need new solid facts before we clearly understand how steroids interact with the brain to facilitate behavior.

Many questions still await clear answers: How different are steroid receptors in the brain and in the uterus? Is the presence of a lower affinity binder of steroids in the brain important in the induction of behavior? What RNA and proteins synthetized in the brain are under the influence of steroids? How does this action relate to the role of monoamines on behavior? How general are the steroid actions at the membrane level? Some of these questions will only be answered after new technical advances have been made. With the available tools, however, much can still be done.

In birds, numerous questions are now clearly standing which do not seem to have an answer in the analysis of the circulating levels of hormones— questions concerning, for example, the biochemical basis of the sexual differentiation, strain differences in behavior, or the hormonal control of the individual differences. Hopefully some answers will soon be found in the study of the interaction of the steroids with the brain.

ACKNOWLEDGMENTS

I am indebted to Professor E. Schoffeniels for his continued interest and support of my work. I also want to thank Dr. J. D. Blaustein and H. H. Feder who allowed me to perform my first experiments on steroid receptors. Through numerous discussions they also taught me the basic information in this field of research. Writing of this paper was supported by grant No. 2.4544.76 from the Fonds de la Recherche Fondamentale Collective to Professor E. Schoffeniels. During the execution of the experiments reported at the end of this paper, J. Balthazart was Chargé de Recherches du Fonds National Belge de la Recherche Scientifique and supported by a Fulbright/ Hays travel grant and a NATO fellowship.

LITERATURE CITED

ABRAHAM, G. E.   1969.   Solid-phase radioimmunoassay of estradiol-17β.   J. Clin. Endocrinol., 29:866-870.
ABRAHAM, G. E., R. SWERDLOFFER, D. TULCHINSKY, AND W. D. ODELL.   1971.   Radioimmunoassay of plasma progesterone.   J. Clin. Endocrinol., 32:619-624.
ADKINS, E. K.   1978.   Sex steroids and the differentiation of avian reproductive behavior.   Amer. Zool., 18:501-509.
———.   1979.   Effect of embryonic treatment with estradiol or testosterone on sexual differentiation of the quail brain.   Neuroendocrinol., 29:178-185.
ADKINS, E. K., AND P. MASON.   1974.   Effects of cyproterone acetate in the male Japanese quail.   Horm. Behav., 5:1-6.
ADKINS, E. K., AND B. L. NOCK.   1976.   The effects of the antiestrogen CI-628 on sexual behavior activated by androgen or estrogen in the quail.   Horm. Behav., 7:417-430.
ADKINS, E. K., J. J. BOOP, D. L. KOUTNIK, J. B. MORRIS, AND E. E. PNIEWSKI.   1980.   Further evidence that androgen aromatization is essential for the activation of copulation in male quail.   Physiol. Behav., 24:441-446.
ARNOLD, A. P.   1980.   Sexual differences in the brain.   Amer. Sci., 68:165-173.
ARNOLD, A. P., F. NOTTEBOHM, AND D. W. PFAFF.   1976.   Hormone concentrating cells in vocal control and other areas of the brain of the Zebra finch.   J. Comp. Neurol., 165:487-512.
ATTARDI, B., AND S. OHNO.   1976.   Androgen and estrogen receptors in the developing mouse brain.   Endocrinology, 99:1279-1290.
BALTHAZART, J.   1976.   Daily variations of behavioral activities and of plasma testosterone levels in the domestic duck, *Anas platyrhynchos*.   J. Zool. (London), 180:155-173.

———. 1978. Behavioral and physiological effects of testosterone propionate and cyproterone acetate in immature male domestic ducks, *Anas platyrhynchos*. Z. Tierpsychol., 47:410-421.

BALTHAZART, J., AND J. C. HENDRICK. 1976. Annual variation in reproductive behavior, testosterone and plasma FSH levels in the Rouen duck, *Anas platyrhynchos*. Gen. Comp. Endocrinol., 28:171-183.

———. 1979. Relationships between the daily variations of social behavior and of plasma FSH, LH and testosterone levels in the domestic duck, *Anas platyrhynchos L.* Behav. Processes, 4:107-128.

BARFIELD, R. J. 1969. Activation of copulatory behavior by androgen implanted into the preoptic area of the male fowl. Horm. Behav., 1:37-52.

———. 1971. Activation of sexual and aggressive behavior by androgen implanted into the male ring dove brain. Endocrinology, 89:1470-1476.

BEYER, C., K. LARSSON, AND M. L. CRUZ. 1979. Neuronal mechanisms probably related to the effect of sex steroids on sexual behavior. Pp. 365-387, *in* Endocrine control of sexual behavior (C. Beyer, ed.), Raven Press, New York, x + 413 pp.

BLAUSTEIN, J. D., AND H. H. FEDER. 1979. Cytoplasmic progestin-receptors in guinea pig brain: characteristics and relationship to the induction of sexual behavior. Brain Res., 169:481-497.

CALLARD, G. V., Z PETRO, AND K. J. RYAN. 1978. Conversion of androgen to estrogen and other steroids in the vertebrate brain. Amer. Zool., 18:511-523.

CHENG, M. F., AND D. S. LEHRMAN. 1975. Gonadal hormone specificity in the sexual behavior of ring doves. Psychoneuroendocrinol., 1:95-102.

DUBE, J. Y., AND R. R. TREMBLAY. 1974. Androgen binding proteins in cock's tissues: properties of ear lobe protein and determination of binding sites in head appendages and other tissues. Endocrinology, 95:1105-1112.

ERICKSON, C. J. 1965. A study of the courtship behavior of male ring doves. Ph.D. Dissertaion, Rutgers University, iv + 173 pp.

FEDER, H. H. 1978. Specificity of steroid hormone activation of sexual behavior in rodents. Pp. 395-424, *in* Biological determinants of sexual behavior (J. B. Hutchison, ed.), John Wiley and Sons, New York, xv + 822 pp.

FEDER, H. H., A. STOREY, D. GOODWIN, C. REBOULLEAU, AND R. SILVER. 1977. Testosterone and "5α-dihydrostestosterone" levels in peripheral plasma of male and female ring doves (*Streptopelia risoria*) during the reproductive cycle. Biol. Reprod., 16:666-667.

FOX, T. O., C. C. VITO, AND S. J. WIELAND. 1978. Estrogen and androgen receptor in embryonic and neonatal brain: hypothesis for roles in sexual differentiation and behavior. Amer. Zool., 18:525-537.

FURUYAMA, A. A., D. M. NAYES, AND C. A. NUGENT. 1970. A radioimmunoassay for plasma testosterone. Steroids, 16:415-428.

GIBSON, M. J., AND M. F. CHENG. 1979. Neural mediation of estrogen-dependent courtship behavior in female ring doves. J. Comp. Physiol. Psychol., 93:855-867.

HARDING, C. F., AND B. K. FOLLETT. 1979. Hormone changes triggered by aggression in a natural population of blackbirds. Science, 203:918-920.

HARLAN, R. E., J. H. GORDON, AND R. A. GORSKI. 1979. Sexual differentiation of the brain: implications for neuroscience. Pp. 31-71, *in* Review of neuroscience, Vol. 4 (D. M. Schnieder, ed.), Raven Press, New York, 000 + 000.

HAYNES, R. L., AND B. GLICK. 1974. Hypothalamic control of sexual behavior in the chicken. Poult. Sci., 53:27-38.

HUTCHISON, J. B. 1971. Effects of hypothalamic implants on gonadal steroids on courtship behavior in Barbary doves (*Streptopelia risoria*). J. Endocrinol., 50:97-113.

———. 1978. Hypothalamic regulation of male sexual responsiveness to androgen. Pp. 277-318, *in* Biological determinants of sexual behavior (J. B. Hutchinson, ed.), John Wiley and Sons, New York, xv + 822 pp.

ISGMAIL, A. A. A., G. D. NISWENDER, AND A. R. MIDGLEY. 1972. Radioimmunoassay of testosterone without chromatography. J. Clin. Endocrinol. Metab., 34:177-184.

KARAVOLAS, H. J. R., AND S. M. HERF. 1971. Conversion of progesterone by rat medial basal hypothalamic tissue to 5α-pregnane-3,20-dione. Endocrinology, 89:940-942.

KAWASHIMA, M., M. KAMIYOSHI, AND K. TANAKA. 1978. A cytoplasmic progesterone receptor in hen pituitary and hypothalamic tissues. Endocrinology, 102:1207-1213.

———. 1979a. Cytoplasmic progesterone receptor concentrations in the hen hypothalamus and pituitary: difference between laying and non laying hens and changes during the ovulatory cycle. Biol. Reprod., 20:581-585.

———. 1979b. Effects of progesterone, estradiol and testosterone on cytoplasmic progesterone receptor concentrations in the hen hypothalamus and pituitary. Biol. Reprod., 21:639-646.

KOMISARUK, B. R. 1967. Effects of local brain implants of progesterone on reproductive behavior in ring doves. J. Comp. Physiol. Psychol., 64:219-224.

KORENBROT, C. C., D. W. SCHONBERG, AND C. J. ERICKSON. 1974. Radioimmunoassay of plasma estradiol during the breeding cycle of ring doves. Endocrinology, 94:1126-1132.

LARSSON, K. 1979. Features of the neuroendocrine regulation of masculine sexual behavior. Pp. 77-163, in Endocrine control of sexual behavior (C. Beyer, ed.), Raven Press, New York, x + 413 pp.

LEHRMAN, D. S. 1958. Effect of female sex hormones on incubation behavior in the ring dove (Streptopelia risoria). J. Comp. Physiol. Psychol., 51:142-145.

LIEBERBURG, I., N. J. MACLUSKY, AND B. S. MC EWEN. 1977. 5α-dihydrotestosterone (DHT) receptors in rat brain and pituitary cell nuclei. Endocrinology, 100:598-607.

LIEBERBURG, I., N. J. MACLUSKY, E. J. ROY, AND B. S. MC EWEN. 1978. Sex steroid receptors in the perinatal brain. Amer. Zool., 18:539-544.

LISK, R. D. 1960. Estrogen-sensitive centers in the hypothalamus of the rat. J. Exp. Zool., 145:197-208.

———. 1962. Diencephalic placement of estradiol and sexual receptivity in the female rat. Amer. J. Physiol., 203:493-496.

LUINE, V., AND B. S. MC EWEN. 1977. Effects of an estrogen antagonist on enzyme activities and (³H) estradiol nuclear binding in uterus, pituitary and brain. Endocrinology, 100:903-910.

LUTTGE, W. G. 1979. Endocrine control of mammalian male sexual behavior: an analysis of the potential role of testosterone metabolites. Pp. 341-364 in Endocrine control and sexual behavior (C. Beyer, ed.), Raven Press, New York, x + 413 pp.

MACLUSKEY, N. J., AND B. S. MC EWEN. 1980. Progestin receptors in rat brain: distribution and properties of cytoplasmic progestin-binding sites. Endocrinology, 106:192-202.

MARTINEZ-VARGAS, M. C., W. E. STUMPF, AND M. SAR. 1976. Anatomical distribution of estrogen target cells in the avian CNS: a comparison with mammalian CNS. J. Comp. Neurol., 167:83-104.

MASHALY, M. N., AND B. GLICK. 1979. Comparison of androgen levels in normal males (Gallus domesticus) and in males made sexually inactive by embryonic exposure to testosterone propionate. Gen. Comp Endocrinol., 38:105-110.

MASSA, R., L. CRESTI, AND L. MARTINI. 1977. Metabolism of testosterone in the pituitary gland and in the central nervous system of the European starling (Sturnus vulgaris). J. Endocrinol., 75:347-354.

MC EWEN, B. S., I. LIEBERBURG, C. CHAPTAL, P. G. DAVIS, L. C. KREY, N. J. MACLUSKY, AND E. J. ROY. 1979. Attenuating the feminization of the neonatal rat brain: mechanisms of action of cyproterone acetate, 1,4,5-androstratriene-3,17-dione and a synthetic progestin, R 5020. Horm. Behav., 13:269-281.

MEYER, C. C. 1973. Testosterone concentration in the male chick brain: an autoradiographic survey. Science, 180:1381-1383.

MORALI, G., AND C. BEYER. 1979. Neuroendocrine control of mammalian estrous behavior. Pp. 33-75, in Endocrine control of sexual behavior (C. Beyer, ed.), Raven Press, New York, x + 413 pp.

NISWENDER, G. D., AND A. R. MIDGLEY.   1970.   Hapten-radioimmunoassay for steroid hormo-
    nes.   Pp. 149-173, in Immunologic methods in steroid determination (F. G. Peron and
    B. V. Caldwell, eds.), Appleton-Century-Crofts, New York.

OTTINGER, M. A., AND H. J. BRINKLEY.   1978.   Testosterone and sex-related behavior and mor-
    phology: relationship during maturation and in the adult Japanese quail.   Horm.
    Behav., 11:175-182.

PECZELY, P., AND G. PETHES.   1979.   Alterations in plasma sexual steroid concentrations in the
    collared dove (Streptopelia decaocto) during sexual maturation and reproduction cycle.
    Acta Physiol. Acad. Sci. Hung., 54:161-170.

PRÖVE, F.   1978.   Quantative untersuchungen zu wechselbeziehungen zwishchen balzaktivitat
    und testosterontitern bei mannlichen Zebrafinken (Taeniopygia guttata castanots
    Gould). Z. Tierpsychol., 48:47-67.

SILVER, R.   1977.   Effects of the antiandrogen cyproterone acetate on reproduction in male and
    female ring doves.   Horm. Behav., 9:371-379.

SILVER, R., C. REBOULLEAU, D. S. LEHRMAN, AND H. H. FEDER.   1974.   Radioimmunoassay of
    plasma progesterone during the reproductive cycle of male and female ring doves
    (Streptopelia risoria).   Endocrinology, 94:1547-1554.

STERN, J. M.   1974.   Estrogen facilitation of progesterone-induced incubation behavior in cas-
    trated male ring doves.   J. Comp. Physiol. Psychol., 87:332-337.

STERN, J. M., AND D. S. LEHRMAN.   1969.   Role of testosterone in progesterone-induced incuba-
    tion behavior in male ring doves (Streptopelia risoria).   J. Endocrinol., 44:13-22.

STUMPF, W. D., AND M. SAR.   1975.   Autoradiographic techniques for localizing steroid hor-
    mones.   Pp. 135-156, in Methods in enzymology, Vol. 36 (B. E. O'Malley and J. G. Hardman,
    eds.), Academic Press, New York.

WILSON, J. A., AND B. GLICK.   1970.   Ontogeny of mating behavior in the chicken.   Amer. J.
    Physiol., 218:951-955.

WOOD-GUSH, D. G. M., G. A. S. LANGLEY, A. F. LEITCH, M. J. GENTLE, and A. G. GIL-
    BERT.   1977. An autoradiographic study of sex steroids in the chicken telencephalon.
    Gen. Comp. Endocrinol., 31:161-168.

ZIGMOND, R. E., R. NOTTENBOHM, AND D. W. PFAFF.   1973.   Androgen-concentrating cells in
    the midbrain of a songbird.   Science, 179:1005-1007.

SECTION III

NUTRITIONAL INFLUENCES UPON GROWTH, REPRODUCTION, AND METABOLISM: ENDOCRINE EFFECTS

INTRODUCTION

M. A. OTTINGER

Department of Poultry Science, University of Maryland, College Park, Maryland 20742

Dietary requirements for birds have been carefully researched, particularly in domestic species such as the chicken. Much of this research has been conducted as an effort to increase the economic return associated with raising the meat-type chicken. In addition, nutritional regimes used in poultry have maximized the growth potential of the bird and minimized the time required for growth. These developments present a number of interesting questions from the standpoint of the physiology of growth as affected by nutrition. Papers presented in this section discuss a number of these questions. Certain portions of the diet appear to be of particular importance (Carew, Engster, and Balzer). Among the hormones involved in these responses, growth hormones and thyroid hormones play important roles in the control of rate of growth and attainment of maximal growth capacity (Scanes and Harvey). Furthermore, dietary administration of thyroid hormones results in alterations in body composition and changes in plasma concentrations of hormones (May).

Detailed studies of nutritional requirements of adult birds have been applied to optimize reproduction potential. Restricted feeding regimes are routinely used in the poultry industry to maintain a body weight that promotes a longer reproductive lifespan. However, if the diet is insufficient, the reproductive performance of the bird will decline. Fasting for short periods of time has been found to result in ovarian decline. Circulating concentrations of gonadotropin and gonadal steroids also decline associated with this level of nutrition (Tanabe, Nakamura, and Tanabe). The actual control of feed intake appears to be under the control of rather specific areas of the brain. The anatomical proximity of these areas remains a complex issue (Robinzon, Snapir, and Lepkovsky), and the exact function of specific areas remains a subject of both current and future research (Kuenzel).

Finally, many chemicals found in the environment are possible contaminants of feed, both for wild or domestic birds. A group of naturally occurring contaminants are mycotoxins produced by molds (Ottinger and Doerr). One mycotoxin, aflatoxin, exerts powerful adverse effects on growth and sexual maturation. Not only is there depression of both, but there is also a residual suppression of sexual maturation. This is reflected in depression of circulating concentrations of gonadal steroids.

These topics encompass basic endocrine events and response to diet. In addition, some practical applications of dietary effects are considered. Therefore, these papers provide information that is of interest for the avian endo-crinologist as well as for application by the poultry industry.

Reprinted from
ASPECTS OF AVIAN ENDOCRINOLOGY:
PRACTICAL AND THEORETICAL IMPLICATIONS (C. G. Scanes *et al.*, eds.)
Grad. Studies, Texas Tech Univ., 1982, 26:1-411.

ESSENTIAL FATTY ACID DEFICIENCY
AND ENDOCRINE FUNCTION IN CHICKENS AND QUAIL

LYNDON B. CAREW, JR., HENRY M. ENGSTER[1], AND PATRICIA M. BALZER

Bioresearch Center, Department of Animal Sciences, Vermont Agricultural Experiment Station, University of Vermont, Burlington, Vermont 05401 USA; [1]Present address: Ralston Purina Co., Checkerboard Square, St. Louis, Missouri 63188 USA

The effects of nutritional deficiencies on specific cellular biochemical changes in animals, including avian species, have been widely studied. On the other hand, changes in endocrine function and hormonal activity in states of nutritional deficit have, until recently, been largely overlooked. This neglect undoubtedly is related to the relative ease for some years now of measuring aspects of cellular biochemistry with well-developed methods compared to the more tedious and inexact biological assays that often had to be used with hormones. But with the important discovery of radioimmunoassay (RIA) procedures by Yalow and Berson (1960), and the application of these procedures to an increasing number of hormones, this situation is changing rapidly.

Nevertheless, research in avian endocrinology is not keeping pace with the need for new information. In particular, investigations of nutrition-endocrine interactions with avian species are lagging behind discoveries with other species, including humans. This issue is critical, in part, because birds serve as important models for animal research in many nutritional studies. But more important is the role of certain birds in supplying a substantial share of the world's food needs. In order for poultry to meet increasing world demands for animal protein, the fund of basic scientific information concerning avian metabolism, including endocrinology, must be enlarged.

Two areas of avian endocrinology, among others, in which information on nutritional effects is deficient, include thyroid function and the hormonal regulation of growth. Few papers are being published with new RIA techniques on the role of nutritional status on avian thyroid function. On the other hand, extensive work has been in progress for several years now on the role of diet composition on thyroid function and energy metabolism in both humans (Danforth *et al.*, 1979) and rats (Young *et al.*, 1980). To date, an acceptable rapid assay for avian thyroid stimulating hormone (TSH) is not available, and most work on thyroid function has involved 3,5,3'-triiodothyronine (T_3) and thyroxine (T_4), and not other thyronine derivatives or T_3 or T_4 receptors. The regulation of energy use for growth and egg production is a topic of paramount importance to the international poultry industry. Similarly, regarding growth aspects, extensive studies are needed on the avian endocrine mechanisms by which nutritional disorders alter the function of pituitary growth hormone, as well as other hormones with related functions such as the somatomedins. Few laboratories in the world are currently able to conduct RIA analyses on these hormones in birds.

Research in our laboratory in the last few years has centered on the role of dietary deficiencies of essential fatty acids and, very recently, of essential amino acids on the function of certain aspects of the avian endocrine systems, especially those controlled by the pituitary gland.

Linoleic, arachidonic, and linolenic acids are considered the essential fatty acids (EFA), although the last is often excluded because its addition to diets does not cure all EFA deficiency symptoms. Linoleic acid is the only EFA found widely in feed ingredients. Since the initial discovery of EFA by Burr and Burr (1929, 1930), dietary exclusion of these polyunsaturated fatty acids (but not other unrelated fatty acids) has been shown to produce a variety of deficiency symptoms in all higher animals studied, including the human infant and the chicken. Reviews concerning EFA requirements in poultry (Balnave, 1972) and other species (Holman, 1968; Guarnieri and Johnson, 1970; Alfin-Slater and Aftergood, 1968, 1971) have been published.

Several consistently observed symptoms of an EFA deficiency in chickens, such as slow growth, slow sexual development, small testes size, and increased water intake suggest that these fatty acids are important for the proper function of many aspects of the avian endocrine system. Furthermore, changes in size of the pituitary and pineal glands have also been reported in EFA-deficient chickens (Menge, 1967). Additional characteristics of an EFA deficiency observed in mammalian species, such as epiphyseal degeneration, adrenal cortical degeneration, and hyperactive thyroid glands, indicate widespread involvement of the endocrine system (other symptoms of EFA deficiency often observed in birds and mammals include enlarged and fatty livers, high liver cholesterol, high 20:3w9/20:4w6 fatty acid ratios, reduced resistance to disease, and dermatitis).

A prominent feature of an EFA deficiency is interruption of sexual development (Alfin-Slater and Aftergood, 1968, 1971). This is evident in the develpment of secondary sexual characteristics of EFA-deficient chickens. Delayed comb and wattle growth give EFA-deficient chickens the appearance of immaturity well into adult age (Edwards, 1967). These symptoms of an EFA deficiency are accompanied by reductions in gonadal size, testicular degeneration, incomplete spermatogenesis, and changes in gonadal metabolism (Edwards, 1967; Lillie and Menge, 1968; Nugara and Edwards, 1970a, 1970b). Although these changes strongly suggest that changes have occurred in pituitary gonadotropic and gonadal hormone activities, little direct evidence of such changes is available. One of the few studies in this area is an extensive report by Roland and Edwards (1971). They showed that injection of ovine luteinizing hormone (oLH) into EFA-deficient chickens increased concentration of sperm, increased packed cell volume of sperm, and reversed testicular degeneration to almost normal. A similar treatment with ovine follicle stimulating hormone and testosterone proprionate did not improve semen quality and only enhanced testicular degeneration. Because pituitary glands from EFA-deficient chickens contained increased amounts of periodic acid Schiff-positive material (secretory), the authors concluded that EFA might operate at the level of the brain, perhaps through release or production of hypothalamic LH-releasing factors.

A marked retardation in body size in EFA-deficient chicks suggests decreases in rate of bone growth and nitrogen retention due to alteration in the functions of growth hormone (GH) and related hormones. Studies of interactions between EFA deficiency and the endocrine mechanism of growth retardation are limited. In one of the few such studies, Deuel and colleagues (1950) reported that injection of GH into EFA-deficient rats would not improve growth rate. However, if weight gain of deficient rats was rehabilitated by feeding EFA, the rats recovered more rapidly when also injected with GH, suggesting that an inadequate level of GH in the deficient rats was indeed a limiting factor.

EFA deficiency effects on thyroid function in avian species have received little attention. The few reports available have relied on anatomical measures rather than hormonal assays. Either slight increases (Menge, 1967) or no change (Edwards, 1967; Lowe et al., 1975; Engster et al., 1978) in avian thyroid gland weight have been reported. These have not been correlated with energy metabolism. In rats, an EFA deficiency is usually accompanied by increased metabolic rate when measured by oxygen consumption (Burr and Beber, 1937; Panos and Finerty, 1954). The thyroid gland in EFA-deficient rats is smaller with a histological appearance of hyperactivity (Alfin-Slater and Bernick, 1958). However, these changes do not correlate with changes in I^{131} uptake, Protein Bound Iodide (PBI), or pituitary TSH concentration (Morris et al., 1957), suggesting that changes in energy metabolism are not a consequence of altered thyroid function.

The objective of our studies, then, was to elaborate on the role of EFA in avian endocrine function.

MATERIALS AND METHODS

A purified EFA-deficient diet was used in all studies. A typical composition is shown in Table 1 (small modifications from this composition were used in some experiments and will be noted). The nutrient levels were increased somewhat in studies with quail to account for their higher nutritional needs. This diet contained 0.03% linoleic acid, compared to the nutritional requirement of the chick of about 1%. This level of linoleic acid readily produced an EFA deficiency in newly hatched chicks after about one week of feeding. The saturated fat, hydrogenated coconut oil (HCO), was used to accelerate the EFA deficiency (Peifer and Holman, 1959). Either HCO or glucose was replaced with isocaloric amounts of corn oil to produce EFA-adequate, positive control diets. With corn oil present, the purified diet was adequate in all nutritional needs, and chickens eating it grew as rapidly as when fed a commercial poultry ration.

TABLE 1.—*Typical composition of an essential fatty acid-deficient diet. For detailed composi-tions see Engster* et al. *(1978) and Balzer (1980).*

Ingredient	%
Glucose monohydrate	55.667
Casein (vitamin-free)	25.000
L-arginine-HCl	1.500
Methionine hydroxy analog	0.700
Glycine	1.000
Cellulose	4.000
Vitamin mixture	0.500
Mineral mixture	6.380
Hydrogenated coconut fat	5.000
Choline Cl (70%)	0.240
Antioxidant	0.013

Chicks were obtained from local commercial hatcheries. All were fed the EFA-deficient diet for seven to ten days and then divided into experimental groups. Japanese quail (*Coturnix coturnix japonica*) were obtained from our own breeding flock. Unless noted otherwise, at least duplicate groups of eight to ten birds were used as the treatment unit.

Except where pair-feeding was employed, birds were allowed *ad libitum* intakes of feed and water. Where pair-feeding was used, it was carried out on a daily basis, and the groups of pair-fed chickens were given a caloric allowance equal to the previous 24-hour intake of the respective control chickens. Feeder space was sufficient to avoid competition. Metabolizable energy values of the diets were measured according to the methods of Hill and Anderson (1958) and Hill *et al*. (1960).

Biochemical evidence for the presence of an EFA deficiency was measured by large increases in the ratios of 18:1w9/18:2w6 and 20:3w9/20:4w6 fatty acids in testes, heart, and/or liver tissue. Evidence for such changes in an EFA deficiency in chickens is well established (Bieri *et al*., 1957; Machlin and Gordon, 1961). Additional measurements such as reduced body size, small comb and testes weights, feather loss and, where applicable, testicular degeneration, were used as evidence of an EFA deficiency. These symptoms of an EFA deficiency were always observed where expected. The analysis for fatty acids is described elsewhere in the literature (Engster *et al*., 1978).

The following methods were used in the assay of tissue hormone levels: gonadotropic activity of the anterior pituitary by ^{32}P uptake of testes of day-old test chicks (Kamiyashi *et al*., 1972); anterior pituitary growth hormone by bioassay using hypothysectomized rats (Greenspan *et al*., 1949); plasma growth hormone by RIA (Harvey and Scanes, 1977); plasma thyroxine and triiodothyronine by double antibody RIA (Beckman, 1976, 1977), using procedures developed and validated for chicken plasma in our laboratory; and plasma LH by RIA (Follett *et al*., 1972).

Analysis of variance was carried out on all data with F-test comparisons made in experiments 1 and 2 (Snedecor, 1956). The method of Scheffé (1953)

TABLE 2.—*Effect of an EFA deficiency on growth and parameters of reproductive development in Japanese quail.*[a]

| | Dietary treatment | | | |
| | Experiment 1[b] | | Experiment 2[b] | |
Variate	EFA def.	+ 10% corn oil	EFA def.	+ 10% corn oil
Body wt gain, g, male	74	87*	65	77*
Body wt gain, g, female	79	108*	71	95*
Testes, mg/g body wt.	13.64	27.12*	16.83	19.99
Testes showing advanced spermatogenesis, %	10	80*	0	55*
Oviduct, mg/g body wt.	15.94	33.85*	13.40	34.27*
Mean ovarian follicle diameter, mm[c]	5.41	10.50*	5.53	11.06*

[a]Experiments 1 and 2 were conducted for 54 and 53 days, respectively. Values are the means from all birds per treatment.

[b]The EFA deficient diet contained 2% HCO. For the positive control, corn oil was added isocalorically at the expense of glucose.

[c]Measures were made on all enlarged, yellow follicles.

*Indicates that the value is significantly different ($P < 0.05$) from the EFA-deficient group.

was used for comparisons in experiments 3-7, and the method of Duncan (1955) in experiment 8.

RESULTS AND DISCUSSION

The results in Table 2 demonstrate the pronounced effects that an EFA deficiency has on growth and reproductive performance. In this experiment, Japanese quail were used because their life cycle is much shorter than that of the chicken, and maturity is attained in six to seven weeks. By 53 or 54 days of age (experiments 1 and 2), consumption of the EFA-deficient diet caused marked retardations in growth of male and female quail. The growth response (per cent weight gain response over the deficient group) to dietary supplementation with corn oil, which contains over 50 per cent linoleic acid, varied from 17-18 per cent in males to 34-36 per cent in female quail. This would suggest either a more rapid metabolism of EFA in the female quail or perhaps a higher requirement as compared to males.

The important effect of EFA on reproduction, reported in the earliest work by Burr and Burr (1929, 1930) in young rats, is also demonstrated for young quail (Table 2). Depressions in both anatomical development and function of the male and female reproductive systems were evident in the EFA-deficient quail. In the male *Coturnix*, evidence of advanced spermatogenesis, based on histological examination, was almost totally lacking in the EFA deficient group. In females, measurement of the size of developing ovarian follicles demonstrated a marked retardation in ovarian function in the EFA deficient group, although we had no certainty that this indicated interruption of ovulation. However, Calvert (1969) reported that when normal, mature quail are fed an EFA-deficient diet, both egg size and number of eggs produced are reduced. The combined evidence suggests that both growth rate of ova and rate of ovulation are reduced.

TABLE 3.—*Effect of an EFA deficiency on growth and parameters of reproduction in White Leghorn cockerels.*[a,b]

| | Dietary treatments[c] | | | | | |
| | Experiment 3 | | Experiment 4 | | Experiment 5 | |
Variate	EFA deficient	+ 5% corn oil	EFA deficient	+ 5% corn oil	EFA deficient	+ 5% corn oil
Body wt gain, g	1405	1843*	1450	1713*	1557	1854*
Comb, mg/g body wt	10.39	25.34*	18.09	32.50*	19.87	35.59*
Testes, mg/g body wt	0.77	9.14*	3.69	7.01*	5.67	9.10*
Plasma LH, ng/ml	13.45	22.92*	6.82	9.53*	17.05	26.91*
Pituitary gonadotropic activity, % increase			155	184*	126	169*

[a]Experiment 3 was for 20 weeks; experiments 4 and 5, 24 weeks.

[b]Values in experiment 3 are the means of 6 observations; in experiments 4 and 5, 10 to 15 observations.

[c]The EFA deficient diet contained 2% hydrogenated coconut oil. Corn oil was added to the positive control diet isocalorically at the expense of glucose.

*Indicates that the value is significantly different ($P < 0.05$) from the EFA-deficient group.

The data in Table 3 show the effect of an extended EFA deficiency on growth and reproductive development in male White Leghorn chickens. As with quail, the EFA deficiency in all three experiments brought about marked depressions in body weight gain.

Avian comb growth represents a secondary sexual characteristic that responds markedly to the presence of androgens. We therefore used comb weight in these studies as an indicator of circulating androgens. Comb growth was sharply reduced in the EFA deficient groups, suggesting interruption of testicular production and release of male hormone. Disruption of testicular development in the EFA-deficient chickens was also in evidence from the smaller size of the testes themselves. The biggest difference in testes size between the deficient and normal cockerels occurred at 20 weeks of age (experiment 3), but this difference diminished by 24 weeks of age (experiments 4 and 5). This suggests that an EFA deficiency slows the rate of growth of testicular tissue, but does not inhibit it entirely. As seen in our experiments with quail, impairment of normal spermatogenesis in EFA-deficiency also occurs in chickens, although the data are not presented here.

In these studies we measured the level of LH circulating in the blood and the total gonadotropin content of the pituitary gland. The latter analysis was used because no reliable assay for separating avian pituitary FSH and LH was available when these experiments were started. Both plasma LH and pituitary gonadotropic activities were reduced in the EFA-deficient cockerels. This is the first time direct evidence has been presented showing reductions in specific gonadotropic activity in blood or pituitary glands in an EFA deficiency. These changes explain, at least in part, the reductions in comb and testes size and impairment of spermatogenesis in the case of an EFA deficiency.

Total pituitary gonadotropic activity probably measured both LH and FSH activity. It should definitely measure FSH, because the test we used is based on the growth of immature chick testes. Based on their measurements of increased Periodic Acid Schiff (PAS)-positive material (secretory) in pituitary glands of EFA-deficient chickens, Roland and Edwards (1971) suggested that LH accumulates in the pituitary gland and that a fault in releasing mechanisms had occurred. We suggest that LH and FSH levels in the pituitary gland of EFA-deficient birds are actually reduced, and the defect is in synthesis or storage of the hormones rather than in release. Both Roland and Edwards' (1971) data and ours agree, however, in that responses of EFA-deficient chickens to injections of ovine LH occur, and our data further show that circulating LH is indeed low in the deficient group. On the other hand, Roland and Edwards (1971) found no positive response of EFA-deficient chickens to ovine FSH, whereas our data, showing smaller testes, testicular degeneration, and reduced pituitary gonadotropin content in the EFA deficiency, suggest lower than normal levels of FSH. Chickens might not respond well to injections of ovine FSH, and further resolution of these observations will depend on the use of specific assays for avian FSH.

In most states of nutritional deficiency, appetite, and food intake, and as a consequence, growth rate, are markedly depressed. It is recognized in many species that reductions in food intake and energy intake can suppress various components of the endocrine system. In chickens, restricted energy intake of nutritionally adequate diets causes reductions in testes size and in the volume and fertilizing capacity of semen in cockerels (Parker and Arscott, 1964) and reduced reproductive capacity in hens (Aukland and Fulton, 1973). Unfavorable effects of underconsumption of feed on egg production and egg size are widely recognized in the poultry industry.

In our studies, the EFA deficiency caused marked reductions in food intake which in turn was reflected in reduced rates of growth. We were concerned that many of the endocrine effects we were observing might be a consequence of reduced food intake rather than a specific effect of the EFA *per se*. Essential fatty acids, therefore, would exert their effects on endocrine mechanisms indirectly by affecting the appetite-regulating mechanisms. To take this effect into account, we pair-fed chicks receiving the totally adequate, positive control diet to those chicks receiving the EFA-deficient diet. By pair-feeding, we mean that chicks fed the adequate diet were allowed to consume, on a daily basis, only the same number of calories of food as the deficient chicks had consumed the previous day. Because all diets were kept in identical nutrient balance, all chicks, deficient or normal, received identical intakes of all nutrients except for EFA during pair-feeding.

Our concerns about the role of energy intake in the etiology of EFA deficiency symptoms were confirmed as shown in Table 4. Experiment 6 represents a four-week study with 20-week old EFA-deficient cockerels. When cockerels fed the diet with 5 per cent corn oil were pair-fed to the EFA-deficient cockerels, they grew at the slower rate of the deficient ones. More significant, however, is the observation that when the EFA-adequate chickens were pair-fed to the EFA-deficient chickens, the sizes of their combs

TABLE 4.—*Growth and parameters of reproduction in EFA-deficient and pair-fed, nondeficient White Leghorn cockerels.*[a]

	Dietary treatments[b]—Experiment 6		
Variate	EFA deficient	+ 5% corn oil pair-fed	+ 5% corn oil *ad libitum*
Body wt gain, g	80	100	203*
Energy intake, kcal	6001	6003	7148*
Comb, mg/g body wt	19.87	22.48	25.23*
Testes, mg/g body wt	5.67	7.86	11.16*
Plasma LH, ng/ml	17.05	17.12	19.91
Pituitary gonadotropic activity, % increase	126	126	143

[a]All chickens were fed the EFA deficient diet to 20 weeks of age, and then fed the respective test diets 20 to 24 weeks of age; values are means from two replicate pens of 10 to 15 chickens each.

[b]The EFA-deficient diet contained 2% hydrogenated coconut oil. Corn oil was added to the positive control diet isocalorically at the expense of glucose.

*Indicates that the value is significantly different ($P < 0.05$) from the EFA-deficient group.

and testes were also more similar to that of the deficient chickens. This means that some of the inhibitory effects on avian development and function attributed to an EFA deficiency might be a consequence of reduced feed or energy intake rather than a direct effect of the EFA. The data in Table 4 also suggest that reductions in plasma LH and pituitary gonadotropic activity in the EFA deficiency occurred as a result of reduced feed intake, because pair-fed and deficient chickens had values that were almost identical and lower than those of the positive controls. However, variability in these measures was sufficiently great that none of the values was significantly different ($P < 0.05$) although the LH values approached significance at ($P = 0.09$). As a result of these observations, we included a pair-fed, EFA-adequate control in all of our subsequent studies on nutrition-endocrine interactions.

In experiment 7 (Table 5), the effect of an EFA deficiency on various mechanisms of growth was examined. Significant reductions in body weight gain, tibia length, tibia width, epiphyseal plate width, and pituitary growth hormone (GH) content were observed in the EFA deficiency. However, similar reductions in all of these values, except for the latter, also occurred in the EFA-adequate pair-fed chicks. This suggests that, although the EFA deficiency has widespread effects on measures of growth, these changes are an indirect effect of the EFA deficiency on food intake. Only the reduction in pituitary GH content occurred in the EFA-deficient chicks and not in the pair-fed normal chickens; this suggests that EFA directly and specifically affects the capacity of the pituitary gland to synthesize or store GH.

Plasma GH was elevated in the EFA-deficient chickens, and a similar elevation occurred when non-deficient chickens were pair-fed. It has been demonstrated that in states of reduced food intake or fasting, the level of circulating GH increases (Harvey *et al.*, 1978), perhaps reflecting its role in internal energy-yielding mechanisms in times of dietary deficit (Turner and Bagnara, 1976; Harvey *et al.*, 1977). Thus, elevated plasma GH is not a specific conse-

TABLE 5.—*Effect of an EFA deficiency on parameters of growth in eight-week-old White Leghorn male chicks.*[a] *Modified from Engster* et al. *(1979).*

Variate	Dietary treatments—Experiment 7		
	EFA deficient	+ 4% corn oil pair-fed	+ 4% corn oil *ad libitum*
Body wt gain, g	489	532	677*
Energy intake, kcal	3928	3906	4896*
Comb, mg/g body wt	7.28	11.81	12.36
Testes, mg/g body wt	36.66	50.62	55.17*
Tibia length, cm	8.09	8.39	8.89*
Tibia width, cm	0.46	0.48	0.52*
Epiphyseal plate width, μ	799	824	890*
Nitrogen retained, mg/day	991	930	1127
Plasma growth hormone, ng/ml	113	98	63*
Pituitary growth hormone activity, relative response	204.5	215.8*	221.4*

[a]Values are means from 3 replicate pens of 10-12 chicks each.
*Indicates that the value is significantly different ($P < 0.05$) from the EFA-deficient group.

quence of the EFA deficiency, but a generalized effect resulting from the reduced intake of dietary energy.

Nitrogen retention was also slightly (but not significantly) reduced in the EFA-deficient and pair-fed normal cockerels. Earlier we reported that nitrogen retention, measured directly by body composition, was decreased in an EFA deficiency (Carew and Foss, 1974). However, this reduction in nitrogen balance was fully explained by the reduced intake of dietary energy and disappeared with pair-feeding. The present results confirm that report.

It should be noted at this point that whereas certain symptoms of an EFA deficiency can be attributable to reduced feed intake, symptoms relating specifically to an identifiable biochemical role of EFA are not. For example, the elevation in eicosatrienoic acid (20:3w9) in an EFA deficiency never occurs in pair-fed chicks fed linoleic acid. In a like fashion, enlargement of the liver, typical of an EFA deficiency and representing alteration in fat metabolism due to a specific lack of EFA, does not occur in pair-fed chicks receiving EFA. We interpret our results with pair-feeding as showing whether or not EFA participate in a direct and specific way in the biological effect being observed. We would conclude that EFA are directly and specifically responsible for the reduction in pituitary GH content because pair-fed, EFA-adequate chickens did not show a similar decrease. On the other hand, elevation of plasma GH in the EFA deficiency occurred regardless of the presence or absence of EFA in the diet so long as there was a reduction (in other words, a restriction) in food intake.

Experiment 8 (Table 6) represents an effort to determine the effect of an EFA deficiency on thyroid function. Two EFA-deficient diets were used. One was the standard EFA-deficient diet (I) that contained 5 per cent HCO (Table 1). To this we added 10 per cent HCO (diet II) to greatly increase the

TABLE 6.—*Effect of an EFA deficiency on growth, testes weight, and parameters of thyroid function of 8-week-old broiler male chicks.*[a]

| | Dietary treatments—Experiment 8[b] | | | | |
| | Diet type I | | | Diet type II | |
Variate	EFA deficient	+4% corn oil pair-fed to I	+4% corn oil *ad libitum*	+10% HCO	+4% corn oil pair-fed to II
Body wt gain, g	1837 b	1748 b	2267 c	1296 a	1414 a
Testes, mg/100g body wt	23.7 a	25.2 a	31.1 b	24.2 a	26.5 a
Thyroid, mg/100g body wt	8.77 b	6.33 a	6.03 a	8.35 b	5.22 a
Thyroid follicle diameter, μ	244 c	180 ab	206 b	185 b	156 a
Plasma T_3 ng/dl, 4 wk	294 b	260 b	333 b	151 a	307 b
Plasma T_3 ng/dl, 8 wk	212 b	211 b	153 a	177 ab	221 b
Plasma T_4 μg/dl, 4 wk	2.61 a	2.38 a	2.51 a	2.72 b	2.45 a
Plasma T_4 μg/dl, 8 wk	2.65 b	2.54 b	2.72 b	2.31 a	2.69 b
Energy intake, kcal	10938	10991	13278	8000	9075
Caloric efficiency[c], %	37.6	39.4		29.3	33.1

[a]Each value represents the mean obtained from three replicate groups totalling 8 to 20 chicks per treatment. All values were obtained at 8 weeks of age except for the 4-week T_3 and T_4 values, and caloric efficiency which were taken at 4 weeks.

[b]Means in the same horizontal line followed by the same on-line lowercase letter are not significantly different ($P > 0.05$).

[c]Caloric efficiency is the ratio of total energy gain in the carcass divided by total metabolizable energy intake from 1 to 4 weeks of age.

level of saturated fat. We expected that this would intensify the EFA deficiency (Peifer and Holman, 1959). Thus we were producing what we considered to be both moderate and severe EFA deficiencies.

The results on body weight gain (Table 6) indicate that moderate and severe EFA deficiencies were indeed produced. The weight depression with 10 per cent HCO added to the diet (diet II) was more severe than in diet I. When chicks fed the EFA-adequate diets containing added corn oil were pair-fed to the respective deficient groups, the growth depressions were similar, showing that the reduced growth was largely explained by the reduction in food intake. Note that pair-feeding between the deficient chicks fed the diet with the added 10 per cent HCO and their respective controls was not exact. This represents an unexpected decrease in the energy value of the diet with 10 per cent HCO. We suspect that the high level of HCO was less absorbed than expected.

Testes weights were smaller in the EFA-deficient birds. Similarly, small testes were also observed in EFA-adequate chickens pair-fed to the EFA-deficient ones. This shows that the reduction in testes size is accounted for by the reduced feed and energy intake, and agrees with our previous conclusions for experiment 6.

Thyroid weights were increased in both the moderate and severe EFA deficiencies. This thyroid enlargement did not occur in the pair-fed EFA-adequate chickens; thus this alteration in thyroid size is a specific effect of the EFA deficiency apart from changes in energy intake. This goiter-like condition was accompanied by an increase in thyroid follicle diameter only in the EFA-deficient birds, but not in their pair-fed controls. These results demonstrate an important role for EFA in the maintenance of normal development and function of the avian thyroid. An enlarged thyroid often represents a defect in the iodine uptake mechanism. In rats, however, it has been reported that iodine uptake by the thyroid is not altered in an EFA deficiency (Morris *et al.*, 1957). Presence of an enlarged thyroid gland also suggests an increase in TSH stimulation. It is unfortunate that an RIA for avian plasma TSH is not available. However, we are proceeding with a biological assay of hypophyseal TSH in this study.

An interesting observation with regard to thyroid follicle diameter is that in the EFA-adequate groups of chickens the diameter was smaller in the pair-fed chickens than in chickens in the *ad libitum* fed group. This probably reflects a reduction thyroid activity because it is well known that starvation and fasting inhibit thyroid activity (Penn and Huston, 1968; May, 1978).

If thyroid function is affected in an EFA deficiency, it might be reflected in changes in circulating levels of T_3 and T_4. The data in Table 6 show that plasma T_3 and T_4 levels were little affected at either four or eight weeks of age in the moderate EFA deficiency. Although T_3 values with diet I were elevated at eight weeks compared to the corn oil diet fed *ad libitum*, they were not different from the pair-fed control, and therefore represent an effect of level of food intake only. However, with the severe EFA deficiency (diet II), there were definite changes in plasma T_4 and T_3 levels. T_4 was significantly increased at four weeks of age and significantly decreased by eight weeks of age. With T_3, plasma levels were lower in the deficient birds at both ages although not statistically significant ($P < 0.05$) at the older age. We believe that the increased thyroid size, increased follicle diameter, and low plasma T_3 in the severe EFA deficiency argue for reduced thyroid activity, especially at eight weeks when plasma T_4 levels are also low. This is strengthened by an earlier observation (Engster *et al.*, 1978) that thyroid follicle cell height is also diminished in the EFA deficiency. An explanation for the lack of change in plasma T_4 and T_3 levels in the moderate EFA deficiency might be that, by enlarging, the thyroid gland is able to compensate sufficiently for the lowered efficiency of hormone synthesis. This compensation might be effective with a mild deficiency but not a severe one. Certainly much more research is needed to unravel the mechanisms involved, and to determine if age and degree of deficiency are factors in the nature of the response as these results suggest. However, if thyroid activity is truly decreased in EFA-deficient chickens, this means that these chickens respond differently than has been reported for EFA-deficient rats (Alfin-Slater and Bernick, 1958).

If thyroid activity is diminished, it should be reflected in a reduction in energy metabolism represented either by lower oxygen consumption or lower thermogenesis. We used the latter measure and determined it by comparing total energy retention in the body with total energy intake. This gave us a measure of caloric efficiency. The relevant comparisons were only between the deficient and pair-fed groups. As shown in Table 6, there were no significant differences in these comparisons. But efficiency of caloric retention was always poorer in the deficient chickens than in the pair-fed controls. This suggests a small wastage of heat. These data on caloric efficiency are more consistent with either normal thyroid activity or overactivity than with underactivity, and do not appear consistent with our other data on thyroid function. However, a goiter-like condition does not necessarily represent hypothyroidism; hyperthyroidism and euthyroidism have been observed in goiter. Also, measures of body composition are not very sensitive, and do not allow partitioning of heat production into its components of basal metabolism, activity, and heat increment. We intend to repeat these studies using a calorimeter.

Summary

An essential fatty acid (EFA) deficiency was produced in chickens and quail by feeding purified casein-glucose diets containing 0.03 per cent linoleic acid and 2 to 15 per cent hydrogenated coconut oil. Control chicks were fed diets containing 4 to 10 per cent corn oil. Adult cockerels and quail showed marked depressions in reproductive performance in the EFA deficiency. This was characterized by slower testicular and comb growth, impaired spermatogenesis, impaired ovarian and oviducal development, low plasma luteinizing hormone (LH) levels, and decreased pituitary gonadotropic activity. Pair-feeding of EFA-adequate cockerels to the lower feed intake of the deficient cockerels brought about reductions in body growth rate, testicular and comb growth, and decreases in plasma LH and pituitary gonadotropic activity similar to that observed in the deficient chickens. This suggests that these symptoms are related to the lower level of feed intake and not a specific and direct effect of the EFA deficiency.

When parameters of growth and thyroid function were examined in an EFA deficiency in young chickens, the following effects were observed: low weight gain; smaller length and width of the tibia, and narrower epiphyseal plate width; reduced nitrogen retention; increased plasma growth hormone (GH) content; decreased pituitary GH content; enlarged thyroid glands with larger follicles; reduced plasma T_3 and T_4 levels in a very severe EFA deficiency (except that T_4 was elevated at an early age); and no significant change in caloric efficiency as a measure of thermogenesis. When EFA-deficient and EFA-adequate cockerels were pair-fed, body weight gain, bone growth, nitrogen retention, and plasma GH were similar. However, changes in pituitary GH content, thyroid development, and plasma T_3 and T_4 in most cases did not occur in the EFA-adequate, pair-fed chickens as they did in the deficient chickens.

These results show that an EFA deficiency has widespread effects on reproductive performance, measures of growth, and thyroid function, and that these effects are mediated in part by changes in plasma and pituitary levels of hormones. Many of these changes are a consequence of the reduction in feed and energy intake in the EFA-deficient chickens, especially the measures of reproductive performance and growth. However, most changes in thyroid function, as well as pituitary growth hormone content, occur only when EFA are deficient and seem to be a direct and specific consequence of the EFA deficit, not just the reduction in feed intake.

ACKNOWLEDGMENTS

We thank Drs. Colin G. Scanes, Steve Harvey, and Frank J. Cunningham for conducting GH and LH radioimmunoassays; to Drs. Donald C. Foss and David E. Bee for their advice; and to Christine Crossman, Geoff Clark, Vera Graham and Kathleen Underhill for their technical assistance. We are grateful to Agway, Inc., Syracuse, New York for financial support, and to Hoffmann-LaRoche, Inc., Monsanto Chemical Co., Commercial Solvents Corp., Proctor and Gamble Co., and Beckman Instruments, Inc., for donating materials. Vermont Agricultural Experiment Station Journal Article No. 454.

LITERATURE CITED

ALFIN-SLATER, R. B., AND L. AFTERGOOD. 1968. Essential fatty acids reinvestigated. Physiol. Rev., 48:758-784.

———. 1971. Physiological functions of essential fatty acids. Prog. Biochem. Pharm., 6:214-241.

ALFIN-SLATER, R. B., AND S. BERNICK. 1958. Changes in tissue lipids and tissue histology resulting from an essential fatty acid deficiency in rats. Amer. J. Clin. Nutr., 6:613-624.

AUCKLAND, J. N., AND R. B. FULTON. 1973. Effects of restricting energy intake in laying hens. Brit. Poult. Sci., 14:579-588.

BALNAVE, D. 1972. Essential fatty acids in poultry nutrition. World's Poult. Sci., 26:442-460.

BALZER, P. M. 1980. Effect of a dietary essential fatty acid deficiency on thyroid function and energy balance in the growing broiler cockerel. M.S. Thesis, Univ. Vermont, vi + 104 pp.

BECKMAN. 1976. T_4 reagent system for the radioimmunoassay of ^{125}I labelled thyroxine. Beckman Instruments, Inc., Fullerton, California, 16 pp.

———. 1977. T_3 reagent system for the radioimmunoassay of triiodothyronine. Beckman Instruments, Inc., Fullerton, California, 15 pp.

BIERI, J. G., C. J. POLLARD, AND G. M. BRIGGS. 1957. Essential fatty acid deficiency in the chick. II. Polyunsaturated fatty acid composition of blood, heart, liver. Arch. Biochem. Biophys., 68:300-307.

BURR, G. O., AND A. J. BEBER. 1937. Metabolism studies with rats suffering from fat deficiency. J. Nutr., 14:553-566.

BURR, G. O., AND M. M. BURR. 1929. A new deficiency disease produced by the rigid exclusion of fat from the diet. J. Biol. Chem., 82:345-367.

———. 1930. On the nature and role of the fatty acids essential in nutrition. J. Biol. Chem., 86:587-621.

CALVERT, C. C. 1969. The performance of adult female Japanese quail on linoleic acid deficient diets. Poultry Sci., 48:975-985.

CAREW, L. B., JR., AND D. C. FOSS. 1974. Effect of linoleic acid on energy utilization by chicks. Fed. Proc., 33:664.

DANFORTH, E., JR., E. S. HORTON, E. A. H. SIMS, A. G. BURGER, A. G. VAGENAKIS, L. E. BRAVERMAN, AND S. H. INGBAR. 1979. Dietary-induced alteration in thyroid hormone metabolism during overnutrition. J. Clin. Invest., 64:1336-1347.

DEUEL, H. J., JR., S. M. GREENBERG, C. E. CALBERT, E. E. SAVAGE, AND T. FUKUI. 1950. The effect of fat level of the diet on general nutrition. V. The relationship of linoleic acid requirement to optimum fat level. J. Nutr., 40:351-366.

DUNCAN, D. R. 1955. Multiple range and multiple F tests. Biometrics, 11:1-42.

EDWARDS, H. M., JR. 1967. Studies of the essential fatty acid deficiency of the growing domestic cock. Poultry Sci., 46:1128-1133.

ENGSTER, H. M., L. B. CAREW, JR., AND F. J. CUNNINGHAM. 1978. Effects of an essential fatty acid deficiency, pair-feeding and level of dietary corn oil on the hypothalamic-pituitary-gonadal axis and other physiological parameters in the male chicken. J. Nutr., 108:889-900.

ENGSTER, H. M., L. B. CAREW, JR., S. HARVEY, AND C. G. SCANES. 1979. Growth hormone metabolism in essential fatty acid-deficient and pair-fed nondeficient chicks. J. Nutr., 109:330-338.

FOLLETT, B. K., C. G. SCANES, AND F. J. CUNNINGHAM. 1972. A radioimmunoassay for avian luteinizing hormone. J. Endocrinol., 52:359-378.

GREENSPAN, F. S., C. H. LI, M. E. SIMPSON, AND H. M. EVANS. 1949. Bioassay of hypophyseal growth hormone: the tibia test. Endocrinology, 45:455-463.

GUARNIERI, M., AND R. M. JOHNSON. 1970. The essential fatty acids. Adv. Lipid Res., 8:115-174.

HARVEY, S., AND C. G. SCANES. 1977. Purification and radioimmunoassay of chicken growth hormone. J. Endocrinol., 73:321-329.

HARVEY, S., C. G. SCANES, A. CHADWICK, AND N. J. BOLTON. 1978. Influence of fasting, glucose and insulin on plasma levels of growth hormone and prolactin in the domestic fowl (Gallus domesticus). J. Endocrinol., 76:501-506.

HARVEY, S., C. G. SCANES, AND T. HOWE. 1977. Growth hormone effects on in vitro metabolism of avian adipose and liver tissue. Gen. Comp. Endocrinol., 33:322-328.

HILL, F. W., AND D. L. ANDERSON. 1958. Comparison of metabolizable energy and productive energy determinations with growing chicks. J. Nutr., 64:587-604.

HILL, F. W., D. L. ANDERSON, R. RENNER, AND L. B. CAREW, JR. 1960. Studies of the metabolizable energy of grain and grain products for chickens. Poult. Sci., 39:573-579.

HOLMAN, R. T. 1968. Essential fatty acid deficiency. Prog. Chem. Fats and Other Lipids, 9:275-348.

KAMIYASHI, M., K. TANAKA, AND Y. TANABE. 1972. A 6 hour bioassay for pituitary gonadotropins based on radioactive phosphorus uptake by chick testes. Endocrinology, 91:385-388.

LILLIE, R. J., AND H. MENGE. 1968. Effect of linoleic acid deficiency on the fertilizing capacity and semen fatty acid profile of the male chicken. J. Nutr., 93:311-315.

LOWE, C. L., L. B. CAREW, JR., D. C. FOSS, AND H. M. ENGSTER. 1975. Essential fatty acid deficiency in growing quail. Poultry Sci. 54:1786 (abstr.).

MACHLIN, L. J., AND R. S. GORDON. 1961. Effect of dietary fatty acids and cholesterol on growth and fatty acid composition of the chicken. J. Nutr., 75:157-164.

MAY, J. D. 1978. Effect of fasting on T_3 and T_4 concentration in chicken serum. Gen. Comp. Endocrinol., 34:323-327.

MENGE, H. 1967. Fatty acid composition and weights of organs from essential fatty acid deficient and non-deficient hens. J. Nutr., 92:148-152.

MORRIS, D. M., T. C. PANOS, J. C. FINERTY, R. L. WALL, AND G. F. KLEIN. 1957. Relation of thyroid activity to increased metabolism induced by fat deficiency. J. Nutr., 62:119-128.

NUGARA, D., AND H. M. EDWARDS, JR. 1970a. Changes in fatty acid composition of cockerel testes due to age and fat deficiency. J. Nutr., 100:156-160.

———. 1970b. In vitro androgen metabolism by fat-deficient cockerel testes and uropygial gland. J. Nutr., 100:539-544.

PANOS, T. C., AND J. C. FINERTY. 1954. Effects of a fat-free diet on growing male rats with special reference to the endocrine system. J. Nutr., 54:315-329.

PARKER, J. E., AND G. H. ARSCOTT. 1964. Energy intake and fertility of male chickens. J. Nutr., 82:183-187.

PEIFER, J. J., AND R. T. HOLMAN. 1959. Effect of saturated fat upon essential fatty acid metabolism of the rat. J. Nutr., 68:155-168.

PENN, R. P., AND T. M. HUSTON. 1968. The influence of feed and water restriction and pitressin administration upon thyroxine secretion in domestic fowl. Poultry Sci., 47:1432-1436.

ROLAND, D. A., AND H. M. EDWARDS, JR. 1971. Evidence for direct effects of essential fatty acids at the hypothalamus-pituitary level in domestic fowl. J. Nutr., 101:1683-1694.

SCHEFFÉ, H. 1953. A method for judging all contrasts in the analysis of variance. Biometrika, 40:87-104.

SNEDECOR, G. W. 1956. Statistical methods. Fifth edition, The Iowa State College Press, Ames, Iowa, xiii + 534 pp.

TURNER, C. D., AND J. T. BAGNARA. 1976. General Endocrinology. Sixth edition, W. B. Saunders Co., Philadelphia, x + 596 pp.

YALOW, R. S., AND S. A. BERSON. 1960. Immunoassay of endogenous plasma insulin in man. J. Clin. Invest., 39:1157-1175.

YOUNG, R. A., O. L. TULP, AND E. S. HORTON. 1980. Thyroid and growth responses of young Zucker obese and lean rats to a low protein-high carbohydrate diet. J. Nutr., 110:1421-1431.

Reprinted from
ASPECTS OF AVIAN ENDOCRINOLOGY:
PRACTICAL AND THEORETICAL IMPLICATIONS (C. G. Scanes *et al.*, eds.)
Grad. Studies, Texas Tech Univ., 1982, 26:1-411.

HORMONES, NUTRITION, AND METABOLISM IN BIRDS

COLIN G. SCANES[1] AND STEVEN HARVEY[2]

[1]*Department of Physiology, Rutgers-The State University, New
Brunswick, New Jersey 08903 USA;* [2]*The Wolfson Institute, The
University of Hull, Hull HU6 7RX, England*

This review deals with three aspects of avian growth and metabolism: 1)
endocrine control; 2) interactions between the two; and 3) influence of nutrition on both. Particular emphasis is placed on growth hormone (GH)
because of recent advances in our knowledge of this hormone, especially in
birds.

Hormones and Growth

The phenomenon of growth in animals is a complex process, and depends
upon the inherent genome of the individual, the cellular metabolism of
nucleic acids, protein, carbohydrates, and lipids, the tissues receiving
optimal concentrations of substrates (which in turn depends on the nutrition
of the animal), the hormonal milieu, and the tissue responsiveness to hormones (see general scheme, Fig. 1). A number of hormones appear to be
required for growth. In both birds and mammals, there is a requirement for
GH, thyroxine (T_4), and probably for insulin, somatomedins, and other circulating growth factors, namely, prolactin and androgen (Fig. 1).

Growth Hormone

There is considerable circumstantial evidence that GH is as important to
growth in birds as it is in mammals, although direct evidence is lacking.
Whereas hypophysectomy has been shown to reduce the growth rate of
young chickens (Nalbandov and Card, 1943), there is no information on the
effect of replacement with GH or other pituitary hormones in growing birds.
In adult hypophysectomized pigeons, administration of mammalian GH
does increase body weight (Bates *et al.*, 1962). Studies on the effect of exogenous mammalian GH on the growth of intact chicks are somewhat equivocal. Growth in young chickens has been reported to be stimulated by a
trypsin-treated bovine GH (Meyers and Peterson, 1974), but not by bovine
GH itself (Libby *et al.*, 1955). In addition, bovine GH stimulates liver RNA
metabolism in young chicks (Scanes *et al.*, 1975).

Among the best evidence that GH exerts an important role in avian
growth is the report of the *in vivo* use of an antisera raised against chicken
GH and longitudinal studies of circulating concentrations of GH with age
in birds. The administration of antisera raised against avian GH to chicks
reduced their growth rate (Scanes *et al.*, 1977), which is similar to the effect

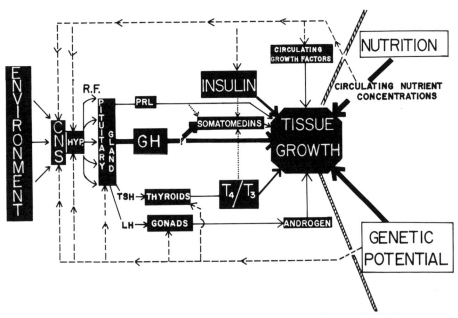

Fig. 1.—Scheme for control of tissue growth by hormones, nutrition, and other factors.

of anti-rat GH in rats (Grindeland *et al.*, 1974). It is particularly interesting that the plasma concentrations of GH vary with stage of growth and development in birds: high in the early rapid stage of post-hatch growth and low during the latter stages of growth (Fig. 2). Data illustrating this can be found for chickens (Harvey *et al.*, 1979), turkeys (Harvey *et al.*, 1977*a*; Proudman and Wentworth, 1980), geese (Scanes *et al.*, 1979), and doves (Scanes and Balthazart, 1981). It may be noted that the plasma concentration of GH appears to be higher at a somewhat earlier stage in faster growing strains of chickens (Harvey *et al.*, 1979), and in turkeys (Proudman and Wentworth, 1980).

Thyroid Hormones

Thyroid hormones appear to be very important in avian growth, particularly in muscle and bone growth. For instance, King and King (1976) reported that chemical thyroidectomy by propylthiouracil administration is followed by decreased bone length (shank-toe), muscle (gastrocnemius and sartorius) weight, and muscle DNA content in growing chicks, and that this trend could be reversed by injections of thyroxine. These authors suggested that the thyroidal influence on growth may be mediated by decreased GH secretion. However, this does not appear to be the case as goitrogen administration has been found to be accompanied by increased circulating concentrations of GH (Chaisson *et al.*, 1979). Although thyroxine is important to growth, it is of note that the plasma concentrations of thyroxine appear to

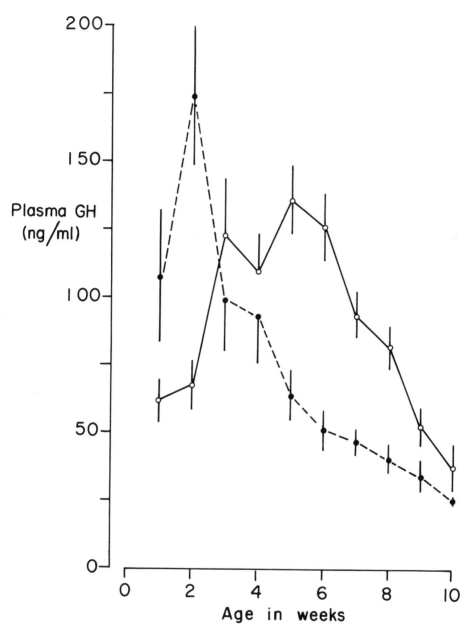

FIG. 2.—Variation in the plasma concentration of GH with age in growing male chickens of either heavy, broiler type (closed circles) or light, layer type strains (open circles). Vertical bars indicate SE. Closed circles are pooled data from Chaisson *et al.* (1979) and Harvey *et al.* (1979), $N = 21$. Open circles represent pooled data from Harvey and Scanes (1978, 1979) and Harvey *et al.* (1979), $N = 36$.

be very constant throughout growth in at least the domestic fowl (Davison, 1976).

Hormones and Metabolism
Pancreatic Hormones

Glucagon and insulin have major roles in the control of carbohydrate and lipid metabolism in both birds and mammals (Hazelwood, 1976). To summarize, glucagon acts to mobilize liver glycogen and increase the already high plasma concentrations of glucose, while insulin exerts an opposite effect (in chickens, Heald et al., 1965; Langslow et al., 1970; Hazelwood, 1976; in geese, Grande, 1969; Sitbon and Mialhe, 1978a, 1978b). Glucagon is the most potent lipolytic and anti-lipogenic hormone in the domestic fowl (Langslow and Hales, 1969; Goodridge, 1973). Whereas insulin stimulates lipogenesis in the chicken (Goodridge, 1973; Kompiang and Gibson, 1973), it has little anti-lipolytic action alone or with glucagon (ducks, Grande, 1969; chickens, Langslow and Hales, 1969). However, insulin appears to potentiate glucagon induced lipolysis in vitro (Langslow and Hales, 1969). These effects on lipolysis and lipogenesis explain the elevated plasma concentrations of free fatty acid found after insulin administration (chickens, Heald et al., 1965; Langslow et al., 1970). The proposed third pancreatic hormone, avian pancreatic polypeptide, has little effect on plasma glucose concentrations or liver carbohydrate metabolism in the chicken, but does reduce the circulating concentration of free fatty acids and partially suppresses glucagon induced lipolysis (Langslow and Hazelwood, 1975; McCumbee and Hazelwood, 1977). Finally, both glucagon and insulin depress the plasma concentration of GH in young chickens (Harvey et al., 1978). The control of pancreatic hormone secretion is as follows: glucagon secretion is stimulated by low (and inhibited by high) ambient glucose concentrations (in chickens, Honey et al., 1980; ducks, Laurent and Mialhe, 1976; geese, Sitbon and Mialhe, 1978a). It is also elevated by arginine (chickens, Honey et al., 1980; ducks, Laurent and Mialhe, 1978), somatostatin (in ducks, Strosser et al., 1980), free fatty acids (ducks, Laurent and Mialhe, 1978), and GH (ducks, Foltzer et al., 1975). Insulin release, in turn, is stimulated by glucose (chickens, Langslow et al., 1970; King and Hazelwood, 1976; ducks, Laurent and Mialhe, 1976; geese, Sitbon and Mialhe, 1978a), arginine (chickens, Honey et al., 1980; ducks, Laurent and Mialhe, 1978), and glucagon (chickens, King and Hazelwood, 1976; Naber and Hazelwood, 1977) and is inhibited by epinephrine (chickens, Langslow et al., 1970), GH (ducks, Foltzer and Mialhe, 1976), and somatostatin (ducks, Strosser et al., 1980). Somatostatin release from the chicken pancreas is increased by glucose, arginine (Honey et al., 1980), and insulin (Honey and Weir, 1979).

Growth Hormone

The following is a discussion of the role of GH in the short-term control of metabolism in birds. All general conclusions reached will be of a somewhat speculative nature, primarily because only limited data is available; studies have been done in only two species (domestic fowl and duck).

GH appears to be involved in the control of carbohydrate, lipid, and protein metabolism. GH appears to stimulate lipolysis and inhibit lipogenesis; the former results in elevated plasma concentrations of fatty acids and the latter tends to increase plasma concentrations of glucose. GH also elevates plasma amino acids, presumably (by analogy with mammals) by decreased catabolism in the liver. GH does not affect liver glycogen levels, but has been reported to increase muscle glycogen (Hazelwood and Lorenz, 1959). Furthermore, GH affects the circulating concentrations of both insulin and glucagon.

Of these proposed effects, the effect of GH on lipid metabolism is best characterized. Hypophysectomized chickens become obese (Nalbandov and Card, 1943). This however, may be due to a deficiency of T_4. The increase in adipose tissue following hypophysectomy can be explained both by a decrease in lipolysis (Gibson and Nalbandov, 1966a) and by an increase in lipogenesis (reported by Kompiang and Gibson, 1973; but not observed by Chandrabose and Bensadoun, 1971). There is also evidence that GH administration produces effects opposite to those following hypophysectomy. For instance, in the domestic fowl, GH stimulates lipolysis *in vitro* (Langslow and Hales, 1969; Harvey *et al.*, 1977b), whereas avian pituitary extracts (containing GH) or preparations of bovine or chicken GH inhibit lipogenesis *in vitro* (Kompiang and Gibson, 1976; Harvey *et al.*, 1977b). Further support for the lipolytic effect of GH comes from *in vivo* studies in the duck where hypophysectomy depresses the plasma concentrations of free fatty acids; this effect is then overcome by GH administration (Foltzer and Mialhe, 1976). Similarly, extracts of avian pituitary glands (containing GH) elevate plasma concentrations of free fatty acids in chickens (Gibson and Nalbandov, 1966b).

The evidence for effects of GH in protein and amino acid metabolism rest largely on analogy with mammals. However, Foltzer and Mialhe (1976) found decreases in circulating levels of non-protein nitrogen (mostly amino acids) following hypophysectomy; this was reversed with GH replacement therapy.

In carbohydrate metabolism studies, hypophysectomized ducks exhibit depressed plasma concentrations of glucose, which are again elevated by GH therapy (Foltzer *et al.*, 1975). It is these changes in glucose levels and not GH that directly influence the circulating glucagon concentration, this being elevated following hypophysectomy and depressed with GH therapy (Foltzer *et al.*, 1975). Circulating concentrations of insulin are also affected by hypophysectomy (decreased) and GH (increased). It is, however, uncertain whether this is a direct effect of GH or not (Foltzer and Mialhe, 1976). Similarly, in the chicken, a more rapid, prolonged, and acute decrease in circulating glucose levels during fasting has been observed in hypophysectomized birds (Koike *et al.*, 1964).

Thus, GH appears to mobilize energy stored in the adipose tissue, to spare glucose and amino acids from utilization, and to direct metabolites to the muscle. While this is of obvious significance for muscle growth, it is also probably very important in the short term regulation of metabolism. For instance, in the chicken, the circulating level of GH is elevated by starvation (Harvey *et al.*, 1978) and by longer term nutritional deprivation (see below). GH could aid the response of the bird to these constraints by acting to maintain circulating glucose levels, and to stimulate breakdown of stored triglyceride, thus releasing the readily useable free fatty acids and glycerol.

Adrenal Hormones

The effects of adrenal hormones on avian metabolism are reviewed extensively elsewhere (for example, Hazelwood, 1976) and hence only will be briefly covered here. Epinephrine increases the plasma concentration of glucose by mobilizing glycogen (Langslow *et al.*, 1970). It has, however, little effect on the plasma concentration of free fatty acids (Langslow *et al.*, 1970), although a high concentration of epinephrine stimulates lipolysis (Langslow and Hales, 1969) and inhibits lipogenesis (Gibson and Nalbondov, 1966*a*).

The glucocorticoids, of which corticosterone is the principal one in birds, appear to elevate both the plasma concentration of glucose and the liver concentration of glycogen (Hazelwood, 1976). This probably is achieved by increased glycogenesis (from amino acids and glycerol, Hazelwood, 1976) and possibly also from stimulation of lipolysis (Harvey *et al.*, 1977*b*). In addition, corticosterone appears to extend some synergistic effects with GH on plasma concentrations of metabolites and pancreatic hormones in hypophysectomized ducks (Foltzer *et al.*, 1975; Foltzer and Mialhe, 1976).

It may be noted that insulin induced hypoglycemia is associated with high plasma concentrations of corticosterone in young chickens (Scanes *et al.*, 1980). The glucocorticoid, in turn, then presumably acts to elevate plasma glucose concentrations. Indeed, in hypophysectomized chickens, the insulin effect is greatly prolonged (Koike *et al.*, 1964).

Other Hormones

It would be facile to list all the effects of hormones on metabolism in birds. Some of the more interesting of these will be considered. It can be noted that adrenocorticotropic hormone has a potent direct lipolytic effect on chicken adipose tissue (Langslow and Hales, 1969). Furthermore, thyroidal hormones affect carbohydrate metabolism (Hazelwood, 1976) in the chicken. For instance, triiodothyronine reduces liver glycogen (Arrondo *et al.*, 1978), and increases glucose uptake by heart muscle cells (Segal *et al.*, 1977).

TABLE 1.—*Effect of fasting in the domestic fowl.*

Variate	Effect[a]	Age in weeks	Body weight[b] (g)	Reference
Plasma Hormones				
Insulin	O[c]	6		Langslow *et al.*, 1970
Corticosterone	+	2.6		Scanes *et al.*, 1980
Growth Hormone	+	2.6		Harvey *et al.*, 1978
Prolactin	−	6		Harvey *et al.*, 1978
Luteinizing Hormone	−	6		Scanes *et al.*, 1976
Follicle Stimulating Hormone	−	6		Scanes *et al.*, 1976
Thyroxine	+	7		May (1978)
Triiodothyronine	−	0-7		May (1978); King *et al.*, 1977
Plasma Metabolite				
Glucose	0		(1200-3000)	Belo *et al.*, 1976; Brady *et al.*, 1978
	−	6		Langslow *et al.*, 1970
Free fatty acid	+	6		Langslow *et al.*, 1970
Amino acids	+		Adult (3000)	Belo *et al.*, 1976
Pyruvate	+		(1200-3000)	Belo *et al.*, 1976; Brady *et al.*, 1978
Lactate	0		(1200-3000)	Belo *et al.*, 1976; Brady *et al.*, 1978
β Hydroxybutyrate	+		(1200-3000)	Belo *et al.*, 1976; Brady *et al.*, 1978
Acetoacetate	0		(1200-3000)	Belo *et al.*, 1976; Brady *et al.*, 1978
Tissue Metabolite				
Liver glycogen	−	10		Hazelwood and Lorenz, 1959
Liver pyruvate	0		(1200)	Brady *et al.*, 1978
Liver lactate	0		(1200)	Brady *et al.*, 1978
Liver β hydroxybutyrate	+		(1200)	Brady *et al.*, 1978
Liver glutamate	+		(1200)	Brady *et al.*, 1978
Liver glutamine	+		(1200)	Brady *et al.*, 1978
Heart glycogen	+	10		Hazelwood and Lorenz, 1959
Other Variates				
Glucose replacement rate	−		Adult (3000)	Belo *et al.*, 1976
Adipocyte responsiveness to glycagon	−	0-1		Krug *et al.*, 1976

[a]A plus sign indicates an increase following fasting; a minus sign, a decrease; a zero, no significant effect of fasting.

[b]If data on age are not available, body weight is included in parentheses.

[c]Simon and Rosselin (1978) reported a reduced insulin response following a glucose load in chicks fasted for 65 hours.

The Nutritional, Metabolic, Endocrine Interface

It is intuitively obvious that diet can affect the circulating concentration of metabolites and also that of hormones. For instance, in the domestic fowl, starvation for as short a period as twenty-four hours is accompanied by changes in many metabolic and endocrine parameters (Table 1). As might be expected, fasting shifts glucose stored in the liver (as glycogen) to the plasma and increases the circulating concentrations of free fatty acids via an effect on lipolysis. The fatty acids are then available as an energy source and their use as such is reflected in increases in the concentration of the ketone, β hydroxybutyrate, in both the liver and the circulation. In addition, muscle protein is likely to be broken down to generate amino acids for catabolism.

TABLE 2.—*Plasma GH and LH in concentrations in young (four-week-old) male chickens on different protein diets.*[a]

Dietary protein (%)	Plasma GH[b] $\bar{X} \pm$ (N) SE	Plasma LH[b] $\bar{X} \pm$ (N) SE
3	334.8±(10)21.7 }**	31.4±(10 3.1
6	210.9±(10)37.0 }*	36.4±(10) 2.9 }***
12	137.0±(10)17.4	110.3±(10)15.0
18	117.4±(0)13.0	132.7±(10)22.2
24	100.0±(10) 8.7	100.0±(9)10.5

[a]Birds were fed *ad libitum* for 2 weeks (from 2 to 4 weeks old) on synthetic isocaloric diet with varying protein concentrations.
[b]Hormone concentration expressed as a percentage of that in birds on the control (24%) protein diet. Data based on Buonomo *et al.*, 1982; Scanes *et al.*, 1981.
*$P<0.05$, **$P<0.01$, ***$P<0.001$

Supporting evidence for this comes from the reported increased concentrations of amino acids in the plasma, and presence of glutamate and glutamine in the liver in fasted birds (see Table 1).

Nutrition and Metabolic Hormones

It would be reasonable to suggest that the metabolic changes accompanying starvation (see above and Table 1) in part result from the action of metabolic hormones. Glucagon would appear to be the most likely hormone to account for the starvation responses as glucagon increases glycogenolysis and lipolysis and inhibits lipogenesis; indeed, pancreatectomized fasted chickens show fatal hypoglycemia (Krug *et al.*, 1976). Although fasting does not affect the circulating concentrations of glucagon-like immunoreactivity in birds (Krug *et al.*, 1976; Sitbon and Mialhe, 1979), it does elevate that of pancreatic glucagon (Sitbon and Mialhe, 1979). Another possible candidate is insulin. Basal circulating concentrations of insulin are little affected by starvation (Table 1). However, elevated insulin concentrations associated with feeding (Sitbon and Mialhe, 1979) are obviously absent in fasted birds. This change in overall insulin level might reduce the synthesis of glycogen and fatty acids.

Fasting also is associated with elevated plasma concentrations of GH (Harvey *et al.*, 1978) and corticosterone (Nir *et al.*, 1975; Scanes *et al.*, 1980; Table 1). In turn, both are likely to increase lipolysis, while GH will depress lipogenesis and corticosterone will increase the catabolism of proteins (see above). It should be noted that elevated plasma levels of GH in young chickens result from fasting and from chronic dietary deprivation, including reduced caloric intake, essential fatty acid deficiency (Engster *et al.*, 1979), and low levels of protein in the diet (Scanes *et al.*, 1980; also see Table 2). These increases in plasma GH concentrations could reflect increased secretion or a lower rate of disappearance. In rats, starvation is reported to reduce the metabolic clearance rate for GH (Mosier *et al.*, 1980); this also might be the case in birds.

Nutrition and Reproductive Hormones

It is of selective value to a species to restrict reproductive activity at times of very low food availability when survival of offspring is not favored. In addition, redirection of metabolites away from the gonads and reproductive tract to the more essential organs should enhance the likelihood of that animal to withstand adverse dietary conditions. In birds, the effect of nutrition on reproduction is best characterized in the domestic fowl. For instance, starved hens do not ovulate because of decreased gonadotropin secretion (Morris and Nalbandov, 1961). Starvation in young chicks reduces the plasma concentrations of luteinizing hormone (LH) and follicle stimulating hormone (FSH) (Table 1; Scanes *et al.*, 1976). In addition, feeding diets deficient in either essential fatty acid or protein result in low plasma concentrations of LH (Engster *et al.*, 1978; Buonomo *et al.*, 1982; also see Table 2).

ACKNOWLEDGMENTS

Financial Support by the New Jersey State Experimental Station and the Upjohn Company for studies described is acknowledged.

LITERATURE CITED

ARRONDO, J. L. R., M. J. SANCHO, AND J. M. MACARULLA. 1978. Effect of acute administration of triodothyronine in chicken. Liver glycogen depletion and amino acid incorporation to proteins. Experientia, 34:1099-1100.

BATES, R. W., R. A. MILLER, AND M. M. GARRISON. 1962. Evidence in the hypophysectomized pigeon of a synergism among prolactin, growth hormone, thyroxine and pregnisone upon weight of the body, digestive tract, kidney and fat stores. Endocrinology, 71:345-360.

BELO, P. S., D. R. ROMSOS, AND G. A. LEVEILLE. 1976. Blood metabolites and glucose metabolism in the fed and fasted chicken. J. Nutr., 106:1135-1143.

BRADY, L. J., D. R. ROMSOS, P. S. BRADY, W. G. BERGEN, AND G. A. LEVEILLE. 1978. The effects of fasting on body composition, glucose turnover, enzymes and metabolites in the chicken. J. Nutr., 108:648-657.

BUONOMO, F. C., P. GRIMINGER, AND C. G. SCANES. 1982. Effects of gradations in protein-calorie malnutrition on the pituitary-gonadal axis in young domestic fowl. Poultry Sci., In press.

CHAISSON, R. B., P. J. SHARP, H. KLANDORF, C. G. SCANES, AND S. HARVEY. 1979. The effect of rapeseed meal and methimazole on levels of plasma hormones in growing broiler cockerels. Poultry Sci., 58:1575-1583.

CHANDRABOSE, K. A., AND A. BENSADOUN. 1971. Effects of hypophysectomy on some enzymes involved in lipid metabolism of the domestic chicken (*Gallus domesticus*). Comp. Biochem. Physiol., 39B:45-54.

DAVIDSON, T. F. 1976. Circulating thyroid hormones in the chicken before and after hatching. Gen. Comp. Endocrinol., 29:21-27.

ENGSTER, H. M., L. B. CAREW, AND F. J. CUNNINGHAM. 1978. Effects of an essential fatty acid deficiency, pairfeeding and levels of dietary control on the hypothalamic-pituitary-gonadal axis and other physiological parameters in the male chicken. J. Nutr. 108:889-900.

ENGSTER, H. M., L. B. CAREW, S. HARVEY, AND C. G. SCANES. 1979. Growth hormone metabolism in essential fatty acid-deficient and pair-fed non-deficient chicks. J. Nutr., 109:330-338.

FOLTZER, C., AND P. MIALHE. 1976. Pituitary and adrenal control of pancreatic endocrine function in the duck. II. Plasma free fatty acids and insulin variations following hypophysectomy and replacement therapy with growth hormone and corticosterone. Diabetes et Metabolisme, 2:101-105.

FOLTZER, C., V. LECLERCQ-MEYER, AND P. MIALHE. 1975. Pituitary and adrenal control of pancreatic endocrine function in the duck. I. Plasma glucose and pancreatic glucagon variations following hypophysectomy and corticosterone. Diabetes et Metabolisme, 1:39-44.

GIBSON, W. R., AND A. V. NALBANDOV. 1966a. Lipolysis and lipogenesis in liver and adipose tissue of hypophysectomized cockerels. Amer. J. Physiol., 211:1352-1356.

————. 1966b. Lipid mobilization in obese hypophysectomized cockerels. Amer. J. Physiol., 211:1345-1351.

GOODRIDGE, A. G. 1973. Regulation of fatty acid synthesis in isolated hepatocytes prepared from the livers of neonatal chicks. J. Biol. Chem., 248:1924-1931.

GRINDELAND, R. E., A. T. SMITH, E. S. EVANS, AND S. ELLIS. 1974. Induction of chronic growth hormone deficiency by anti-GH serum. Endocrinology, 95:793-798.

GRANDE, F. 1969. Lack of insulin effect on free fatty acid mobilization produced by glucagon in birds. Proc. Soc. Exp. Biol. Med., 130:711-713.

HARVEY, S., AND C. G. SCANES. 1978. Plasma concentrations of growth hormone during growth in normal and testosterone-treated chickens. J. Endocrinol., 79:145-146.

————. 1979. Plasma growth hormone concentrations in growth-retarded, cortisone treated chickens. Brit. Poul. Sci., 20:331-335.

HARVEY, S., P. M. M. GODDEN, AND C. G. SCANES. 1977a. Plasma growth in turkeys. Brit. Poult. Sci., 18:547-551.

HARVEY, S., C. G. SCANES, AND T. HOWE. 1977b. Growth hormone effects on in vitro metabolism of avian adipose and liver tissue. Gen. Comp. Endocrinol., 33:322-328.

HARVEY, S., C. G. SCANES, A. CHADWICK, AND N. J. BOLTON. 1978. Influence of fasting, glucose and insulin on the levels of growth hormone and prolactin in the plasma of the domestic fowl (Gallus domesticus). J. Endocrinol., 76:501-506.

HARVEY, S., C. G. SCANES, A. CHADWICK, AND N. J. BOLTON. 1979. Growth hormone and prolactin secretion in growing domestic fowl: influence of sex and breed. Brit. Poult. Sci., 20:9-17.

HAZELWOOD, R. L. 1976. Carbohydrate metabolism. Pp. 210-232, in Avian physiology (P. D. Sturkie, ed.), Springer-Verlag, New York, xiii + 399 pp.

HAZELWOOD, R. L., AND F. W. LORENZ. 1959. Effects of fasting and insulin on carbohydrate metabolism in the domestic fowl. Amer. J. Physiol., 197:47-51.

HEALD, P. J., P. M. McLACHLAN, AND K. A. ROOKLEDGE. 1965. The effects of insulin glucagon and adrenocorticotrophic hormones on the plasma glucose and free fatty acids of the domestic fowl. J. Endocrinol., 33:83-95.

HONEY, R. N., AND G. C. WEIR. 1979. Insulin stimulates samotostatin and inhibits glucagon secretion from the perfused chicken pancreas-duodenum. Life Sci., 24:1747-1750.

HONEY, R. N., J. A. SCHWARTZ, C. J. MALHE, AND G. C. WEIR. 1980. Insulin, glucagon and somatostatin secretion from isolated perfused rat and chicken pancreas-duodenum. Amer. J. Physiol., 238:E150-E156.

KING, D. L., AND R. L. HAZELWOOD. 1976. Regulation of avian insulin secretion by isolated perfused chicken pancreas. Amer. J. Physiol., 231:1830-1839.

KING, D. B., AND C. R. KING. 1976. Thyroidal influence on gastroeniumius and sartorius muscle growth in young white leghorn cockerels. Gen. Comp. Endocrinol., 29:473-479.

KING, D. B., C. R. KING, AND J. R. ESHLEMAN. 1977. Serum triiodo thyronine levels in the embryonic and post hatching chickens with particular reference to feeding-induced changes. Gen. Comp. Endocrinol., 31:216-223.

KOIKE, T. I., A. V. NALBANDOV, M. K. DIMICK, T. MATSUMURA, AND S. LEPKOVSKY. 1964. Action of insulin upon blood glucose levels in fasted hypophysectomized, depancreatized and normal chickens. Endocrinology, 74:944-948.

KOMPIANG, I. P., AND W. R. GIBSON. 1973. Effect of hypophysectomy on lipogenesis and glycogenesis in cockerels. Amer. J. Physiol., 224:362-366.

———. 1976. Effect of hypophysectomy and insulin on lipogenesis in cockerels. Horm. Metab. Res., 8:340-345.

KRUG, E., G. GROSS, AND P. MIALHE. 1976. The contribution of the pancreas and the intestine to the regulation of lipolysis in birds. 2. Impaired lipolytic activity of pancreatic glucagon in the absence of either the pancreas or the intestine in the chicken. Horm. Metab. Res., 8:345-350.

LANGSLOW, D. R., AND C. N. HALES. 1969. Lipolysis in chicken adipose tissue *in vitro*. J. Endocrinol., 43:285-294.

LANGSLOW, D. R., AND R. L. HAZELWOOD. 1975. Physiological action of avian pancreatic polypeptide (APP). Diabetologia, 11:357.

LANGSLOW, D. R., E. J. BUTLER, C. N. HALES, AND A. W. PEARSON. 1970. The response of plasma insulin, glucose and non-esterified fatty acids to various hormones, nutrients and drugs in the domestic fowl. J. Endocrinol., 46:243-260.

LAURENT, F., AND P. MIALHE. 1976. Insulin and glucose-glucagon feedback mechanism in the duck. Diabetolgia, 12:23-33.

———. 1978. Effect of free fatty acids on glucagon and insulin secretions in normal and diabetic ducks. Diabetologia, 15:313-321.

LIBBY, D. A., J. MEITES, AND J. SCHAIBLE. 1955. Growth hormone in chickens. Poultry Sci., 34:1329-1331.

MAY, J. D. 1978. Effects of fasting on T_3 and T_4 concentrations in chicken serum. Gen. Comp. Endocrinol., 34:323-327.

MCCUMBEE, W. D., AND R. L. HAZELWOOD. 1977. Biological evaluation of the third pancreatic hormone (APP); hepatocyte and adipocyte effects. Gen. Comp. Endocrinol., 33:518-525.

MORRIS, T. R., AND A. V. NALBANDOV. 1961. The induction of ovulation in starving pullets using mammalian and avian gonadotropins. Endocrinology, 68:687-697.

MOSIER, H. D., R. A. JANSONS, C. S. BIGGS, S. M. TANNER, AND L. C. DEARDEN. 1980. Metabolic clearance rate of growth hormone during experimental growth arrest and subsequent recovery in rats. Endocrinology, 107:744-748.

MYERS, W. R., AND R. A. PETERSON. 1974. Responses of six and ten-week-old broilers to a trytic digest of bovine growth hormone. Poultry Sci., 53:508-514.

NABER, S. P., AND R. L. HAZELWOOD. 1977. *In vitro* insulin release from chicken pancreas. Gen. Comp. Endocrinol., 2:495-504.

NALBANDOV, A. V., AND L. E. CARD. 1943. Effect of hypophysectomy on growing chicks. J. Exper. Zool., 94:387-415.

NIR, I. D. YAM, AND M. PEREK. 1975. Effect of stress on the corticosterone content of the blood plasma and adrenal gland of intact and bursectomized *Gallus domesticus*. Poultry Sci., 54:2102-2110.

PROUDMAN, J. A., AND B. C. WENTWORTH. 1980. Ontogenesis of plasma growth in large and midget white strains of turkeys. Poultry Sci., 59:906-913.

SCANES, C. G., AND J. BALTHAZART. 1981. Circulatory concentrations of growth hormone during growth, maturation and reproductive cycles in ring doves (*Strepopelia risoria*). Gen. Comp. Endocrinol., 45:381-385.

SCANES, C. G., S. B. TELFER, A. F. HACKETT, R. NIGHTINGALE, AND B. A. K. SHARIFUDDEN. 1975. Effects of growth hormone on tissue metabolism in broiler chicks. Brit. Poult. Sci., 16:405-408.

SCANES, C. G., S. HARVEY, AND A. CHADWICK. 1976. Plasma luteinizing hormone concentration in fasted immature male chickens. IRCS Med. Sci., 4:371.

———. 1977. Hormones and growth in poultry. Pp. 79-85, *in* Growth and poultry meat production (K. N. Boorman and B. J. Wilson, eds.), Brit. Poult. Sci., Edinburgh, 350 pp.

SCANES, C. G., G. PETHES, P. RUDAS, AND T. MURAY. 1979. Changes in plasma growth hormone concentration during growth in domesticated geese. Acta Vet. Acad. Sci. Hung., 27:183-184.

SCANES, C. G., G. F. MERRILL, R. FORD, P. MAUSER, AND C. HOROWTIZ. 1980. Effects of stress (hypoglycemia, endotoxin and ether) on the peripheral circulatory concentration of corticosterone in the domestic fowl (*Gallus domesticus*). Comp. Biochem. Physiol., 66C:183-186.

SCANES, C. G., P. GRIMINGER, AND F. C. BUONOMO. 1981. Effects of dietary protein restriction on circulating concentration of growth hormone in growing domestic fowl (*Gallus domesticus*). Proc. Soc. Exper. Biol., 168:334-337.

SEGAL, J., H. SCHWARTZ, AND A. GORDON. 1977. The effect of triiodothyronine on 2-deoxy-D-(1-^{3}H) glucose uptake in altered chick embryo heart cells. Endocrinology, 101:143-149.

SIMON, J., AND G. ROSSELIN. 1978. Effect of fasting, glucose, amino acids and food intake on *in vivo* insulin release in chickens. Horm. Metab. Res., 10:93-98.

SITBON, G., AND P. MIALHE. 1978a. Pancreatic hormones and plasma glucose; regulation mechanism in the goose under physiological conditions. 2. Glucose-glucagon and glucose-insulin feedback mechanisms. Horm. Metab. Res., 10:117-123.

⸺. 1978b. Pancreatic hormones and plasma glucose: regulation mechanism in the goose under physiological conditions; 3. Inhibitory effect of insulin on glucagon secretion. Horm. Metab. Res., 10:473-477.

⸺. 1979. Pancreatic hormones and plasma glucose; regulation mechanisms in the goose under physiological conditions; 4. Effects of food and fasting on pancreatic hormones and gut GLI. Horm. Metab. Res., 11:123-129.

STROESSER, M., T. L. COHEN, S. HARVEY, AND P. MIALHE. 1980. Somatostatin stimulates glucagon secretion in ducks. Diabetologia, 18:319-322.

Reprinted from
ASPECTS OF AVIAN ENDOCRINOLOGY:
PRACTICAL AND THEORETICAL IMPLICATIONS (C. G. Scanes *et al.*, eds.)
Grad. Studies, Texas Tech Univ., 1982, 26:1-411.

EFFECT OF DIETARY THYROID HORMONES ON SERUM HORMONE CONCENTRATION, GROWTH, AND BODY COMPOSITION OF CHICKENS

J. D. MAY

U.S. Department of Agriculture, Science and Education Administration, Agriculture Research, Poultry Research Laboratory, R.D. 2, Box 600, Georgetown, Delaware 19947 USA

In mammals, 3,5,3'-triiodothyronine (T_3) and 3,3',5'-triiodothyronine (rT_3, or reverse T_3) arise primarily from peripheral degradation of thyroxine (T_4), and a molecule of T_3 or rT_3 can be produced from each molecule of T_4. The serum levels of T_3 and rT_3 in humans are altered by illness and nutritional state (Chopra *et al.*, 1975). rT_3 is generally believed to be metabolically inactive (Pittman and Pittman, 1974), but is an inhibitor of thyroxine (T_4) in some bioassays (Van Middlesworth, 1974). Chickens have much lower levels of circulating rT_3 than mammals have (Premachandra *et al.*, 1977; May, 1980); thus, rT_3 may not be an intermediate in T_4 catabolism in chickens.

Thyroid hormone preparations have long been used in experiments to improve productivity of poultry. In most experiments, iodinated casein or thyroid extracts have been used, and the reported effects on growth rate and feed efficiency have been contradictory (Andrews and Schnetzler, 1946; Glazener and Jull, 1946). These contradictions may be explained by the presence of both T_3 and T_4 at varying concentrations in the preparation used. In some experiments, the effect of purified hormones has been investigated. Singh *et al.* (1968) reported that daily T_4 injections of 2.0, 3.0, or 4.0 $\mu g/100$ g of body weight resulted in improved growth rate for White Leghorn cockerels. May (1980) reported that 0.10 or 1.00 ppm of T_4 or 0.10 ppm of T_3 in the diet did not affect growth rate of broilers, but 1.00 ppm of T_3 adversely affected growth rate and feed efficiency.

Thyroidectomy or treatments to induce hypothyroidism in chickens result in increased fat deposition (Ringer, 1976). Such fat deposition was desired by commercial poultry producers early in the development of the poultry industry in the U.S. but is now considered a detriment.

The objectives of the experiments reported herein were to investigate the productivity, serum hormone concentration, and body composition of broilers fed diets containing T_3 or T_4.

MATERIALS AND METHODS

Broiler chickens were purchased from a commercial hatchery and fed corn-soybean meal diets in mash form. The diets met all known nutritional requirements. The chickens were housed in metal batteries with continuous lighting and continuous access to feed and water. Crystalline T_3 and T_4 as the free acids were incorporated into the feed at various levels ranging from 0.25 to 10 ppm.

Serum T_3 and T_4 were assayed by the double-antibody radioimmunoassay described by May (1978). Antibody to T_3 and rabbit gamma globulin were those prepared by May (1978), and antibody to T_4 was purchased [Antibodies, Inc., P.O. Box 442, Davis, California 95616. Mention of a trade name is for information only and does not imply guarantee, warranty, or approval by the U.S. Department of Agriculture]. Specific activity for both T_3 ^{125}I and T_4 ^{125}I was 1200 $\mu Ci/\mu g$ [Amersham Corporation, 2636 S. Clearbrook Dr., Arlington Heights, Illinois 60005]. Serum rT_3 was assayed with a single antibody radioimmunoassay kit [Serono, Inc., 607 Boylston Street, Boston, Massachusetts 02116]; polyethylene glycol was used in the precipitation step.

In the carcass composition experiments, female chickens were killed when they were 54 days old in one trial, and both sexes were killed in another trial when they were 51 days old. The feathers, head, and feet were removed and the carcass, including the viscera, was then ground in a meat grinder. Aliquots of the ground meat were analyzed for protein by Kjeldahl analysis, for moisture by drying at 95°C, and for fat by ether extraction in a Goldfisch apparatus. Other chickens of both sexes were killed in both trials and their abdominal fat pads were removed and weighed. The fat removed was that which surrounded the cloaca and viscera and extended to but did not include the fat adhering to the gizzard.

The data were analyzed by using analysis of variance (ANOVA), and significant differences between treatment means were determined by Duncan's Multiple Range Test as given by Steel and Torrie (1960).

RESULTS AND DISCUSSION

T_4 in the diet at 10 ppm resulted in significant increases in serum rT_3 and T_4 concentrations after initiation of T_4 feeding (Fig. 1). Serum T_3 concentration was not significantly increased. These results show that excess T_4 can be degraded to rT_3 and that the chicken is able to maintain stable serum T_3 concentrations even when given a very high dose of T_4. These very low serum rT_3 concentrations have been previously reported (May, 1980) and confirm the report of Premachandra et al. (1977). Serum rT_3 concentration is increased when 10 ppm T_4 is fed, but no data are available to show that rT_3 is normally an intermediate in T_4 catabolism in chickens. The very low serum rT_3 concentrations in the control chickens suggest that rT_3 normally may not be an intermediate in T_4 degradation.

T_3 in the diet at 1 ppm resulted in increased serum T_3 concentration and decreased serum T_4 concentration (Fig. 2). The decrease of serum T_4 concentration may be the result of feedback inhibition of the thyroid. Dietary T_4 at 1.00 ppm resulted in increased serum T_4 concentration, but, although serum T_3 concentration was slightly lower, it was not significantly affected.

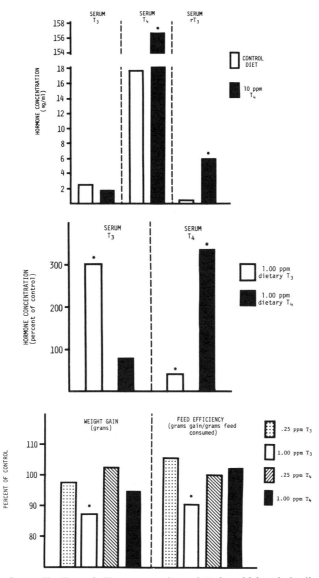

FIG. 1 (top).—Serum T_3, T_4, and rT_3 concentrations of 43-day-old female broilers fed 10 ppm T_4 for two days (asterisk indicates a statistically significant difference from the control, $P<0.05$).

FIG, 2 (middle).—Serum T_3 and T_4 concentrations of 28-day-old female broilers fed thyroid hormones for the 0-28-day period (asterisk indicates a statistically significant difference from the control, $P<0.05$).

FIG. 3 (bottom).—Effect of dietary thyroid hormones on weight gain and feed efficiency of 28-day-old female chickens fed the diets for the 0-28-day period (asterisk indicates a statistically significant difference from the control, $P<0.05$).

TABLE 1.—*Carcass moisture, protein, and fat of 54-day-old female broilers fed thyroid hormones for the 0-54-day period.*

Diet	Moisture (%)	Protein (%)	Fat (%)
Control	61.5^a	17.4^a	16.5^a
0.25 ppm T_3	66.4^b	18.4^a	12.2^b
0.25 ppm T_4	63.4^{ab}	18.3^a	16.8^a

[a,b] Within columns, values not having a common letter are significantly different ($P<0.05$).

Neither T_3 at 0.25 ppm nor T_4 at 0.25 or 1.00 ppm significantly affected growth rate or feed efficiency, but T_3 at 1.00 ppm was detrimental to both (Fig. 3). This difference in the effects of T_3 and T_4 must be considered in the context of the serum hormone concentrations previously observed. Serum rT_3 was not measured at 28 days, but the previous results (Fig. 1) show that serum rT_3 is increased when T_4 is in the diet. This T_4 treatment, which caused increased serum rT_3 and T_4, did not affect growth and feed efficiency, but dietary T_3 resulted in increased serum T_3 and poorer performance as measured by growth and feed efficiency.

T_3 in the diet at 0.25 ppm resulted in significantly decreased carcass fat with a concomitant slight increase in carcass moisture (Table 1). These parameters were not significantly affected by 0.25 ppm of T_4 in the diet, and percentage of carcass protein was not significantly affected by either T_3 or T_4 in the diet.

Percentage of abdominal fat was reduced in male broilers by 0.25 ppm T_3 and 0.25 ppm T_4, but T_4 did not significantly affect that of females (Table 2). This finding suggests that females are less sensitive than males to the effect of T_4 on fat deposition. Male and female chickens respond similarly to T_3 and T_4 when weight gain and feed efficiency are the parameters (May, 1980).

Neither carcass percent moisture, protein, nor fat was significantly affected by 0.25 ppm T_3 or 0.30 ppm T_4 in the diet for the 14-51 day period (Table 3). These results show thyroid hormones have a much greater effect on body composition if thyroid hormone feeding is begun immediately after hatching than if hormone feeding is delayed for two weeks. Hormone feeding for the 14-51 day period had little effect on abdominal fat; only females fed 0.25 ppm T_3 were significantly different from the corresponding controls.

TABLE 2.—*Abdominal fat of 54-day-old broilers fed thyroid hormones for the 0-54-day period.*

Diet	Sex	Abdominal fat (%)
Control	F	2.23^a
	M	1.76^{ab}
0.25 ppm T_3	F	1.61^b
	M	0.59^c
0.25 ppm T_4	F	1.99^{ab}
	M	1.07^c

[a,b,c] Values not having a common letter are significantly different ($P<0.05$).

TABLE 3.—*Carcass moisture, protein, and fat of 51-day-old broilers fed thyroid hormones for the 14-51-day period.*

Diet	Sex	Moisture (%)	Protein (%)	Fat (%)	Abdominal fat (%)
Control	F	63.4[a]	16.9[a]	13.0[a]	3.08[ab]
	M	64.2[a]	17.6[a]	12.2[a]	2.31[cd]
0.25 ppm T_3	F	64.3[a]	16.9[a]	14.5[a]	2.46[cd]
	M	65.6[a]	17.4[a]	11.2[a]	2.21[d]
0.30 ppm T_4	F	65.0[a]	17.0[a]	13.4[a]	3.26[a]
	M	65.0[a]	17.0[a]	13.3[a]	2.75[bc]

[a,b,c,d]Within columns, values not having a common letter are significantly different ($P<0.05$).

The data presented here show that chickens are more responsive to T_3 than to T_4. Excess T_4 in the diet can be degraded to rT_3, and circulating levels of T_3 remain remarkably constant, even with extremely high levels of T_4.

LITERATURE CITED

ANDREWS, F. N., AND E. E. SCHNETZLER. 1946. Influence of thiouracil on growth and fattening of broilers. Poultry Sci., 25:124-129.

CHOPRA, I. J., V. CHOPRA, S. R. SMITH, M. REZA, AND D. H. SOLOMON. 1975. Reciprocal changes in serum concentrations of 3,3′,5′-triiodothyronine (Reverse T_3) and 3,5,3′-triiodothyronine (T_3) in systemic illnesses. J. Clin. Endocrinol. Metab., 41:1043-1049.

GLAZENER, E. W., AND M. A. JULL. 1946. Effects of thiouracil, desiccated thyroid, and stilbestrol derivatives on various glands, body weight, and dressing appearance in the chicken. Poultry Sci., 25:236-241.

MAY, J. D. 1978. A radioimmunoassay for 3,5,3′-triiodothyronine in chicken serum. Poultry Sci., 57:1740-1745.

———. 1980. Effect of dietary thyroid hormone on growth and feed efficiency of broilers. Poultry Sci., 59:888-892.

PITTMAN, C. S., AND J. A. PITTMAN. 1974. Relation of chemical structure to the action and metabolism of thyroactive substances. Pp. 233-253, *in* Handbook of physiology, Section 7, Vol. III (R. O. Greep and E. B. Astwood, eds.), Williams and Wilkins Co., Baltimore, vii+491 pp.

PREMACHANDRA, B. N., S. LANG, J. A. ANDRADA, AND J. H. KITE, JR. 1977. Reverse triiodothyronine in the chicken. Life Sci., 21:205-212.

RINGER, R. K. 1976. Thyroids. Pp. 348-358, *in* Avian physiology (P. D. Sturkie, ed.), Springer-Verlag, New York, xiii+400 pp.

SINGH, A., E. P. REINEKE, AND R. K. RINGER. 1968. Influence of thyroid status of the chick on growth and metabolism, with observations on several parameters of thyroid function. Poultry Sci., 47:212-219.

STEEL, R. G. D., AND J. H. TORRIE. 1960. Principle and procedures of statistics. McGraw Hill Book Co., New York, 481 pp.

VAN MIDDLESWORTH, L. 1974. Metabolism and excretion of thyroid hormones. Pp. 215-231, *in* Handbook of physiology, Section 7, Vol. III (R. O. Greep and E. B. Astwood, eds.), Williams and Wilkins Co., Baltimore, vii+491 pp.

Reprinted from
Aspects of Avian Endocrinology:
Practical and Theoretical Implications (C. G. Scanes *et al.*, eds.)
Grad. Studies, Texas Tech Univ., 1982, 26:1-411.

ENDOCRINE MECHANISM OF OVARIAN FOLLICULAR ATRESIA CAUSED BY FASTING OF THE HEN

Hisako Tanabe[1], Takao Nakamura[2], and Yuichi Tanabe[2]

[1]*Department of Human Nutrition and Food Science, Gifu Women's University, Taromaru, Gifu 501-25, Japan; and* [2]*Department of Poultry and Animal Sciences, Faculty of Agriculture, Gifu University, Kakamigahara, Gifu 504, Japan*

It is well known that the ovarian follicular atresia of the hen is induced by starvation, and the change can be prevented by daily administration of gonadotropic hormone (Hosoda *et al.*, 1955, 1956; Morris and Nalbandov, 1961).

Starvation of laying hens for seven to ten days with or without a combination of water withdrawal for two to three days is used as a practical management technique in poultry husbandry to induce forced molting which will then be followed by increased egg production when feeding is resumed. Forced molting might be induced by ovarian atrophy (atresia) because starvation causes the cessation of egg production and decreased ovarian function in the hen (Tanabe *et al.*, 1957; Tanabe and Katsuragi, 1962; Brake and Thaxton, 1979).

The present work was undertaken to elucidate the endocrine mechanism of supression of ovarian function caused by starvation in the laying hen. Some parts of the work have been published elsewhere (Tanabe *et al.*, 1981).

Effect of Starvation on Plasma Luteinizing Hormone (LH), Estradiol, and Progesterone Concentrations

Fourteen 250-day-old White Leghorn laying hens weighing 1.8-2.0 kilograms were housed in individual cages with feed (a commercial ration) and water available *ad libitum*, under 14 hours of light (from 5:00-19:00) and 10 hours of darkness per day. These birds were divided into three groups. Five hens were deprived of food for seven days and refed *ad libitum* after the seventh day of food withdrawal. Another five hens were deprived of food for seven days and water for the first three days, and refed *ad libitum* after the seventh day of food and water withdrawal. The third group of four hens were given food and water *ad libitum* during the experimental period. Daily blood samples (1.2 ml) were taken at 13:00 from the hen's wing vein using a heparinized syringe. These samples were taken from the time of food withdrawal to the seventh day of refeeding. The blood samples were centrifuged at 3,000 rpm for 10 minutes to separate the plasma, which was then frozen and kept at -20°C until use. One tenth milliliter of plasma for LH or progesterone determinations and 0.2 milliliter plasma for estradiol determination

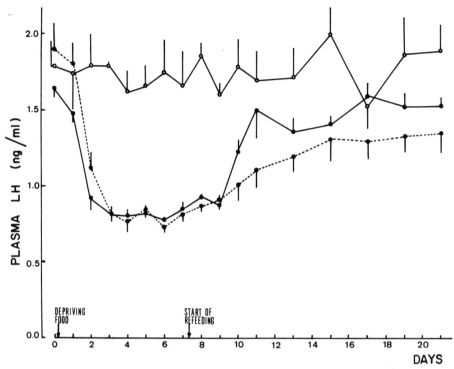

FIG. 1.—Changes in plasma LH in hens fed *ad libitum* (open circles), deprived of food for seven days and refed (closed circles connected by an unbroken line), and hens deprived of food for seven days and water for the first three days and refed (closed circles connected by a broken line). Each open point and vertical line represents a mean of four birds and SE, and each solid point and vertical line represents a mean of five birds and SE, respectively.

was used. For the assay of estradiol and progesterone, plasma samples were extracted four times with ethyl ether. Radioimmunoassays (RIAs) without chromatography of estradiol and progesterone and RIAs for avian LH were carried out using assay methods described previously (Shodono *et al.*, 1975; Tanabe *et al.*, 1979). Rabbit anti-estradiol-17β-6-(*O*-carboxymethyl)-oxime-BSA serum and anti-progesterone-3-(*O*-carboxymethyl)-oxime-BSA serum were obtained from Teikokuzoki Pharmaceutical Company, Tokyo. The crossreactions of the antisera to other steroids were low as described by Tanabe *et al.* (1979). Rabbit anti-avian LH serum (3027/5), avian LH preparations (AEI/B2) for radioiodination, and standard preparation (IRC2) were kindly supplied by Drs. Follett and Scanes. Estradiol-17β (2,4,6,7-^{3}H; 105 Ci/m mol) and (1.2,6,7-^{3}H) progesterone were used as antigens in radioimmunoassay for the steroids. Na^{131}I was used for radioiodination of avian LH. Radioactivity was measured with a liquid scintillation spectrometer (Beckman LS 9000) or with an autogamma counter (Beckman Gamma 9000).

The effects of food deprivation alone or food deprivation in combination with water deprivation on plasma LH are given in Fig. 1. Plasma LH sig-

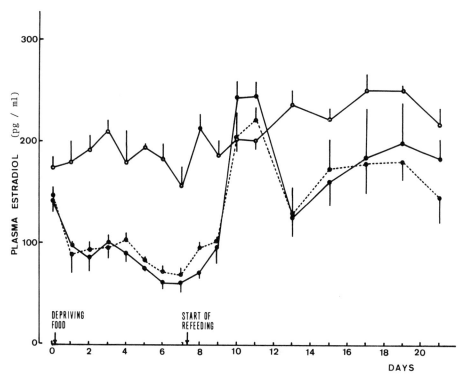

FIG. 2.—Changes in plasma estradiol for hens fed *ad libitum* (open circles), hens deprived of food for seven days and refed (closed circles connected by an unbroken line), and hens deprived of food for seven days and water for the first three days and refed (closed circles connected by a broken line). Each open point and vertical line represents a mean of four birds and SE, and each solid point and vertical line represents a mean of five birds and SE, respectively.

nificantly decreased two days after both food withdrawal and food plus water withdrawal. It remained at low levels until three days after refeeding in the food deprived group, and for a longer period in the food and water deprived group, as compared with the group fed *ad libitum*. All hens deprived of food or food and water stopped egg production at the third day of food withdrawal. No noticeable change in plasma LH was observed in hens fed *ad libitum*.

The effects of food deprivation alone and food and water deprivation on plasma estradiol are given in Fig. 2. In both treatment groups, plasma estradiol significantly decreased one day after food deprivation and reached the minimal level on the sixth and seventh days of food withdrawal. Two days after resumption of feeding, plasma estradiol levels increased rapidly and reached a significantly higher level than that in the group fed *ad libitum*; levels then declined on the fourth through sixth days of refeeding. The only difference between the two treatment groups was a slower recovery of estradiol level after refeeding in the groups deprived of both food and water. No noticeable change in plasma estradiol was observed in hens fed *ad libitum*.

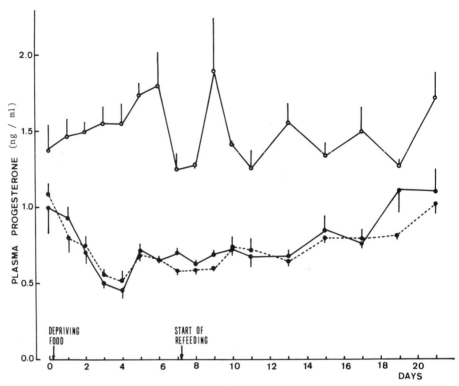

Fig. 3.—Changes in plasma progesterone in hens fed *ad libitum* (open circles), hens deprived of food for seven days and refed (closed circles connected by an unbroken line), and hens deprived of food for seven days and water for the first three days and refed (closed circles connected by a broken line). Each open point and vertical line represents a mean of four birds and sE, and each solid point and vertical line represents a mean of five birds and sE, respectively.

The effects of food deprivation and food and water deprivation on plasma progesterone are given in Fig. 3. Plasma progesterone significantly decreased one day after food withdrawal or food and water withdrawal and remained at a low level during the food deprivation period. It increased gradually after the resumption of feeding and reached its initial level on the twelfth day of refeeding. Rather variable changes in plasma progesterone were observed in hens fed *ad libitum*, but the level was always higher than those of hens deprived of food and food and water.

Effect of Starvation on Pituitary LH Concentration

Twenty-seven 250-day-old White Leghorn hens weighing 2.0 kilograms maintained under similar conditions to the first experiment were divided into three groups. Nine, six, and 12 hens were killed before starvation, on the third day of food withdrawal, and on the seventh day of food withdrawal, respectively. A five milliliter blood sample was taken from the wing veins, and the ovary, oviducts, and pituitary were removed. The pituitary gland was homogenized with one milliliter 0.1 M phosphate buffer using a

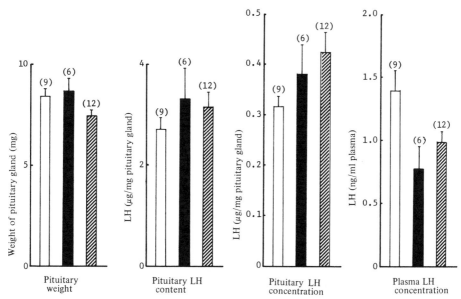

FIG. 4.—Effect of starvation on pituitary weight, pituitary LH content and concentration, and plasma LH concentration of the hen. Columns with lines represent means ± SE; open columns, laying hens; solid columns, hens starved for three days; and hatched columns, hens starved for seven days. Numbers in parentheses at the top of each column indicate the number of birds determined.

loose-fitting Teflon-glass homogenizer for LH assay. The following procedure is the same as those described previously (Doi *et al.*, 1980).

When compared with laying hens, statistically significant decreases in ovary and oviduct weights were observed in the hens starved for three days and seven days. Follicular atrophy (atresia) of the ovary already had occurred on the third day of starvation, and the whole ovary became atretic on the seventh day of starvation. Average ovarian weights of the hen before, after three days, and after seven days of starvation were 40.1, 20.2, and 12.0 grams, respectively. Pituitary weights, pituitary LH content and concentration, and plasma LH concentration in the hens starved for three and seven days as compared with those in intact laying hens are illustrated in Fig. 4. Small and statistically nonsignificant increases, either in LH contents or concentrations in the pituitary were observed in hens starved for three and seven days as compared with those of the control group, while a statistically significant decrease in plasma LH concentration was observed in hens starved for three and seven days as compared with the control group.

Effect of Starvation on Pituitary and Ovarian Responses to Luteinizing Hormone Releasing Hormone (LHRH)

Seven 250-day-old White Leghorn laying hens weighing 1.8-2.0 kilograms and maintained under similar conditions to the previous experiments were injected intravenously with 20 μg synthetic LHRH (Eizai Pharmaceutical

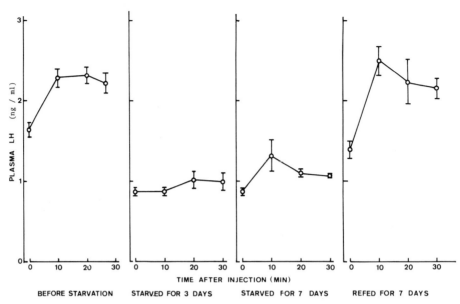

F<small>IG</small>. 5.—Changes in plasma LH after a single intravenous injection of LHRH. Each point with vertical line represents a mean of seven birds injected with 20 µg LHRH and s<small>E</small>.

Company, Tokyo) dissolved in 0.2 milliliter isotonic saline immediately before the food withdrawal, on the third day of food withdrawal, on the seventh day of food withdrawal, and seven days after refeeding. Blood samples (1.2 ml) were taken from the wing vein at the time of each injection, and 10, 20, and 30 minutes after each injection.

All hens stopped egg laying within the third day of food withdrawal, and resumed laying 30 days after the resumption of feeding. Most of the hens did not molt because they were too young to force molt. The effects of 20 µg LHRH injections prior to starvation, on the third and seventh day of food withdrawal, and on the seventh day after refeeding on plasma LH concentration, are illustrated in Fig. 5: a single injection of 20 µg LHRH into the intact laying hens before starvation significantly increased plasma LH 10, 20, and 30 minutes after the injection; a dose of 20 µg injected LHRH failed to increase plasma LH in the hen starved for three days; the injection of LHRH at a 20 µg dose induced a small but statistically significant ($P<0.05$) increase in plasma LH in the hen starved for seven days; and the hens recovered initial sensitivity to LHRH in increasing LH secretion after seven days of refeeding. The injection of LHRH at 20 µg dose did not cause a significant increase either in plasma estradiol or in progesterone of the hen at any stage of the feeding regimens in this experiment.

Mechanism of Supression of Ovarian Function Caused by Starvation of the Hen

The present study demonstrates that whereas deprivation of food for seven days in the laying hens prevented egg laying and significantly decreased plasma LH, estradiol, and progesterone concentrations, it did not significantly change pituitary LH content or concentration. In addition, a single injection of 20 μg LHRH into the intact laying hen before starvation significantly increased plasma LH, but failed to increase plasma LH on the third day of food withdrawal, and increased it to a lesser degree on the seventh day of food withdrawal. Bonney et al. (1974) reported that a dose of 20 μg synthetic LHRH into the laying hen induced a significant increase of plasma LH, and the time of injection of LHRH in relation to ovulation did not affect the response unless it was administered immediately after the preovulatory LH peak.

The present study also demonstrates that resumption of feeding raises plasma LH and progesterone back to initial levels and raises plasma estradiol to twice its initial level within three to four days of refeeding; the value declines to the initial level thereafter. Senior (1974) reported that plasma levels of estradiol were three times higher before the onset of laying compared with those after the initiation of egg laying. The higher than normal estradiol levels found before the onset of laying in the young hen and before the resumption of egg laying in the starved hen might be essential for the synthesis of yolk protein precursors by the liver.

It is noteworthy that slower recoveries of LH and progesterone levels in the plasma after the resumption of feeding were observed in the hens deprived of both food and water than in hens deprived of food only. Because food deprivation combined with water deprivation causes a stronger suppression of the pituitary-gonadal function of the female chicken than simple food deprivation alone, the former technique may be more effective in forcing molt in the hen (Tanabe and Katsuragi, 1962; Brake and Thaxton, 1979).

In the female rat, it is well known that starvation significantly reduces ovarian weight and plasma LH concentration, but does not change significantly the pituitary LH concentration or plasma FSH concentration (Walker and Frawley, 1977). These authors conclude that reduced plasma LH is the major factor contributing to reproductive dormancy in the starved animal.

In immature male chickens, it is demonstrated that plasma LH concentration is significantly reduced within 12 hours of food withdrawal, and continues to fall over the period of fasting, whereas no significant changes in plasma FSH concentration are observed until after 48 hours of food withdrawal (Scanes et al., 1976). A marked effect of fasting on the secretion of LH on the domestic fowl with smaller and later reduction on the plasma FSH concentration (demonstrated by Scanes et al., 1976) agrees generally with data obtained for the rat (Walker and Frawley, 1977). Further, in the chicken, fasting rapidly induces cessation of lay and atresia of the ovary and oviduct, but these changes can be prevented by the administration of exogenous gonadotropins (Hosoda et al., 1955, 1956; Morris and Nalbandov, 1961). Hosoda et al. (1956) indicated exogenous mammalian LH was more effective than FSH in preventing atresia of the ovary in fasting hens.

Taken all together, these results suggest that suppression of gonadal function in the fasting chicken is caused by depression of gonadotropin (mainly

LH) secretion from the pituitary gland, although more studies on chicken FSH are needed. Mori and Masaki (1976) showed that both the serum level of LH and its increased rate following the administration of LHRH were significantly smaller in starved male rats than in normally fed male rats, indicating that anterior pituitary response to LHRH is reduced in the fasting animal. In the female chicken, the present study also demonstrated that starvation caused follicular atrophy due mainly to the decrease of LH secretion from the pituitary, and that starvation reduced the sensitivity of pituitary to LHRH.

Interestingly, Luck and Scanes (1977) showed that, using an *in vitro* laying hen's pituitary cell incubation system with a low calcium concentration (0.9 mM) in the medium, LH secretion from the cell was reduced and could not be restored using LHRH (10 ng ml^{-1}). This suggests that low plasma calcium may affect LH secretion directly at the pituitary level. The present results also suggest that fasting in laying hens may affect LH secretion directly at the pituitary level rather than at the hypothalamus level. It is possible that reduced plasma glucose level in fasting hens reduces the sensitivity of the pituitary to LHRH in increasing LH secretion.

SUMMARY

The effect of starvation on pituitary and plasma LH, plasma estradiol and progesterone levels, and pituitary and ovarian responses to LHRH in the laying hen were studied. Deprivation of food for seven days or additional deprivation of water for the first three days prevented egg laying and significantly decreased plasma LH, estradiol, and progesterone concentrations, but did not significantly change pituitary LH content or concentration. Resumption of feeding increased plasma LH and progesterone to their initial levels, but increased plasma estradiol to twice its initial level within three to four days of refeeding. A single injection of 20 µg LHRH to intact laying hens before starvation significantly increased plasma LH, but did not increase either plasma estradiol or progesterone. LHRH at 20 µg dose failed to increase plasma LH on the third day of food withdrawal, but increased it to some extent on the seventh day. The hen recovered initial sensitivity to LHRH in increasing LH secretion either after seven days of refeeding or after the resumption of egg laying. These results indicate that starvation caused follicular atresia in the hen mainly due to the decrease of LH secretion from the pituitary and to the reduction of plasma LH, and that starvation might reduce the sensitivity of pituitary gonadotrop—in cells to LHRH in the hen.

ACKNOWLEDGMENTS

We wish to thank Professors B. K. Follett, University of Bristol, and C. G. Scanes, Rutgers University, New Brunswick, for their donation of avian LH and its antiserum; Professor Katuhide Tanaka, Gifu University, for his donation of synthetic LHRH; and Mr. Tamotsu Ogawa, Gifu University, for his technical assistance.

LITERATURE CITED

BONNEY, R. C., F. J. CUNNINGHAM, AND B. J. A. FURR. 1974. Effect of synthetic luteinizing hormone releasing hormone on plasma luteinizing hormone in the female domestic fowl, *Gallus domesticus*. J. Endocrinol., 63:539-547.

BRAKE, J., AND P. THAXTON. 1979. Physiological changes in caged layers during a forced molt. 2. Gross changes in organs. Poultry Sci., 58:707-716.

DOI, O., T. TAKAI, T. NAKAMURA, AND Y. TANABE. 1980. Changes in the pituitary and plasma LH, plasma and follicular progesterone and estradiol, and plasma testosterone and estrone concentrations during the ovulatory cycle of the quail (*Coturnix coturnix japonica*). Gen. Comp. Endocrinol., 41:156-163.

HOSODA, T., T. KANEKO, K. MOGI, AND T. ABE. 1955. Effect of gonadotropic hormone on ovarian follicles and serum vitellin of fasting hens. Proc. Soc. Exp. Biol. Med., 88:502-504.

————. 1956. Forced ovulation in gonadrotropin treated fasting hens. Proc. Soc. Exp. Biol. Med., 92:360-362.

LUCK, M. R., and C. G. SCANES. 1977. Gonadotrophin secretion in the domestic fowl during calcium deficiency. 9th Conf. Eur. Comp. Endocrinol., Giessen, Abstr., p 41.

MORI, J., AND J. MASAKI. 1976. Effect of fasting on pituitary response to luteinizing hormone releasing hormone in male rats. Japanese J. Zootech. Sci., 47:602-603.

MORRIS, T. R., AND A. V. NALBANDOV. 1961. The induction of ovulation in starving pullets using mammalian and avian gonadotropins. Endocrinology, 68:687-697.

SCANES, C. G., S. HARVEY, AND A. CHADWICK. 1976. Plasma luteinizing hormone and follicle stimulating hormone concentration in fasting immature male chickens. IRCS Med. Sci., 4:371.

SENIOR, B. E. 1974. Oestradiol concentration in the peripheral plasma of the domestic hen from 7 weeks of age until the time of sexual maturity. Reprod. Fert., 41:107-112.

SHODONO, M., T. NAKAMURA, Y. TANABE, AND K. WAKABAYASHI. 1975. Simultaneous determinations of oestradiol-17β,progesterone and luteinizing hormone in the plasma during the ovulatory cycle of the hen. Acta Endocrinol., 78:565-573.

TANABE, Y., AND T. KATSURAGI. 1962. Thyroxine secretion rates of molting and laying hens, and general discussion of molting in hens. Bull. Nat. Inst. Agric. Sci., G21:49-59.

TANABE, Y., K. HIMENO, AND H. NOZAKI. 1957. Thyroid and ovarian function in relation to molting in the hen. Endocrinology, 61:661-666.

TANABE, Y., T. NAKAMURA, K. FUJIOKA, AND O. DOI. 1979. Production and secretion of sex steroid hormones by the testes, the ovary, and the adrenal glands of embryonic and young chickens (*Gallus domesticus*). Gen. Comp. Endocrinol., 39:26-33.

TANABE, Y., T. OGAWA, AND T. NAKAMURA. 1981. The effect of short-term starvation on pituitary and plasma LH, plasma estradiol and progesterone, and on pituitary response to LH-RH in the laying hen (*Gallus domesticus*). Gen. Comp. Endocrinol., 43:392-398.

WALKER, R. F., AND L. S. FRAWLEY. 1977. Gonadal function in underfed rats: II. Effect of estrogen on plasma gonadotropins after pinealectomy or constant light exposure. Biol. Reprod., 17:630-634.

Reprinted from
ASPECTS OF AVIAN ENDOCRINOLOGY:
PRACTICAL AND THEORETICAL IMPLICATIONS (C. G. Scanes *et al.*, eds.)
Grad. Studies, Texas Tech Univ., 1982, 26:1-411.

HYPOTHALAMIC HYPERPHAGIA, OBESITY, AND GONADAL DYSFUNCTION: ABSENCE OF CONSISTENT RELATIONSHIP BETWEEN LESION SITE AND PHYSIOLOGICAL CONSEQUENCES

B. ROBINZON[1], N. SNAPIR[1], AND S. LEPKOVSKY[2]

[1]*Department of Animal Sciences, Faculty of Agriculture, Hebrew University of Jerusalem, P.O.B. 12, Rehovot, Israel; and* [2]*Department of Poultry Husbandry, University of California, Berkeley, California 94720 USA*

Hypothalamic hyperphagia and obesity in the White Leghorn cock were first reported by Lepkovsky and Yasuda (1966). In their study, large bilateral electrolytic lesions were introduced to the basomedial hypothalamus. The whole area from the rostral part of the ventromedial hypothalamic nucleus (VMH) up to the caudal part of the mammillary nuclei (MN) was destroyed. The cockerels developed hyperphagia following the classical pattern of dynamic and static phases and became markedly obese. These lesions resulted also in involution of both testes and comb and a phenotype similar to that achieved after surgical castration. This type of gonadal dysfunction has been termed as functional castration (FC).

The involvement of the hypothalamus in the regulation of gonadal functions, food intake, and adiposity in the White Leghorn male was studied further (Perek *et al.*, 1973; Ravona *et al.*, 1973*a*, 1973*b*; Snapir *et al.*, 1969, 1973, 1974*a*; Robinzon *et al.*, 1977*a*, 1977*b*). It was again found that lesions in the basomedial hypothalamus (BMH) which damaged both the VMH and the MN induced the FC syndrome. The latter was defined by involution of the testes from 15-20 grams, to about 0.5 grams in weight, reduction of comb weight from 80-90 grams to about 3 grams, and involution of the adenohypophysis to about half of its usual weight. Blood level of testosterone diminished to almost zero. Histologically, only primary spermatocytes could be observed in the seminiferous tubules, and no dehydrogenase activity was found in the interstitial tissue. In the adenohypophyses of the FC cocks, no gonadotropes could be found. All the FC cocks were hyperphagic and became obese. Lesions only in the mammillary nuclei (MN) produced a second type of gonadal dysfunction in which the testes were involuted but a large comb remained. This syndrome was termed functional castration large comb (FCLC). The FCLC cocks had few gonadotropes in their adenohypophyses, somewhat larger testes than the FC ones, both primary and secondary spermatocytes in the seminiferous tubules, and dehydrogenase activity in the interstitial tissue. Plasma testosterone levels of the FCLC cocks were between levels found in the FC and normal levels. All the FCLC cocks were hyperphagic and developed obesity. Lesions limited only to the VMH also induced hyperphagia and adiposity, but there was no gonadal dysfunction.

In the Red-winged blackbird male, involution of the testes, a marked decrease in plasma testosterone, hyperphagia, and obesity were found to follow placement of lesions in the BMH (Robinzon and Katz, 1980). This syndrome in the blackbird was very similar to the FC syndrome in the chicken with lesions at the same site.

From these results it could be apparently concluded that destruction of both VMH and MN brings in the FC phenotype, ablation of the MN only causes the FCLC syndrome, and lesions in the VMH alone are followed by hyperphagia and obesity only. However, the results of the following experiment show that there is not necessarily a consistent relationship between hypothalamic lesion sites and their physiological consequences.

MATERIALS AND METHODS

Seventy White Leghorn cocks, three months of age, were kept in individual cages, fed a commercial breeder mash *ad libitum*, and subjected to 14 hours of light daily.

Bilateral electrolytic lesions aimed to the BMH and neighboring nuclei were placed in 60 of the cocks by a method described by Feldman *et al.* (1973) and Snapir *et al.* (1974c). Eight of the lesioned cocks died during surgery or immediately thereafter. The remainder of the cocks were sham-operated and served as controls.

The entire experimental period lasted five months during which individual daily food intake was recorded. Body weights were measured at 10-day intervals. At the termination of the experiment, heparinized blood samples were drawn from the brachial vein and hematocrit was determined. The cocks were killed and the brains were removed and fixed in 10 per cent neutral formalin. The following organs or tissues were immediately removed, cleaned from adhering tissues, and weighed: comb, abdominal adipose tissue, liver, testes, adrenals, thyroids, and adenohypophyses.

Frozen frontal sections of 25 μm thickness each were prepared from the brains and stained with thionine. The sections were examined microscopically for localization of lesion sites which were defined according to several atlases of the chicken brain (van Tienhoven and Juhasz, 1962; Junghar, 1969; Feldman *et al.*, 1973; Snapir *et al.*, 1974c; Youngren and Phillips, 1978). The lesion site of each cock was reconstructed on a schematic drawing of frontal sections of the chicken brain (Snapir *et al.*, 1974c) as in Fig. 1.

RESULTS

No consistent relationship between lesion sites and physiological consequences was observed. However, considering both the lesion site and its physiological consequences, the results from the lesioned cocks reasonably could be put into seven categories: 1) BMH-lesioned cocks with FC syndrome

TABLE 1.—*Food intake and adiposity criteria at time of autopsy for lesioned and control cocks, mean values ± SE. Means followed by unshared letters are statistically different (P<0.05).*

Group of cocks	N	Food intake in first month (g)	Body weight (g)	Abdominal adipose tissue weight (g)	Liver weight (g)
Control	10	2737±101 c	2223± 65 b	9± 3 c	36± 2 b
BMH-FC	20	4023±185 a	2696±168 ab	188±18 a	60± 7 a
BMH-FCLC	8	3886±252 ab	2840±154 a	164±26 a	48± 6 ab
BMH-NT	3	3072± 60 bc	2771±120 ab	43±10 bc	31± 5 ab
VMH-FCLC	3	3992±345 abc	2936±244 ab	127±33 abc	54±15 ab
VMH-NT	11	3672±267 ab	2794±106 a	119±21 ab	37± 5 ab
POA-FCLC	4	3880±458 abc	2481± 79 ab	77± 6 b	60±17 ab
MN-FCLC	3	4417±212 a	2550±172 ab	137±17 ab	45± 3 ab

(BMH-FC) were hyperphagic and obese with hypertrophied livers (Table 1) and had low hematocrits, involuted testes, combs, and adenohypophyses (Table 2). The smallest lesion in the BMH-FC group was at the caudoventral part of the VMH and the ventromedial MN, whereas the largest one occupied most of the VMH and most of the MN (Fig. 2). 2) BMH-lesioned cocks with the FCLC syndrome (BMH-FCLC) were also hyperphagic and obese but their livers were less hypertrophied (Table 1). The BMH-FCLC cocks had involuted testes but weighed about three times more then those of the BMH-FC cocks (Table 2). These cocks had low hematocrits, but their comb and adenohypophyseal weights were only moderately lower then those of the controls (Table 2). No difference in the lesion site could be found between the BMH-FC and the BMH-FCLC groups (Fig. 3). 3) BMH-lesioned cocks with almost normal testes (BMH-NT) were similar to the controls in being only slightly obese (Tables 1 and 2). However, the lesion site of the BMH-NT cocks was similar to those of the BMH-FC and the BMH-FCLC ones (Fig. 4). 4) VMH-lesioned cocks with the FCLC syndrome (VMH-FCLC) were indistinguishable from the BMH-FCLC ones (Tables 1 and 2) but their lesions did not extend beyond the VMH area so that their MN were left intact (Fig. 5). 5) VMH-lesioned cocks with normal testes (VMH-NT) were hyperphagic and obese (Table 1) but had normal gonadal criteria

TABLE 2.—*Gonadal and some other endocrine criteria at time of autopsy for lesioned and control cocks, mean values ± SE. Means followed by unshared letters are statistically different (P<0.05).*

Group of cocks	N	Hematocrit (%)	Comb weight (g)	Testes weight (g)	Adenohypophysis weight (g)	Thyroid weight (g)	Adrenal weight (g)
Control	10	49.9±1.3 ab	85± 8 a	22.8±3.2 a	16.4±1.6 a	238±12 a	238±19 a
BMH-FC	20	32.8±0.8 c	3± 1 c	0.5±0.1 c	8.6±1.3 b	234±13 a	205±12 a
BMH-FCLC	8	35.2±1.1 c	52± 6 b	1.7±0.3 b	14.5±2.1 ab	243±30 a	222±29 a
BMH-NT	3	43.3±0.2 a	125±14 ab	12.6±1.3 a	17.0±2.3 a	251±12 a	177±12 a
VMH-FCLC	3	34.8±0.5 c	60± 4 ab	2.1±0.4 bc	9.7±1.2 ab	213±16 a	211±16 a
VMH-NT	11	37.6±2.0 bc	84±11 ab	19.4±3.2 a	14.6±1.1 a	228±30 a	236±29 a
POA-FCLC	4	32.1±3.1 bc	53± 4 b	1.3±0.1 b	10.3±1.3 ab	276±49 a	264±13 a
MN-FCLC	3	30.0±3.9 abc	43± 8 ab	2.3±0.8 bc	9.2±1.6 ab	220±75 a	316±57 a

(Table 2). The lesions were similar to those of the VMH-FCLC cocks and completely covered the smallest lesion of the VMH-FCLC group (Fig. 6). 6) Cocks with lesions in the preoptic area (POA) which showed the FCLC symptoms (POA-FCLC; Table 2) were less obese than the BMH-FCLC or the VMH-FCLC ones but had hypertrophied livers (Table 1). The smallest POA-lesion damaged the medial preoptic nucleus, the supraoptic nucleus, and the medial portion of the suprachiasmatic nucleus; the largest lesion extended also to the lateral parts of the suprachiasmatic nucleus and the rostral apex of the VMH (Fig. 7). 7) Cocks with lesions in the MN (Fig. 8) and the FCLC symptoms (MN-FCLC) showed similar syptoms to those of the BMH-FCLC and the VMH-FCLC ones (Tables 1 and 2).

DISCUSSION

From the results it can be concluded that in order to obtain the FC syndrome in the White Leghorn cock, the BMH should be destroyed. This is in agreement with previous reports (Lepkovksy and Yasuda, 1966; Snapir et al., 1969; Perek et al., 1973; Ravona et al., 1973a, 1973b; Snapir et al., 1973, 1974a, 1974b; Robinzon et al., 1977a, 1977b). However, although most of the BMH-lesioned cocks did manifest the complete FC syndrome, there were some with only a partial gonadal disfunction (BMH-FCLC cocks) and a few with no gonadal disturbances at all (BMH-NT cocks). Thus, although the destruction of the BMH complex is a must for obtaining the FC syndrome, it does not guarantee the whole syndrome in each individual cock.

The induction of the FCLC syndrome by lesions in the mammillary nuclei (MN) found in the present study confirms previous reports (Ravona et al., 1973a, 1973b; Snapir et al., 1973, 1974b). However, the present study points out that this syndrome may also be a result of lesions at other hypothalamic sites such as the VMH and the POA. The involvement of the POA in the maintenance of normal gonadal activity and sexual behaviors in birds is well documented (Gardner and Fisher, 1968; Hutchison, 1974; Davies and Follett, 1975; Oliver et al., 1978), and lesions in the POA were found to suppress testicular growth in quail (Davies and Follett, 1975; Oliver and Baylé, 1976) and pigeon (Bouille and Baylé, 1973). In the White-throated sparrow, the hyperphagia and obesity that followed the destruction of the VMH were found to be accompanied by gonadal involution in some of the birds (Kuenzel, 1974).

The two types of gonadal dysfunction observed in the present study (FC and FCLC) seem to differ by the presence or absence of LH activity. In both FC and FCLC cocks the seminiferous tubules showed no spermatogenesis (Ravona et al., 1973a, 1973b; Robinzon et al., 1977a), suggesting that both syndromes involve a lack in FSH activity (Brown et al., 1975). However, whereas the FC cocks had involuted comb, no dehydrogenase activity in the interstitium (Ravona et al., 1973a), almost no plasma testosterone (Snapir et al., 1974b), and no gonadotropes in the adenohypophysis (Ravona et al., 1973b; Robinzon et al., 1977a, 1977b), this was not the case in the FCLC

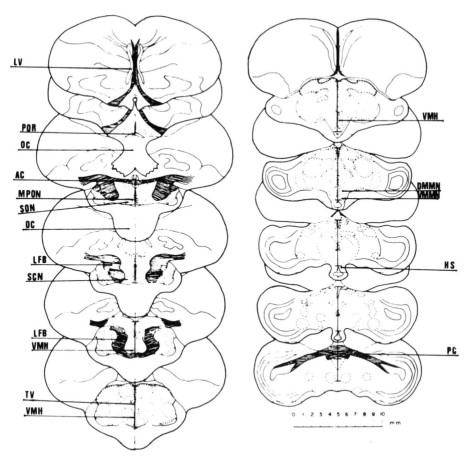

FIG. 1.—Schematic drawing of the chicken brain (Snapir *et al.*, 1974c). AC, anterior commissure; DMMN, dorsomedial mammillary nucleus; HS, hypophyseal stalk; LFB, lateral forebrain bundles; LV, lateral ventricle; MPON, medial preoptic nucleus; OC, optic chiasms; PC, posterior commissure; POR, preoptic recess; SCN, suprachiasmatic nucleus; SON, supraoptic nucleus; TV, third ventricle; VMN, ventromedial hypothalamic nucleus; VMMN ventromedial mammillary nucleus.

cocks. The latter had large comb, considerable levels of plasma testosterone (Snapir *et al.*, 1974b), interstitial dehydrogenase activity (Ravona *et al.*, 1973a), and gonadotropes in their adenohypophyses (Ravona *et al.*, 1973b). Thus, it may be suggested that while FC cocks had neither LH nor FSH activities, the FCLC ones did have a certain degree of LH activity (Brown *et al.*, 1975). It was suggested that in mammals, LH and FSH share a common releasing hormone of which a much higher level is needed to release FSH than for the release of LH (Greep, 1973; Debeljuk *et al.*, 1974; Franchimont *et al.*, 1974). In the duck hypothalamus, two LHRH secreting systems have been found (Bons *et al.*, 1978), one in the BMH and the other in the POA. It may be suggested that damage to one of these systems or to their connections with the adenohypophyseal portal system will reduce the amount of

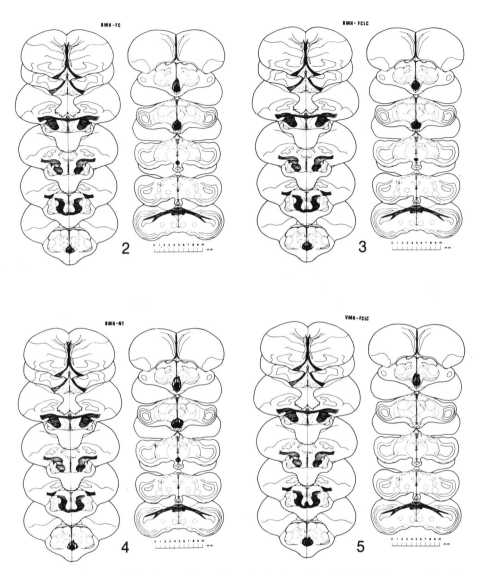

FIGS. 2-5.—Schematic drawings of the chicken brain (Snapir *et al.*, 1974*c*) demonstrating the location of the hypothalamic lesion. Blackened area covers the site of the smallest lesion; heavy barred area, the largest lesion. Fig. 2, BMH-FC cocks; 3, BMH-FCLC cocks; 4, BMH-NT cocks; 5, VMH-FCLC cocks.

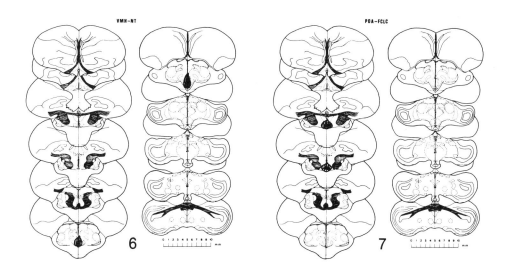

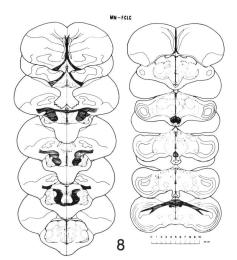

FIGS. 6-8.—Schematic drawings of the chicken brain (Snapir *et al.*, 1974c) demonstrating the location of the hypothalamic lesion. Blackened area covers the site of the smallest lesion; heavy barred area, the largest lesion. Fig. 6, VMH-NT cocks; 7, POC-FCLC cocks; 8, MN-FCLC cocks.

LHRH reaching the pituitary to a level insufficient for FSH release but adequate for LH release that may or may not be at a normal level. This could result in the FCLC syndrome that was observed in cocks lesioned in either POA, MN, VMH, or BMH. In accordance with this suggestion, only lesions that would destroy both LHRH secreting systems (or their efferents to the portal system) could totally deprive the pituitary of LHRH and thus cause the FC syndrome. In this study, the only cocks with the latter syndrome were among those with the BMH-lesion which destroyed both MN and VMH but did not damage the POA. It can be assumed that in the BMH-FC cocks, the BMH LHRH secreting system was destroyed while the anterior-POA system was only disconnected from the portal system. The absence of a consistent relationship between the lesion site and the degree of gonadal disfunction may then be a result of multiplicity and dispersion of the LHRH secreting systems and their efferent pathways to the portal system. One could suggest that in those VMH-lesioned cocks with the FCLC syndrome the lesion destroyed efferents from the POA to the median eminence.

A certain degree of hyperphagia and obesity was observed in all cocks with lesions in the medial hypothalamus from as rostral as the POA to as caudal as the MN. In Red-winged blackbirds, mesencephalic lesions in the interpeduncular nucleus as well as BNH-lesions were found to induce overeating and adiposity (Robinzon and Katz, 1980). The concept of "satiety center" located in the VMH has already been questioned (Gold, 1973). The above analysis of experimental results suggests that a dispersed neuronal system lying along the ventromedial brainstem is involved in the inhibition of food consumption and thus presents an additional challenge to the current emphasis on the VMH as the "satiety center."

ACKNOWLEDGMENT

This study was partially supported by the Charles Revson Foundation.

LITERATURE CITED

BONS, N., B. KERDELHUÉ, AND I. ASSENMACHER. 1978. Mise en évidence d'un deuxième systeme neurosécretoire à LH-RH dans l'hypothalamus du canard. C. R. Acad. Sci. (Paris), Ser. D.-Sci. Natur., 287:145-148.

BOUILLE, C., AND J.-D. BAYLÉ. Experimental studies on the adrenocortictropic area in pigeon brain. Neuroendocrinol., 11:73-91.

BROWN, N., J.-D. BAYLÉ, C. G. SCANES, AND B. K. FOLLETT. 1975. Chicken gonadotrophins: their effects on testes of immature and hypophysectomized Japanese quail. Cell Tiss. Res., 156:499-520.

DAVIES, D. T., AND B. K. FOLLETT. 1975. The neuroendocrine control of gonadotropin release in the Japanese quail. II. The role of the anterior hypothalamus. Proc. Roy. Soc. (London), 191:303-315.

DEBELJUK, L., J. A. VILCHEZ-MARTINEZ, A. ARIMURA, AND A. V. SCHALLY. 1974. Effect of gonadal steroids on the response to LH-RH in intact and castrated male rat. Endocrinology, 94:1519-1524.

FELDMAN, S. E., N. SNAPIR, F. YASUDA, F. TREUTING, AND S. LEPKOVSKY. 1973. Physiological and nutritional consequences of brain lesions: a functional atlas of the chicken hypothalamus. Hilgardia, 41:605-623.

Franchimont, P., H. Becher, C. Ernould, C. Thys, A. Demoulin, J. P. Bourguignon, J. J. Legros, and J. C. Valche. 1974. The effect of hypothalamic luteinizing hormone (LH-RH) on plasma gonadotrophin levels in normal subjects. Clin. Endocrinol., 3:27-39.

Gardner, J. E., and A. E. Fisher. 1968. Induction of mating in male chicks following preoptic implantation of androgen. Physiol. Behav., 3:709-712.

Gold, R. M. 1973. Hypothalamic obesity: the myth of the ventromedial nucleus. Science, 182:488-490.

Greep, R. O. 1973. The gonadotrophins and their releasing factors. J. Reprod. Fert., Suppl., 20:1-9.

Hutchison, J. B. 1974. Post castration decline in behavioral responsiveness to intrahypothalamic androgen in dove. Brain Res., 81:169-181.

Junghar, E. L. 1969. The neuroanatomy of the domestic fowl. Avian Dis., Special Issue, April, 1-159 pp.

Kuenzel, W. J. 1974. Multiple effects of ventromedial lesions in White-throated sparrow, Zonotrichia albicollis. J. Comp. Physiol., 90:169-182.

Lepkovsky, S., and M. Yasuda. 1966. Hypothalamic lesions, growth and body composition of male chicken. Poultry Sci., 45:582-588.

Oliver, J., and J.-D. Baylé. 1976. The involvement of the preoptic-suprachiasmatic region in the photosexual reflex in quail: effects of selective lesions and photic stimulation. J. Physiol. (Paris), 72:627-637.

Oliver, J., S. Herbuté, and J.-D. Baylé. 1978. Étude du rôle joué par le complexe infundibular et ses afferences nerveuses dans le réflexe photosexual chez la Caille. C. R. Acad. Sci. (Paris), Ser. D.-Sci. Natur., 287:825-827.

Perek, M., H. Ravona, N. Snapir, and D. Luxemburg. 1973. Sexual behavior of White Leghorn cocks bearing basal hypothalamic lesions in various locations. Physiol. Behav., 10:479-484.

Ravona, H., N. Snapir, and M. Perek. 1973a. The effect on gonadal axis in cockerels with electrolytic lesions in various regions of the basal hypothalamus. Gen. Comp. Endocrinol., 20:112-124.

———. 1973b. Histological changes of pituitary gland in cockerels bearing basal hypothalamic lesions and demonstration of its anti-HCG immunofluorescence reacting cells. Gen. Comp. Endocrinol., 20:490-497.

Robinzon, B., and Y. Katz. 1980. The effects of hypothalamic and mesencephalic lesions on food and water intakes, adiposity and some endocrine criteria in the Red-winged blackbird (Agelaius phoeniceus). Physiol. Behav., 24:347-356.

Robinzon, B., N. Snapir, and M. Perek. 1977a. The interrelationship between the olfactory bulbs and the basomedial hypothalamus in controlling food intake, obesity and endocrine functions in the chicken. Brain Res. Bull., 2:465-479.

———. 1977b. Histological changes in adenohypophysis and thyroid gland in propylthiouracil treated chicken following placement of basomedial hypothalamic lesion. Gen. Comp. Endocrinol., 33:365-370.

Snapir, N., I. Nir, F. Furuta, and S. Lepkovsky. 1969. Effect of administered testosterone propionate on cocks functionally castrated by hypothalamic lesions. Endocrinology, 8:611-618.

Snapir, N., H. Ravona, and M. Perek. 1973. Effect of electrolytic lesions in various regions of basal hypothalamus in White Leghorn cockerels upon food intake, obesity, blood plasma triglycerides and proteins. Poultry Sci., 52:629-636.

Snapir, N., I. Nir, F. Furuta, and S. Lepkovsky. 1974a. Effect of functional castration and of surgical castration upon body weight, food intake, selected bodily tissues and reproductive function. Gen. Comp. Endocrinol., 24:53-64.

Snapir, N., S. Lepkovsky, H. Ravona, and M. Perek. 1974b. Plasma testosterone in functionally castrated cockerels with hypothalamic lesions. Brit. Poult. Sci., 15:441-448.

SNAPIR, N., I. M. SHARON, F. FURUTA, S. LEPKOVSKY, H. RAVONA, AND B. ROBINZON. 1974c. An X-ray atlas of the saggital plane of the chicken diencephalon and its use in the precise localization of brain sites. Physiol. Behav., 12:419-424.

VAN TIENHOVEN, A., AND L. P. JUHASZ. 1962. The chicken telencephalon, diencephalon and mesencephalon in stereotaxic coordinates. J. Comp. Neurol., 118:185-198.

YOUNGREN, O. M., AND R. E. PHILLIPS. 1978. A stereotaxic atlas of the brain of three-day-old chick. J. Comp. Neurol., 181:567-600.

Reprinted from
ASPECTS OF AVIAN ENDOCRINOLOGY:
PRACTICAL AND THEORETICAL IMPLICATIONS (C. G. Scanes *et al.*, eds.)
Grad. Studies, Texas Tech Univ., 1982, 26:1-411.

CENTRAL NEURAL STRUCTURES AFFECTING FOOD INTAKE IN BIRDS: THE LATERAL AND VENTRAL HYPOTHALAMIC AREAS

WAYNE J. KUENZEL

Department of Poultry Science, University of Maryland, College, Park, Maryland 20742 USA

The central neural control of food intake in mammals was, until recently, thought to reside in two loci of the brain, the ventromedial and lateral hypothalamic areas. The concept had its beginnings after Hetherington and Ranson (1940) lesioned the ventromedial hypothalamic area (VMH) in rats. Experimentals became hyperphagic and obese. A decade later Anand and Brobeck (1951) bilaterally lesioned the lateral hypothalamic area (LHA) which produced aphagia, adipsia, and eventual death in all operated animals. Anand and Brobeck (1951) and Stellar (1954) proposed a dual hypothalamic feeding model and suggested that the VMH served as a center inhibiting a second center located within the LHA. The two-center model remained viable for about twenty years and is still emphasized in textbooks that touch upon the physiology and psychology of feeding behavior. More recently, the entire medial region of the diencephalon, mesencephalon, and rhombencephalon has been viewed as a sympathetic zone (Ban, 1964). It follows that adrenergic receptors within the medial brain region should play a major role in the regulation of fundamental physiological and behavioral processes such as feeding. Support for the idea is Gold's (1973) finding that the ventromedial hypothalamic nucleus serves as an important locus for the ventral noradrenergic bundle. Cutting or destroying the bundle rather than the VMH results in obesity (Gold, 1973). Additional empirical evidence was provided by Leibowitz (1978) who made a significant discovery that a dorsomedial hypothalamic structure, the paraventricular nucleus (Fig. 1), contains both alpha and beta adrenergic receptors. Applying neuropharmacological agents, she showed that chemical stimulation of the former produced feeding responses while stimulation of the latter terminated the behavior (Leibowitz, 1978).

The LHA has been subjected to considerable scrutiny over the years and the past concept of a "feeding center" is inappropriate. Morgane (1961) showed that the region is quite complex and, in addition to cell bodies, has many fibers of passage which may affect feeding. Morgane (1964) defined the LHA in mammals as shown in Fig. 1A. The media-lateral range of the LHA extends from the fornix (F) to the internal capsule (ICap). It therefore is comprised of some hypothalamic structures (the medial forebrain bundle and cell bodies within the medial LHA) but has a major contribution of thal-

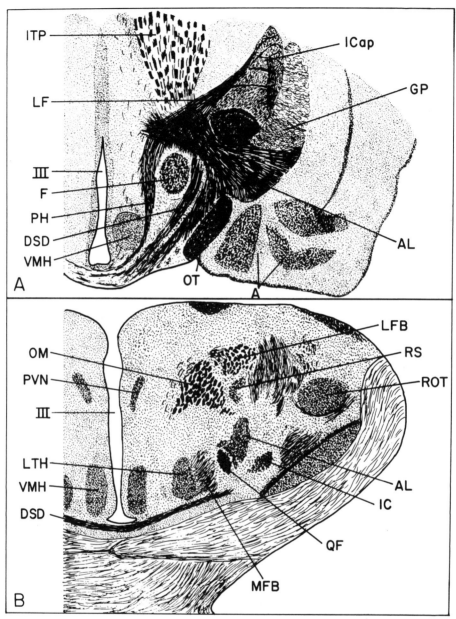

Fig. 1.—Cross-sections of a mammalian and avian brain (hypothalamus) showing areas controlling food intake.   A. Rat brain (after Morgane, 1964): A, Amygdala; AL, Ansa lenticularis; DSD, Dorsal supraoptic decussation; F, Fornix; GP, Globus pallidus; ICap, Internal capsule; ITP, Inferior thalamic peduncle; LF, Lenticular fasciculus; OT, Optic tract; PH, Pallido-hypothalamic tract; VMH, Ventromedial hypothalamic nucleus; III, Third ventricle.   B. Chick brain (8-week-old broiler): AL, Ansa lenticularis; DSD, Dorsal supraoptic decussation; IC, Nucleus intercalatus; LFB, Lateral forebrain bundle; LTH, Lateral hypothalamic nucleus; MFB, Medial forebrain bundle; OM, Occipital mesencephalic tract; PVN, Paraventricular nucleus; QF, Quinto-frontal tract; ROT, Nucleus rotundus; RS, Nucleus reticularis superior; VMH, Ventromedial hypothalamic nucleus; III, Third ventricle.

amic components, including the ansa lenticularis, lenticular fasciculus, lateral portions of Ganser's commissure, and projections of the inferior thalamic peduncle. More recently it has been shown that lateral hypothalamic and most extrahypothalamic lesions that have produced aphagia and adipsia in rats interrupt dopamine containing neurons of the nigrostriatal bundle as they course through the ventral diencephalon (Ungerstedt, 1971). It is clear that the LHA is not a homogeneous structure and not surprising that other persistent deficits, besides feeding and drinking, have been associated with lesions to this region of the diencephalon.

REVIEW: AVIAN NEURAL STRUCTURES AFFECTING FEEDING

The purpose of this mini review is to define anatomically the LHA and VMH in birds and include avian lesion studies that affected food intake after destruction of one or more components of these complex brain regions.

Avian Lateral Hypothalamic Area

Figure 1B shows a cross section of the brain of an eight-week-old broiler, Gallus domesticus. It is taken from a level comparable to that of the rat (Fig. 1A) as evidenced by the presence of the ventromedial hypothalamic nucleus (VMH). The LHA in birds extends from the lateral hypothalamic nucleus (LTH) to the lateral forebrain bundle (LFB). Specific neural structures included in the region are the LTH, the medial forebrain bundle (MFB), the occipital mesencephalic tract (OM), quinto-frontal tract (QF), ansa lenticularis (AL), the nucleus reticularis superior (RS), the nucleus intercalatus (IC), and LFB. Only the first two structures are considered truly hypothalamic. More caudal to the section shown in Fig. 1B is an additional structure called the stratum cellulare externum (SCE). Part of it resides within the LHA whereas part is considered thalamic. Note that, just as in mammals, the major components of the LHA are thalamic rather than hypothalamic structures.

The first experimenters to report aphagia in birds following bilateral lesions to the diencephalon were Feldman et al. (1957). Although they described the lesions as being placed in the lateral hypothalamus, it can be seen by their published photomicrographs (p. 260) that at least three of their cases had primarily thalamic and mesencephalic tissue destroyed. On the other hand, they did claim that some chickens in their study sustained bilateral damage to the anterior and mid-hypothalamus and showed persistent aphagia. No brain photomicrographs were presented for the birds. It is not clear which specific neural structures were damaged except to generalize that bilateral lesions directed to the diencephalon and mesencephalon of fowl resulted in various degrees of aphagia as well as other behavioral and physiological disturbances. Smith (1969) reported aphagia in chickens following bilateral damage to the lateral hypothalamic nucleus. It was not stated, however, how many days the chicks were aphagic after surgery and what other neural structures were damaged by the lesions.

Zeigler and his colleagues (Zeigler *et al.*, 1969; Zeigler and Karten, 1973, 1974) have concentrated on a system of fibers and nuclei involved with the fifth cranial nerve in pigeons, *Columba livia*. The structures investigated include the trigeminal nucleus, nucleus basalis, and the quinto-frontal tract (QF). The latter structure is of interest as it passes through the lateral hypothalamic region of birds (Fig. 1B). Pigeons having bilateral lesions to quinto-frontal structures have persistent aphagia (Zeigler and Karten, 1973). The aphagia in lesioned pigeons is due to sensory and motor deficits to the buccal region such that use of the mandibles to manipulate seeds is drastically affected.

Wright (1975) analyzed feeding deficits and several behavioral changes after bilateral lesions were directed to various diencephalic sites in the Barbary dove (*Streptopelia risoria*). He concluded that destruction of an area extending from the medial borders of the lateral forebrain bundle through the stratum cellulare externum resulted in aphagia for an average of nine days (range 2-35 days, N=13) but little or no adipsia.

In summary, the following structures located in the diencephalon of birds have been implicated in effecting consistent feeding deficits: lateral hypothalamic area, lateral hypothalamic nucleus, quinto-frontal tract, lateral forebrain bundle, and stratum cellulare externum.

Avian Medial Hypothalamic Area

The medial hypothalamus at the level shown in Fig. 1B includes basically two major structures: the ventromedial hypothalamic nucleus (also known as the nucleus hypothalamicus posterior medialis by van Tienhoven and Juhasz, 1962) and the paraventricular nucleus (PVN). The PVN is one of the most clearly defined nuclei within the avian hypothalamus. It is of interest to note that the comparable nucleus in mammals has been subdivided into 10 discrete functional zones (Koh and Ricardo, 1980). The avian PVN may soon be shown to be as complex.

The first experimenters to report the development of obesity in chickens following ventromedial hypothalamic (VMH) lesions were Lepkovsky and Yasuda (1966). Shortly thereafter, VMH lesions were shown to effect hyperphagia and obesity in a migratory bird, the White-throated sparrow, *Zonotrichia albicollis* (Kuenzel and Helms, 1967). Further research, involving lesions directed to the basal hypothalamus of fowl, caused intestinal constriction (Lepkovsky and Dimick, 1969) and several metabolic changes (Snapir *et al.*, 1973, 1974).

In all the avian studies mentioned above, anatomical detail of neural structures destroyed was not given. The omissions were perhaps due to the diffuse nature of avian hypothalamic nuclei, thus making it difficult to identify boundaries and to apply the appropriate nomenclature for each structure. In contrast to mammalian investigations to date, it is not known for any avian species whether the ventromedial hypothalamic nucleus, the ventral noradrenergic bundle, and/or the paraventricular nucleus play a major role in effecting hyperphagia when lesioned.

Future studies employing electrolytic lesions, electrical stimulation, knife cuts, chemical lesions, and/or pharmacological agents to specific loci in the avian brain should include appropriate histological techniques to identify specific neural substrates affected by each neurosurgical procedure. It is most important in order to identify a structure that alters feeding behavior and systems of neurons and fiber tracts which ultimately regulate that behavior.

ACKNOWLEDGMENTS

The author wishes to acknowledge Ms. Manju Masson for her excellent technical assistance and for drawing Fig. 1. This is Scientific Article No. A2912, Contribution No. 5949, of the Maryland Agricultural Experiment Station (Department of Poultry Science).

LITERATURE CITED

ANAND, B. K., AND J. R. BROBECK. 1951. Hypothalamic control of food intake in rats and cats. Yale J. Biol. Med., 24:123-140.

BAN, T. 1964. The hypothalamus, especially on its fiber connections, and the septo-preoptico-hypothalamic system. Med. J. Osaka Univ., 15:1-83.

FELDMAN, S., S. LARSSON, M. DIMICK, AND S. LEPKOVSKY. 1957. Aphagia in the chicken. Amer. J. Physiol., 191:259-261.

GOLD, R. M. 1973. Hypothalamic obesity: the myth of the ventromedial nucleus. Science, 182:488-490.

HETHERINGTON, A., AND S. RANSON. 1940. Hypothalamic lesions and adiposity in the rat. Anat. Rec., 78:149-172.

KOH, E. T., AND J. A. RICARDO. 1980. Paraventricular nucleus of the hypothalamus; anatomical evidence of ten functionally discrete subdivisions. Soc. Neurosci., 6:521.

KUENZEL, W. J., AND C. W. HELMS. 1967. Obesity produced in a migratory bird by hypothalamic lesions. Bioscience, 17:395-396.

LEIBOWITZ, S. F. 1978. Paraventricular nucleus: a primary site mediating adrenergic stimulation of feeding and drinking. Pharm. Biochem. Behav., 8:163-175.

LEPKOVSKY, S., AND M. K. DIMICK. 1969. The hypothalamus and pancreas in intestinal function. Ann. New York Acad. Sci., 157:1062-1068.

LEPKOVSKY, S., AND M. YASUDA. 1966. Hypothalamic lesions, growth, and body composition of male chickens. Poultry Sci., 45:582-588.

MORGANE, P. J. 1961. Medial forebrain bundle and "feeding centers" of the hypothalamus. J. Comp. Neurol., 117:1-25.

————. 1964. Limbic-hypothalamic-midbrain interaction in thirst and thirst motivated behavior. Pp. 429-455, in Thirst (M. J. Wayner, ed.), MacMillan Co., New York, viii+570 pp.

SMITH, C. J. V. 1969. Alterations in the food intake of chickens as a result of hypothalamic lesions. Poultry Sci., 48:475-477.

SNAPIR, N., H. RAVONA, AND M. PEREK. 1973. Effect of electrolytic lesions in various regions of the basal hypothalamus in White Leghorn cockerels upon food intake, obesity, blood plasma triglycerides and proteins. Poultry Sci., 52:629-636.

SNAPIR, N., I. NIR, F. FURUTA, AND S. LEPKOVSKY. 1974. Effects of functional and surgical castration of White Leghorn cockerels, and replacement therapy, on food intake, obesity, reproductive traits, and certain components of blood, liver, muscle and bone. Gen. Comp. Endocrinol., 24:53-64.

STELLAR, E.   1954.   The physiology of motivation.   Psychol. Rev., 61:5-22.

UNGERSTEDT, U.   1971.   Adipsia and aphagia after 6-hydroxydopamine induced degeneration of the nigro-striatal dopamine system.   Acta Physiol. Scand., 82(Suppl. 567):95-122.

VAN TIENHOVEN, A., AND L. P. JUHÁSZ.   1962.   The chicken telencephalon, diencephalon and mesencephalon in stereotaxic coordinates.   J. Comp. Neurol., 118:185-198.

WRIGHT, P.   1975.   The neural substrate of feeding behavior in birds.   Pp. 319-349, in Neural and endocrine aspects of behavior in birds (P. Wright, P. G. Caryl, and D. M. Vowles, eds.), Elsevier, Amsterdam, x+408 pp.

ZEIGLER, H. P., AND H. J. KARTEN.   1973.   Brain mechanisms and feeding behavior in the pigeon (Columba livia). I. Quinto-frontal structures.   J. Comp. Neurol., 152:59-82.

———.   1974.   Central trigeminal structures and the lateral hypothalamic syndrome in the rat.   Science, 186:636-638.

ZEIGLER, H. P., H. J. KARTEN, AND H. L. GREEN.   1969.   Neural control of feeding in the pigeon.   Psychon. Sci., 15:156-157.

Reprinted from
ASPECTS OF AVIAN ENDOCRINOLOGY:
PRACTICAL AND THEORETICAL IMPLICATIONS (C. G. Scanes *et al.*, eds.)
Grad. Studies, Texas Tech Univ., 1982, 26:1-411.

EFFECTS OF MYCOTOXINS ON AVIAN REPRODUCTION

MARY ANN OTTINGER AND JOHN A. DOERR

*Department of Poultry Science, University of Maryland, College
Park, Maryland 20742 USA*

Mycotoxins are toxic metabolites produced by a large number of fungi under a wide range of environmental conditions. Many of these fungi are particularly well adapted to invasion of cereals, nuts, and grains that eventually are used in the manufacture of animal feeds. Poultry and livestock that subsequently ingest mycotoxin contaminated feed can exhibit numerous, diverse pathological manifestations that are dependent upon the particular mycotoxin(s), the amount of toxin, the length of exposure, and the age, sex, health, and susceptibility of the animal. Therefore, the occurrence of mycotoxins in standing or stored crops poses a threat to domesticated species that might be exposed to them. It is further possible, and perhaps probable, that toxigenic fungi may be a factor in the health of wild species in that environmental conditions are often appropriate for fungal invasion of the plants, seeds, and grains that such species rely on for food.

The descriptions of pathologies associated with mycotoxicoses are still incomplete and often confusing, resulting in only limited ability to predict sequence and importance of various symptoms for any specific mycotoxin and difficulty in determining precise causal associations. Laboratory evidence shows that relatively high doses of some mycotoxins elicit extensive liver damage and lead to aberrations in various metabolic pathways. Chronic exposure to very low levels of these mycotoxins may result in ill-defined, subclinical symptoms. However, laboratory results can not be directly related to commercial or field environments because additional factors also affect animals in these environments. Table 1 indicates some problems commonly associated with experimental aflatoxicosis in poultry. Reproductive failure has not been associated routinely with aflatoxin although some other toxins such as zearalenone and T-2 toxin have been demonstrated to have deleterious effects on reproduction in mammals. These results suggest that mycotoxins may have substantial effects on endocrine function and reproductive capacity sufficient to warrant careful consideration of this aspect of animal health.

This paper will highlight some experimental evidence and field observations concerning several mycotoxins that are recognized as serious physiological and economic threats to birds. Although emphasis will be centered on avian species, allusion to mammalian problems will be made where appropriate.

TABLE 1.—*Principal effects of aflatoxin in poultry.*

1. Growth inhibition
2. Increased mortality
3. Increased susceptibility to pathogens, stressors, and nutritional deficits
4. Malabsorption syndrome
5. Depression of humoral and cellular immunity
6. "Fatty liver" syndrome
7. Hemolytic anemia
8. Blood coagulopathy
9. Impaired lipid metabolism

Aflatoxin

Aflatoxins, secondary metabolites of the *flavus-parasiticus* group of the genus *Aspergillus*, are likely to occur in warm, humid climates and are readily found in corn, peanuts, and other feed ingredients (Ford *et al.*, 1978; Pier *et al.*, 1980; Shotwell *et al.*, 1980). Reproductive problems were suggested by Kratzer *et al.* (1969) who found that whereas hens fed 2.7 ppm aflatoxin for 34 days did not experience a decline in egg production, hatchability of eggs was reduced. Hamilton and Garlich (1971) demonstrated a dose relationship between aflatoxin and decreased egg production and egg weights. It was further demonstrated (Garlich *et al.*, 1973) that when 20 ppm aflatoxin was incorporated into a standard layer diet for seven days, egg production declined approximately 26 per cent during toxin feeding, with the nadir (approximately 53% reduction) occurring one week later and the period of reduced egg production extending to 18 days after administration of aflatoxin. The authors suggested the lag effect could be related to the time of rapid growth of follicles committed to maturation. Thus, those follicles already undergoing maturation continued to completion, while initiation of new follicle growth was inhibited for about 10 days after withdrawal of toxin (Garlich *et al.*, 1973). Sims *et al.* (1970) reported production declines but no loss in egg weight in aflatoxin-incubated hens, while Huff *et al.* (1975*a*, 1975*b*) found that aflatoxin depressed egg production, egg weight, and yolk weight. Dietary aflatoxin produced a dose-related decrease in egg production, egg weight, hatchability, and egg quality in Japanese quail (Table 2; Sawhney *et al.*, 1973).

Only little consideration has been given to male gonadal function during aflatoxicosis. Wyatt *et al.* (1973) reported a slight but significant ($P<0.05$) depression in body weight and enlargement of liver and spleen in adult broiler breeder males receiving 20 ppm aflatoxin. In related work, Briggs *et al.* (1974) reported sperm counts, semen volume, and semen DNA, RNA, and protein to be unaffected during aflatoxicosis in breeder males. However, both studies used retired breeders suggesting a possible complication from the variables of rooster age and fertility. Jayaraj *et al.* (1969) reported that during chronic aflatoxicosis in rats, gonadal aberrations ranged from mild degeneration of seminiferous tubules with spermatozoan agglutination to

TABLE 2.—*Effects of mycotoxins on reproduction in Japanese quail.*

Mycotoxin	Variate	Response	Reference
Aflatoxin	Egg Production	Decreased	Sawhney *et al.*, 1973
,,	Egg Weight	Decreased	,,
Ochratoxin	Egg Production	Increased	Prior *et al.*, 1978
,,	Egg Weight	Decreased	,,
,,	Hatchability	Decreased	,,

advanced degeneration characterized by hyalinization, fibrosed tubules, formation of polynuclear giant cells, fatty infiltration of spermatogenic epithelium, and encysted sperm. Richir *et al.* (1964) reported testicular atrophy during aflatoxicosis in rats. Administration of istotopic alfatoxin to mice revealed detectable localization of ^{14}C-metabolites in the epididymus of males, in the vagina of females, and in fetal tissue *in utero* (Arora *et al.*, 1978).

Effects of Aflatoxin During Maturation

Dietary aflatoxin in growing animals produces a suppression in somatic development as well as a delay in sexual maturation. Doerr and Ottinger (1980) have reported inhibition of ovarian development and of follicular maturation in quail with adverse effects extending several weeks beyond the administration of aflatoxin. Additionally, egg weights and production were found to be reduced four to five weeks post-aflatoxin treatment (Doerr and Ottinger, 1979).

Dietary aflatoxin (0, 5.0, and 10 μg per gram diet) fed to prepubertal Japanese quail produced a reduction in the efficiency of mating behavior (Table 3; Ottinger and Doerr, 1980). Control males began courtship behavior at 31 days of age as compared to 35 days in toxin-treated males; the per cent of quail exhibiting courtship patterns remained significantly below controls until 56 days of age. In the toxin-treated groups, the total number of mating attempts (mounts) was highly elevated as compared to controls; however, the number of cloacal contacts completed was the same for all groups. Thus, the attempt/success ratio for birds that had received aflatoxin was higher than controls, indicating a reduced efficiency of mating. Testosterone-dependent cloacal gland development was delayed, consistent with depressed serum testosterone values.

Doerr and Ottinger (1979) reported a significant reduction in weight of testes relative to body weight of quail receiving either 5.0 or 10 μg aflatoxin per gram diet. There was a lag of two to three weeks in achieving normal rates of testicular growth even after recovery of normal somatic growth rate and body weight. Serum testosterone concentrations declined during toxin feeding. Quail exposed to the toxin before the normal onset of testicular function (7 to 21 days of age) recovered normal circulating testosterone levels within two weeks of withdrawal of aflatoxin diet. However, those birds

TABLE 3.—*Age (in days) at which changes in variables occurred as estimated by the join point technique of Hudson (1966).*[1]

	Treatment[2]		
Variable	A	B	C
Mean cloacal gland area	37	37	34
Mean testosterone concentration	27	37	26
First incidence of sexual behavior	35	35	31
Mean testes weight	29	36	27

[1]Printed with permission of Poultry Science; Ottinger and Doerr, 1980.
[2]Treatment A, aflatoxin (10 ppm) between 7 and 14 days of age; treatment B, aflatoxin (10 ppm) between 11 and 28 days of age, treatment C, control, (0 ppm aflatoxin).

whose treatment period overlapped onset of maturation (days 14 to day 28) showed a continued depression of testosterone after two weeks on an uncontaminated diet. These data provide evidence of residual effects by aflatoxin on testicular function and emphasize the significance of age at exposure when assessing toxic responses in juvenile quail.

In a subsequent study, males exposed to 10 ppm aflatoxin in the diet for one week (either 14-21 or 21-28 days of age) also showed residual effects of the treatment (Ottinger and Doerr, unpubl. data). Testes weight (Fig. 1) began to rapidly increase in control males associated with sexual maturation. Earlier treatment depressed testes weight; however, males fed aflatoxin between 21 and 28 days of age showed a depressed testes weight after treatment had ended. Possibly, the tissue components of the testes, such as seminiferous tubules, were already developing by the time of treatment and there was a lag time involved in observation of the toxin-produced regression of the tissue. Plasma testosterone (Fig. 2) followed a similar pattern; earlier treated males exhibited a faster recovery than males treated between 21 and 28 days of age. In addition, there appeared to be an altered androgen metabolism involved in aflatoxicoses (Fig. 3). This was most pronounced in the males exposed between 14 and 21 days of age in that the ratio (in per cent) of dihydrotestosterone (DHT) to testosterone was high and was reduced as the plasma testosterone increased to control concentrations. These data might further indicate differential sensitivity of the testes relative to age of exposure. Finally, relative weights of ovaries were not altered when aflatoxin was withdrawn by day 21. Aflatoxin administered from day of age 14 to day 28 inhibited development of ovaries. Furthermore, a count of follicles undergoing maturation revealed that, three to four weeks following withdrawal of aflatoxin diets, there were 60 percent fewer maturing follicles in females that had received toxin than in control females. Preliminary egg data indicated that the onset of lay was delayed four to eight days and that initial egg weights were lower among quail receiving aflatoxin prepubertally (Doerr and Ottinger, 1980).

Possible Mechanisms of Action

The mechanism of action of aflatoxin appears to be through both direct and indirect effects. Aflatoxin B_1, injected directly into the testes of rats,

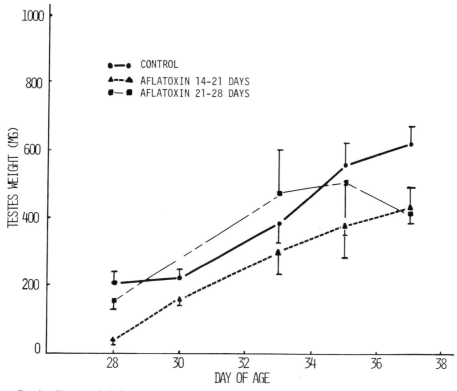

FIG. 1.—Testes weight in male Japanese quail fed aflatoxin during sexual maturation. Points with vertical lines represent means with SE.

resulted in a dose-dependent damage to the tissue as well as altered plasma steroid concentrations (Gopal *et al.*, 1980). Plasma estrogen concentrations were reduced, suggesting a direct effect on the testes rather than an indirect effect through altered liver activity. It had been previously shown that intraperitoneal injection of aflatoxin B_1 in rats had no effect on the reproductive system (Egbunike, 1979), whereas testicular atrophy and azoospermia were found in male rats fed aflatoxin-contaminated feed (Richir *et al.*, 1965). There was additional effect on the liver cells that resulted in destruction of high affinity binding sites for steroids on the endoplasmic reticulum in an *in vitro* liver preparation (Blyth *et al.*, 1971). The severity of toxin effects also appeared to be sexually dimorphic in that the longevity of females exposed to continual high doses of aflatoxin was greater than that of males (Jerrold *et al.*, 1975). Finally, aflatoxin appeared to exert effects on a cellular level and, in particular, it affected RNA polymerase probably by binding to the DNA (Clifford and Rees, 1967; Pong and Wogan, 1969). These cellular effects might be reflected in brain metabolism in that the concentration of catecholamines in specific areas were altered in birds exposed to aflatoxin (El Halawani, pers. comm.). Moreover, aflatoxin has been found to interfere with tryptophan metabolism (Wogan and Freidman, 1965). There were

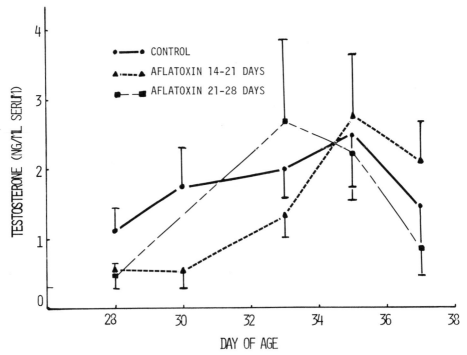

F IG. 2.—Plasma testosterone in male Japanese quail fed aflatoxin during sexual maturation.

increased amounts of serotonin in duodenal tissue in response to aflatoxico-sis (Lalor *et al.*, 1978). These data indicate the possibility that alterations in catecholamines may occur in the brain, thereby affecting pituitary release of gonadotropins which would then alter reproduction.

Finally, because liver tissues are the primary target during aflatoxicosis, alteration or interruption of liver-dependent synthetic pathways, which normally provide precursor materials to extrahepatic synthesis sites, could be expected to decrease macromolecular species originating in other organs such as the adrenal glands or gonads. Donaldson *et al.* (1972) demonstrated impaired lipid metabolism in birds during aflatoxicosis. In addition, trans-port of lipids from the liver is known to be reduced (Tung *et al.*, 1972). As a result, lipid fractions, such as free and esterified cholesterol, are reduced in the sera. Fig. 4 shows similar results obtained in Japanese quail (Doerr and Ottinger, unpubl. data). Whereas testicular cholesterol remains to be deter-mined, reduction of cholesterol in the testes would probably limit steroido-genesis and could contribute to the observed decline in serum testosterone.

Ochratoxin

Munro *et al.* (1973) have reported fetal deaths in pregnant rats adminis-tered either cultures of *Aspergillus ochraceus* or purified ochratoxin A. Och-ratoxin A administered intravenously to pregnant sheep caused 100 per cent

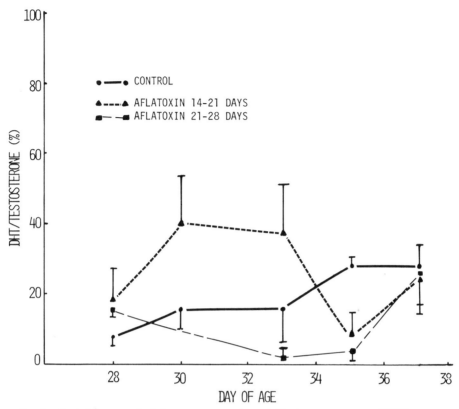

FIG. 3.—Ratio (in per cent) of plasma concentrations of dihydroxytestosterone (DHT) to testosterone in male Japanese quail fed aflatoxin during sexual maturation.

mortality within 24 hours; however, while high levels of ochratoxin could be detected in the blood of dams, only trace levels were found in fetal blood, which the authors suggested implied only limited placental transfer (Still *et al.*, 1971). Comparison of toxigenic isolates of several fungal species obtained from a field outbreak of idiopathic abortion in dairy cows, swine, and cats revealed that *A. ochraceus* evoked 100 per cent fetal death in rats. Szczech and Hood (1978) have described necrosis of brain and retina cells of mice following transplacental exposure to ochratoxin. These teratogenic lesions were dependent on the gestational age at which the toxin was introduced. Teratogeny has been reported for rats, mice, and hamsters (Hayes *et al.*, 1974; Brown *et al.*, 1976; Hood *et al.*, 1976a). In poultry, ochratoxin A is primarily nephrotoxic and produces severe aberrations in kidney size and function, although hepatotoxicity occurs to a lesser extent (Huff *et al.*, 1974, 1975a, 1975b). Literature is lacking concerning specific action of this mycotoxin on the reproductive system of male birds. In mammals, available data suggest little, if any, effect of dietary ochratoxin on testes, but severe, dose-dependent differential pathology when ochratoxin is injected into the testes

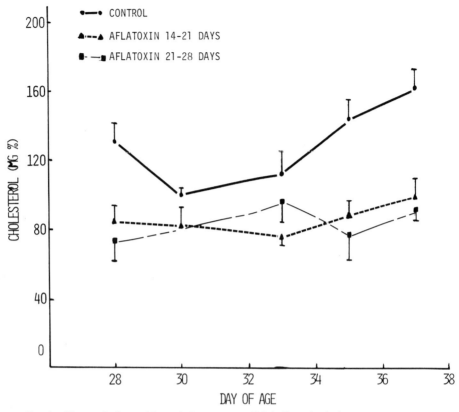

F<small>IG</small>. 4.—Plasma cholesterol in male Japanese quail fed aflatoxin during sexual maturation.

(More and Camguilhem, 1979). In females, metabolites of ochratoxin have been localized in the milk of goats (Nip and Chu, 1979) and rabbits (Galtier *et al.*, 1977), but not of cows (Galtier *et al.*, 1977). It should be noted, however, that these mammalian studies were performed on small numbers of animals and, with the exception of the cow data, represent effects following single dose incubation or injection. In laying hens, ochratoxin was found to decrease egg production and egg weight (Prior and Sisodia, 1978). In that study, there was no loss in fertility or hatchability in contrast to work by Choudhury *et al.* (1971). Prior *et al.* (1978) found significantly reduced egg production and egg weight when ochratoxin was fed for three weeks to Japanese quail (Table 2). Additionally, they reported nearly a total loss in fertility and hatchability. However, quail fed the same diets for six weeks did not differ in fertility from controls although total egg production was lower. Apparently the lengthened exposure period permitted development of compensation or resistance to ochratoxin (Prior *et al.*, 1978).

Other Mycotoxins

Citrinin

Citrinin, a nephrotoxic mycotoxin elaborated by *Penicillium citrinum* and several other species of *Penicillium* and *Aspergillus* (Heatherington and Raistrick, 1931), has been found to contaminate cereal grains alone and in combination with other mycotoxins such as ochratoxin (Krough *et al.*, 1973). Phillips and Hayes (1978) described citrinin-induced reduction of kidney weight and kidney protein in mice; however, major effects involving DNA, RNA, protein, glycogen, and lipid were noted in the livers of affected mice suggesting hepatotoxic actions. Hood *et al.* (1976*b*) injected citrinin intraperitoneally in pregnant mice and found a reduction in fetal survival and fetal weight but no external evidence of teratogenicity. Citrinin produced substantive kidney pathology in rats (Jordan and Carlton, 1978*a*). Urinalysis in affected rats demonstrated altered urine output and abnormal concentrations of electrolytes, glucose, and protein (Jordan and Carlton, 1978*b*). Similar lesions were found in hamsters receiving citrinin either intraperitoneally or orally (Jordan and Carlton, 1978*c*). Ames *et al.* (1976) found no effects of dietary citrinin on production parameters in adult laying hens. In young broilers, body weight, feed and water consumption, liver and kidney weight, and serum sodium levels were affected by this mycotoxin (Ames *et al.*, 1976). However, investigation of citrinin effects on specific reproduction parameters involving young birds has not been performed.

Zearalenone

Zearalenone, a metabolite of several species of the genus *Fusarium*, has been shown to be an anabolic, estrogenic compound, both physiologically and mechanistically (Pier *et al.*, 1980). Indeed, there is commercial interest in the use of zearalenone derivatives as anabolic agents in cattle to improve growth rate and feed efficiency (Katzenellenbogen *et al.*, 1979). Zearalenone interacts with uterine cytoplasmic receptors and is translocated to the nucleus in a fashion similar to other estrogens (Katzenellenbogen *et al.*, 1979; Greenman *et al.*, 1979). *In vitro*, zearalenone competitively inhibits estradiol binding to estrogen receptors in mammary glands (Boyd and Wittliff, 1978).

Sherwood and Peberdy (1973) studied the effects of dietary zearalenone in birds. They reported that weight gain was not affected; however, testes and comb weights were increased, an observation consistent with expected effects of estrogenic compounds (Sherwood and Peberdy, 1973). Zearalenone fed to turkey poults also produced swelling of the vent, oviduct enlargement, cloacal eversion, and cystic development of the right genital tract. Zearalenone has been shown to elicit sperm degeneration in turkeys (Palyuskik *et al.*, 1971), whereas egg production and fertility were not affected in laying geese (Palyusik and Karlic-Kovacs, 1975). Marks and Bacon (1976), however, reported that dietary zearalenone did not adversely affect production or fertility of hens.

Because species of *Fusarium* produce a variety of mycotoxins, some zearalenone research might be complicated by the presence of other compounds; however, the estrogenic manifestations of this mycotoxin do seem to occur in birds. Nonetheless, further research will be required to assess the actual risks of field zearalenone intoxications to reproductive efficiency in avian species.

T-2 Toxin

T-2 toxin, one of the many naturally occurring 12,13-epoxytrichothecenes, has been implicated in field mycotoxicosis of poultry (Wyatt *et al.*, 1972*a*, 1973). Although best known for causing oral lesions (Wyatt *et al.*, 1972*b*) and neural disturbances (Wyatt *et al.*, 1973) in chickens, T-2 toxin has been shown to affect both egg production and egg shell thickness in laying hens (Wyatt *et al.*, 1975). Chi *et al.* (1977) reported lowered egg production, thin egg shells, and reduced hatchability but no adverse effects on fertility when T-2 toxin was fed to laying hens.

Reproductive problems associated with T-2 toxin have been reported in domestic livestock. Weaver *et al.* (1978*b*) administered T-2 toxin intravenously to pregnant sows and reported all fetuses aborted by 80 hours postinjection. In a separate study, Weaver *et al.* (1978*a*) fed 12 ppm purified T-2 toxin in a standard ration to sows. Among their findings were infertility, reduced litter size, and small piglets. Robinson *et al.* (1979) found T-2 toxin in the milk from a Holstein cow and a sow fed 50 and 12 ppm T-2 toxin, respectively. However, the toxin concentrations found in analyzed samples were small, considering the high levels in the diets.

Concluding Remarks

Mycotoxins constitute an important natural contaminant in grains that comprise the diet of birds. This is best recognized by the recurrent mycotoxin problems encountered by the commercial poultry industry. The research described in the foregoing citations gives evidence for mycotoxic interference with normal reproductive and endocrine function. These influences are not limited to the duration of an acute toxic episode, but may be residual effects from prior intoxication. Therefore the capacity to optimize reproductive performance in intensive programs, such as poultry breeder operations, might be restricted by mycotoxin challenges during the maturation period. Furthermore, there is evidence that the mechanism of mycotoxin action is highly complex and could involve both direct and indirect effects. In any event, the reproductive endocrinology of domestic bird species is affected by mycotoxins.

An unexplored area of this problem is the impact of mycotoxins on wild bird populations. Since these mycotoxins have been demonstrated in such a wide variety of grains, nuts, and grasses, and under such a wide range of environmental conditions, it is highly likely that wild birds risk exposure to these toxins as well. It follows then that wild birds could suffer physiological and reproductive consequences from mycotoxin exposure.

ACKNOWLEDGMENTS

The authors wish to thank Ms. Laura Kallay and Ms. Sandi Dougherty for their technical assistance. The assistance of Ms. Louise Palmer with graphs is also appreciated. This is Scientific Article No. A2912, Contribution No. 5969, of the Maryland Agricultural Experiment Station (Department of Poultry Science).

LITERATURE CITED

AMES, D. D., R. D. WYATT, H. L. MARKS, AND K. W. WASHBURN. 1976. Effect of citrinin, a mycotoxin produced by *Penicillium citrinum*, on laying hens and young broiler chickens. Poultry Sci., 55:1294-1301.

ARORA, R. G., L. E. APPELGREU, AND A. BERGMAN. 1978. Distribution of (^{14}C)-labelled aflatoxin B_1 in mice. Acta Pharmacol. Toxicol., 43:273-279.

BLYTH, C. A., R. B. FREEDMAN, AND B. R. RABIN. 1971. The effects of aflatoxin B_1 on sex-specific binding of steroid hormones to microsomal membranes of rat liver. Eur. J. Biochem., 20:580-586.

BOYD, P. A., AND J. L. WITTLIFF. 1978. Mechanism of *Fusarium* mycotoxin action in mammary gland. J. Tox. Envir. Health, 4:1-8.

BRIGGS, D. M., R. D. WYATT, AND P. B. HAMILTON. 1974. The effect of dietary aflatoxin on semen characteristics of mature broiler breeder males. Poultry Sci., 53:2115-2119.

BROWN, M.H., G. N. SZCZECH, AND B. P. PURMALIS. 1976. Teratogenic and toxic effects of ochratoxin A in rats. Toxicol. Appl. Pharmacol., 37:331-338.

CHI, M. S., C. J. MIROCHA, H. J. KURTZ, G. WEAVER, F. BATES, AND W. SHIMODA. 1977. Effects of T-2 toxin on reproductive performance and health of laying hens. Poultry Sci., 56:628-637.

CHOUDHURY, H., C. W. CARLSON, AND G. SEMENIUKZ. 1971. A study of ochratoxin toxicity in hens. Poultry Sci., 50:1855-1859.

CLIFFORD, J. I., AND K. R. REES. 1967. The action of aflatoxin B_1 on rat liver. Biochem. J., 102:65-75.

DOERR, J. A., AND M. A. OTTINGER. 1979. Impaired reproductive capacity during aflatoxicosis in young Japanese quail. Poultry Sci., 58:1051.

———. 1980. Delayed reproductive development resulting from aflatoxicosis in juvenile Japanese quail. Poultry Sci., 59:1995-2001.

DONALDSON, W. E., H. T. TUNG, AND P. B. HAMILTON. 1972. Depression of fatty acid synthesis in chick liver (*Gallus domesticus*) by aflatoxin. Comp. Biochem. Physiol., 41B:843-847.

EGBUNIKE, G. N. 1979. The effects of micro doses of aflatoxin B_1 on sperm production rates, epididymal sperm abnormality and fertility. Zentralblatt Vet. Med., 26:66-72.

FORD, R. E., B. J. JACOBSEN, AND D. G. WHITE. 1978. Mycotoxins-environmental contaminants in nature. Illinois Res., 20:10-11.

GALTIER, P., C. BARADAT, AND M. ALVINERIE. 1977. De L'elimination d'ochratoxine A par le lait chez la lapine. Ann. Nutr. Aliment., 31:911-918.

GARLICH, J. D., H. T. TUNG, AND P. B. HAMILTON. 1973. The effects of short term feeding aflatoxin on egg production and some plasma constituents of the laying hen. Poultry Sci., 52:2206-2211.

GOPAL, T., F. W. OEHME, T. F. LIAO, AND C. L. CHEN. 1980. Effects of intratesticular aflatoxin B_1 on rat testes and blood estrogens. Tox. Letters., 5:263-267.

GREENMAN, D. L., R. G. MEHTA, AND J. L. WITTLIFF. 1979. Nuclear interaction of *Fusarium* mycotoxins with estradiol binding sites in the mouse uterus. J. Tox. Envir. Health, 5:593-598.

HAMILTON, P. B., AND J. D. GARLICH. 1971. Aflatoxin as possible cause of fatty liver syndrome in laying hens. Poultry Sci., 50:800-804.

HAYES, A. W., R. D. HOOD, AND H. L. LEE. 1974. Teratogenic effects of ochratoxin A in mice. Teratology, 9:93-98.

HEATHERINGTON, A. C., AND H. RAISTRICK. 1931. Biochemistry of microorganisms. XIV. Production and chemical constitution of a new yellow coloring matter, citrinin, produced from dextrose by *Penicillium citrinum* Thom. Trans. Roy. Soc. London, B220:269-295.

HOOD, R. D., M. J. NAUGHTON, AND A. W. HAYES. 1976a. Prenatal effects of ochratoxin in hamsters. Teratology, 13:11-14.

HOOD, R. D., A. W. HAYES, AND J. G. SCAMMELL. 1976b. Effects of prenatal administration of citrinin and viriditoxin to mice. Food Cosmet. Toxicol., 14:175-178.

HUDSON, D. 1966. Fitting segmented curves whose points have to be estimated. J. Amer. Stat. Assoc., 61:316.

HUFF, W. E., R. D. WYATT, T. L. TUCKER, AND P. B. HAMILTON. 1974. Ochratoxicosis in the broiler chicken. Poultry Sci., 53:1585-1591.

HUFF, W. E., R. D. WYATT, AND P. B. HAMILTON. 1975a. Effects of dietary aflatoxin on certain egg yolk parameters. Poultry Sci., 54:2014-2018.

———. 1975b. Nephrotoxicity of dietary ochratoxin A in broiler chickens. Appl. Microbiol., 30:48-51.

JAYARAJ, A. P., A. MURTI, V. SREENIVASAMURTHY, AND H. A. B. PARPIA. 1969. Toxic effects of aflatoxin on the testes of albino rats. J. Anat. Soc. India, 17:101-104.

JERROLD, M., J. WARD, M. SONTAG, E. K. WEISBURGER, AND C. BROWN. 1975. Effects of life time exposure to aflatoxin in rats. J. Nat. Cancer Inst., 55:107-113.

JORDON, W. H., AND W. W. CARLTON. 1978a. Citrinin mycotoxicosis in the rat. I. Toxicology and pathology. Food Cosmet. Toxicol., 16:431-439.

———. 1978b. Citrinin mycotoxicosis in the rat. II. Clinicopathological observations. Food Cosmet. Toxicol., 16:441-447.

———. 1978c. Citrinin mycotoxicosis in the Syrian hamster. Food Cosmet. Toxicol., 16:355-363.

KATZENELLENBOGEN, B. S., J. A. KATZENELLENBOGEN, AND D. MORDECAI. 1979. Zearalenones: characterization of the estrogenic potencies and receptor interactions of a series of fungal β-resorcyclic acid lactones. Endocrinology, 105:33-40.

KRATZER, F. H., D. BANDY, M. WILEY, AND A. N. BOOTH. 1969. Aflatoxin effects in poultry. Proc. Soc. Exp. Biol. Med., 131:1281-1284.

KROUGH, P., B. HALD, AND C. J. PEDERSEN. 1973. Occurrence of ochratoxin A and citrinin in cereals associated with mycotoxic porcine nephropathy. Act. Path. Microbiol. Scand., B81:689-695.

LALOR, J. H., T. D. KIMBROUGH, AND G. C. LLEWELLYN. 1978. Induction of duodenal serotonin production by dietary sodium selenite and aflatoxin B_1. Food Cosmet. Toxicol., 16:611-613.

MARKS, H. L., AND C. W. BACON. 1976. Influence of *Fusarium*-infected corn and F-2 on laying hens. Poultry Sci., 55:1864-1870.

MORE, F., AND R. CAMGUILHEM. 1979. Effect of low doses of ochratoxin A after intratesticular injection in the rat. Experentia, 35:890-892.

MUNRO, I. C., P. M. SCOTT, C. A. MOODIE, AND R. F. WILLES. 1973. Ochratoxin A— occurrence and toxicity. J. Amer. Vet. Med. Assoc., 163:1269-1273.

NIP, W. K., AND F. S. CHU. 1979. Fate of ochratoxin A in goats. J. Envrion. Sci. Health, B14:319-333.

OTTINGER, M. A., AND J. A. DOERR. 1980. Early influence of aflatoxin upon sexual maturation in the male Japanese quail. Poultry Sci., 59:1750-1754.

PALYUSIK, M., AND E. KARLIC-KOVACS. 1975. Effect on laying geese of feeds containing the fusariotoxins T-2 and F-2. Acta Vet. Acad. Sci. Hung., 25:363-368.

PALYUSIK, M., E. KOPLINE-KOVACS, AND E. GUZSAL. 1971. A *Fusarium graminearum* Latasa a gunarok es a polykakakasok ondotermelzesene. Magyar Allatorv. Lapja., 26:300-303.

PHILLIPS, R. D., AND A. W. HAYES. 1978. Effect of the mycotoxin citrinin on composition of mouse liver and kidney. Toxicol., 16:351-359.

PIER, A. C., J. L. RICHARD, AND S. J. CYSEWSKI. 1980. Implications of mycotoxins in animal disease. J. Amer. Vet. Med. Assoc., 178:719-724.

PONG, R. S., AND G. N. WOGAN. 1969. Time course of alterations of rat liver polysomes induced by aflatoxin B_1. Biochem. Pharmacol., 18:2357-2361.

PRIOR, M. G., AND C. S. SISODIA. 1978. Ochratoxicosis in White Leghorn hens. Poultry Sci., 57:619-623.

PRIOR, M. G., J. B. O'NEILL, AND C. S. SISODIA. 1978. Effects of ochratoxin A on production characteristics and hatchability of Japanese quail. Can. J. Anim. Sci., 58:29-33.

RICHIR, C., M. MARTINEAUD, J. TOURY, AND H. DUPIN. 1964. Observations sur des accidents de toxicoses fungiques survenue des élevages de canetoues. Nutr. Dieta, 6:229-233.

RICHIR, C., J. TOURY, M. MARTINEAUD, AND H. DIPIN. 1965. Sur les effets cancerigenes de régimes contenant des arachides contaminées. C. R. Soc. Biol., 158:1375-1378.

ROBINSON, T. S., C. J. MIROCHA, H. J. KUTZ, J. C. BEHRENS, M. S. CHI, G. A. WEAVER, AND S. D. NYSTROM. 1979. Transmission of T2 toxin into bovine and porcine milk. J. Dairy Sci., 62:637-641.

SAWHNEY, D. S., D. V. VADEHRA, AND R. C. BAKER. 1973. Aflatoxicosis in the laying Japanese quail (*Coturnix coturnix japonica*). Poultry Sci., 52:465-473.

SHERWOOD, R. F., AND J. F. PEBERDY. 1973. Effects of zearalenone on the developing male chick. Brit. Poult. Sci., 14:127-129.

SHOTWELL, O. L., M. L. GOULDEN, C. W. HESSELTINE, J. W. DICKENS, AND W. F. KWOLEK. 1980. Aflatoxin: distribution in contaminated corn plants. Cereal Chem., 57:206-208.

SIMS, W. M., JR., D. C. KELLEY, AND P. E. SANFORD. 1970. A study of aflatoxicosis in laying hens. Poultry Sci., 49:1082-1084.

STILL, P. E., A. W. MACKLIN, W. E. RIBELIN, AND E. B. SMALLEY. 1971. Relationship of ochratoxin A to foetal death in laboratory and domestic animals. Nature, 234:563-564.

SZCZECH, G. M., AND R. D. HOOD. 1978. Animal model of human disease: ochratoxicosis in dogs, mice, pigs, and rats. Amer. J. Pathol., 91:689-692.

TUNG, H. T., W. E. DONALDSON, AND P. B. HAMILTON. 1972. Altered lipid transport during aflatoxicosis. Toxicol. Appl. Pharmacol., 22:97-104.

WEAVER, G. A., H. J. KURTZ, C. J. MIROCHA, F. Y. BATES, J. C. BEHRENS, T. S. ROBINSON, AND W. F. GIPP. 1978a. Mycotoxin-induced abortions in swine. Can. Vet. J., 19:72-74.

WEAVER, G. A., H. J. KURTZ, F. Y. BATES, M. S. CHI, C. J. MIROCHA, J. C. BEHRENS, AND T. S. ROBINSON. 1978b. Acute and chronic toxicity of T-2 mycotoxin in swine. Vet. Rec., 103:531-535.

WOGAN, G. N., AND M. A. FRIEDMAN. 1965. Effects of pyrrolase induction in rat liver. Proc. Fed. Amer. Soc. Exp. Biol., 24:627.

WYATT, R. D., J. R. HARRIS, P. B. HAMILTON, AND H. R. BURMEISTER. 1972a. Possible field outbreaks of furariotoxicoses in avians. Avian Diseases, 16:1123-1129.

WYATT, R. D., B. A. WEEKS, P. B. HAMILTON, AND H. R. BURMEISTER. 1972b. Severe oral lesions in chickens caused by ingestion of dietary fusariotoxin T-2. Appl. Microbiol., 24:251-257.

WYATT, R. D., W. M. COLWELL, P. B. HAMILTON, AND H. R. BURMEISTER. 1973. Neural disturbances in chickens caused by dietary T-2 toxin. Appl. Microbiol., 26:757-761.

WYATT, R. D., J. A. DOERR, P. B. HAMILTON, AND H. R. BURMEISTER. 1975. Egg production, shell thickness, and other physiological parameters of laying hens affected by T-2 toxin. Appl. Microbiol., 29:641-645.

SECTION IV

Endocrinology of Calcium Metabolism and Its Clinical Relevance

INTRODUCTION

ALEXANDER D. KENNY

Department of Pharmacology and Therapeutics, Texas Tech University Health Sciences Center, Lubbock, Texas 79430 USA

Our current concept of the major endocrine systems involved in mammalian and therefore in human calcium homeostasis embraces three major hormones and three main target tissues. Two of the hormones are hypercalcemic in action (parathyroid hormone and 1,25-dihydroxyvitamin D_3); the third, calcitonin, is hypocalcemic in action. The chief target organs of these calcemic hormones are bone, gut, and kidney. The most important question that we wish to address is: How can avian studies further our understanding of the endocrine control of calcium homeostasis in the human?

Calcium metabolism in birds is characterized by several aspects which distinguish it from that found in humans. First and foremost, certain avian species, particularly those which have been domesticated, such as the chicken and the Japanese quail, have the potential for laying an egg daily for extended periods. This represents a significant daily loss of approximately 10 per cent of the total body stores of calcium and points to the fact that such species must have efficient endocrine systems for coping with such a daily loss. Perturbations of calcium metabolism in the human might call into play homeostatic responses which, because of their subtlety, thus far remain undiscovered. On the other hand, such perturbations in avian species may bring forth homeostatic responses that are more overt and therefore more readily discovered. Armed with such discoveries, it then becomes simpler to seek such homeostatic responses in mammalian species, including the human.

In the four papers to follow, several endocrine aspects of calcium metabolism are explored in two avian species, namely the chicken and Japanese quail. Much of the work described is broadly related to avian reproductive function; some of the findings have direct relevance to the problem of post-menopausal osteoporosis in women.

Luck and Scanes describe studies in which the important relationship between dietary calcium and ovulation in the chicken was investigated. They confirmed the fact that egg production is regulated by calcium availability, and further suggest that there is a direct interaction between the blood ionized calcium concentration and the endocrine and neuroendocrine functions of the pituitary and hypothalamus as well as between vitamin D metabolism and the production of the calcemic hormones. The requirements of bone appear to have priority over those of reproduction. When chickens are subjected to calcium deprivation, ovulation ceases before structural bone becomes seriously depleted. These authors suggest that future studies should focus on the relationships between the physiological fluctuations in plasma LH, 1,25-dihydroxyvitamin D_3, and plasma ionized calcium.

Dacke and Kenny studied the possible relationship between the prostaglandins and vitamin D metabolism in the Japanese quail. None of the prostaglandins tested (PGE_2, PGE_1, or $PGF_{2\alpha}$) showed any effect on plasma calcium levels within a four-hour period following injection. Similarly, there were no consistent effects on vitamin D metabolism. Administration of the prostaglandin synthetase inhibitor, indomethacin, also led to no significant effects on vitamin D metabolism.

Kenny discusses the suitability of the Japanese quail for studying the endocrine aspects of calcium metabolism, and cites two studies from his own laboratory to indicate this point. First was the discovery that ovulation leads to increased production of the hormonal form of vitamin D_3, 1,25-dihydroxyvitamin D_3, and the observation that injection of an estrogen also leads to increased production of the hormone. Secondly, egg-laying Japanese quail may be used as a source of preparing the gut cytosolic receptor for 1,25-dihydroxyvitamin D_3. This receptor preparation was used both for kinetic studies and for the development of a radioreceptor assay for the determination of 1,25-dihydroxyvitamin D levels in plasma.

Turner and associates describe an avian cell culture model that has proved useful for studies of vitamin D metabolism, particularly with respect to the metabolic fate of 25-hydroxyvitamin D_3 and tissue distribution of the 1- and 24-hydroxylases.

Reprinted from
ASPECTS OF AVIAN ENDOCRINOLOGY:
PRACTICAL AND THEORETICAL IMPLICATIONS (C. G. Scanes *et al.*, eds.)
Grad. Studies, Texas Tech Univ., 1982, 26:1-411.

AVIAN MODEL FOR STUDIES OF THE VITAMIN D ENDOCRINE SYSTEM

ALEXANDER D. KENNY

Department of Pharmacology and Therapeutics, Texas Tech University Health Sciences Center, Lubbock, Texas 79430 USA

CURRENT CONCEPTS OF CALCIUM HOMEOSTASIS

Mammalian Calcium Metabolism

The salient points of calcium metabolism in the human adult are summarized in Fig. 1 in which it is assumed that the dietary intake of calcium is 800 mg/day. It is additionally assumed that the subject is in calcium balance, that is, the daily dietary intake of calcium equals the daily urinary and fecal loss of calcium. Under normal circumstances, less than 50 per cent of the dietary calcium passes from the intestinal contents into the systemic circulation. The major endocrine factor regulating the intestinal absorption of calcium is the hormonal form of vitamin D, 1,25-dihydroxyvitamin D_3 [1,25-$(OH)_2D_3$. The total and ionic calcium concentrations in the blood are held within very narrow limits: approximately 10.0 and 4.5 mg/dl, respectively. The skeleton, which contains over 99 per cent of the body calcium, serves as a reservoir. Approximately 500 mg of calcium enter and leave this reservoir per day. All three calcemic hormones, namely parathyroid hormone (PTH), calcitonin (CT), and 1,25-$(OH)_2D_3$, serve regulatory functions in this process. There is an endogenous loss of calcium over which there is little endocrine control. This is represented by the calcium loss into the gastrointestinal tract (secretions, cell loss, and the like) and the filtered calcium that is not reabsorbed by the renal tubules. Additional losses can occur during pregnancy into the fetus and in lactation into the milk but these amount to no more than 2.5 and 7 per cent, respectively, of the total calcium in the maternal skeleton. Hence the total loss of calcium by these routes amounts to no more than 10 per cent of the total maternal calcium stores over an 18-month pregnancy and lactation period. This is in sharp contrast to the daily loss occurring in the egg-laying female bird.

Our current concept of the major endocrine systems involved in mammalian calcium homeostasis is summarized in Fig. 2, in which the roles of the two hypercalcemic hormones [PTH and 1,25-$(OH)_2D_3$] and the single hypocalcemic hormone (CT) are depicted. The latter hormone in mammals has its origin in the thyroid C cells. The major target organs of these calcemic hormones are bone, gut, and kidney. Whereas bone is considered the main target organ of PTH, this hormone also has important effects on the kidney. PTH increases renal tubular reabsorption of calcium and thereby conserves calcium for the body. In addition, PTH acts on the kidney to increase the

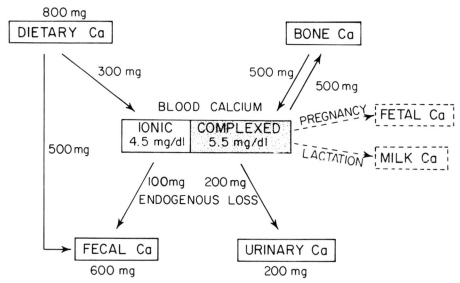

Fig. 1.—Calcium metabolism in the human adult. The quantities of calcium are turnover rates per day. The total calcium loss to the fetus and in the milk during an 18-month pregnancy and lactation period amounts to no more than 10% of the maternal body stores of calcium.

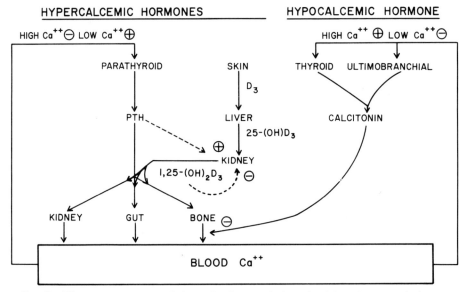

Fig. 2.—The current concept of the endocrine regulation of calcium homeostasis in mammals. Parathyroid hormone (PTH) enhance the renal production of 1,25-(OH)₂D₃, the hormonal form of vitamin D₃.

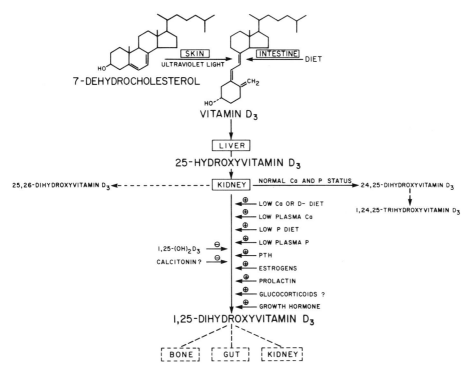

FIG. 3.—Vitamin D metabolism and its regulation. Various nutritional, physiological, and pharmacological factors have been shown to regulate the renal synthesis of 1,25-(OH)$_2$D$_3$.

production of 1,25-(OH)$_2$D$_3$, the hormonal form of vitamin D. Thus, indirectly, PTH can influence gut calcium transport. The liver metabolite, 25-hydroxyvitamin D[25-(OH)D], circulates in the plasma at high concentrations (30 ng/ml) relative to those of 1,25-(OH)$_2$D (30 pg/ml). At physiological concentrations the liver metabolite is considered inactive. Thus, it is poised in the plasma waiting for the signal to activate it to the hormonal form. PTH appears to be one of the main, if not the only, physiological regulators of the renal vitamin D endocrine system, although many other nutritional, physiological, and pharmacological factors have been proposed (Fig. 3). The current concept of the interrelationship of PTH and 1,25-(OH)$_2$D$_3$ in response to the imposition of a low calcium diet is presented in Fig. 4. Calcitonin's main physiological role appears to be inhibition of bone resorption; its actions on other target tissues involved in calcium homeostasis are considered negligible or non-existent. The kidney is rapidly becoming recognized as playing a pivotal role in calcium homeostasis both as a target organ and as an endocrine gland itself.

Avian Calcium Metabolism

Calcium metabolism in birds is characterized by several aspects that distinguish it from that found in many other vertebrates. Firstly, the physiolog-

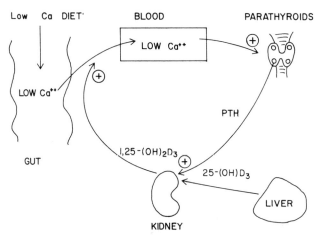

Low Ca DIET·   BLOOD   PARATHYROIDS

LOW Ca⁺⁺

LOW Ca⁺⁺

PTH

1,25-(OH)₂D₃

GUT

25-(OH)D₃

LIVER

KIDNEY

Fig. 4.—Interrelationships between parathyroid hormone (PTH) and vitamin D in response to a low calcium diet. The low calcium diet leads to less calcium absorbed and a consequent fall in plasma ionic calcium. The latter leads to stimulation of the secretion of PTH which in turn causes an increase in the conversion of 25-hydroxyvitamin D_3 to its active metabolite, 1,25-$(OH)_2D_3$. The latter enters the circulation and increases the efficiency of gut calcium absorption which leads to a rise in plasma ionic calcium and a suppression of PTH secretion.

ical role of CT in avian species is unknown (Kenny, 1972). No overt effects of CT on calcium metabolism have been demonstrated in avian species. Nevertheless, the avian ultimobranchial gland is capable of secreting CT *in vitro* (Feinblatt *et al.*, 1973), and high concentrations of bioassayable CT are found in the plasmas of several avian species (Kenny, 1971; Dacke *et al.*, 1972; Boelkins and Kenny, 1973). Calcium challenge increases the plasma CT concentration several-fold within 30 minutes after administration of the calcium to both mature and immature male Japanese quail (Boelkins and Kenny, 1973). Secondly, birds, as do other lower vertebrates, exhibit a hypercalcemic response to estrogen (Baksi and Kenny, 1977a, 1977b, 1977c) and to reproductive activity in the female (Dacke *et al.*, 1973; Boelkins and Kenny, 1973). This extra calcium in the plasma is complexed; the ionic calcium concentrations remain normal (Dacke *et al.*, 1973). It is assumed that the calcium is complexed to yolk precursor proteins released by the liver in response to estrogen. Thirdly, avian species develop a specialized bone in the medullary cavities of the long bones prior to the onset of egg-laying. This medullary bone may be induced pharmacologically by the injection of both androgen and estrogen to immature birds (Dacke, 1979). Fourthly, certain avian species, particularly those which have been domesticated, have the potential for laying an egg daily for extended periods. This represents a daily loss of approximately 10 per cent of the total body calcium stores. Many of these facets peculiar to avian calcium metabolism are illustrated in Fig. 5.

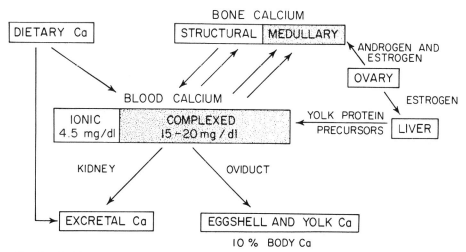

FIG. 5.—Calcium metabolism in egg-laying birds. Avian species have at least four characteristics of calcium metabolism not found in mammals. 1) The egg-laying bird loses as much as 10% of its body stores of calcium every day in the form of the egg. 2) Yolk protein precursors, synthesized by the liver under the influence of estrogen, complex plasma calcium so that the total calcium can rise to very high levels leaving the plasma ionic calcium at normal levels. 3) Under the influence of both estrogen and androgen, medullary bone is formed in the cavities of the long bones. 4) Calcitonin, although circulating at high concentrations in the plasma, has no overt effects on avian calcium metabolism. The increase in total plasma calcium and the formation of medullary bone occur physiologically in the egg-laying bird but these effects may be induced pharmacologically in immature birds or in adult male birds.

THE JAPANESE QUAIL AS AN AVIAN MODEL FOR STUDYING CALCIUM METABOLISM

The author has maintained a breeding colony of Japanese quail (*Coturnix coturnix japonica*) for use in his laboratory since 1968. The breeding colony is normally programmed to produce between 100 and 200 Japanese quail chicks per week. This entails the setting of 250 to 400 eggs per week in the incubator (assuming a hatchability of approximately 50 per cent). Recent breeding data over a 12-week period is given in Table 1. During the past decade we have used Japanese quail to study various aspects of calcium metabolism associated with the three major calcemic hormones, namely CT, PTH, and vitamin D (Abel de la Cruz *et al.*, 1980*a*, 1980*b*, 1981; Baksi and Kenny, 1977*a*, 1977*b*, 1977*c*, 1978*a*, 1978*b*, 1978*c*, 1979*a*, 1979*b*, 1979*c*, 1979*d*, 1980; Baksi *et al.*, 1978; Boelkins and Kenny, 1973; Dacke and Kenny, 1973; Dacke *et al.*, 1972, 1973, 1976; Kenny, 1971, 1972, 1975, 1976, 1979; Kenny and Baksi, 1976, 1978; Kenny and Dacke, 1974; Kenny and Pang, 1980; Kenny *et al.*, 1972, 1974, 1976; Losty *et al.*, 1974; Pang *et al.*, 1980; Pollard *et al.*, 1980). What are some of the advantages of the Japanese quail as a model for studies of avian calcium metabolism?

The Japanese quail, like the chicken, is a prolific egg-layer. This means that the reproductively active female is subjected to daily calcium losses

TABLE 1.—*Hatching records for a Japanese quail breeding colony over a 12-week period.*

Hatch date	No. of eggs incubated	No. of quail chicks hatched	Hatchability
8/21/80	375	190	51%
8/28/80	388	220	57%
9/4/80	382	200	52%
9/11/80	311	191	61%
9/18/80	310	183	59%
9/25/80	305	148	49%
10/2/80	327	175	54%
10/9/80	287	178	62%
10/16/80	300	140	47%
10/23/80	365	199	55%
10/30/80	350	183	52%
11/6/80	350	184	53%
Mean	338	183	54%

representing 10 per cent of her body stores of calcium. It is to be expected that calcium losses of this magnitude must be associated with efficient homeostatic mechanisms for coping with this stress. It can be assumed that these mechanisms are more pronounced and hopefully more easily discovered in this model than in mammalian species. Nevertheless, armed with the knowledge of their existence in the avian model, the investigator can search for their presence in mammals where they may exist covertly. This is the main rationale for using such a model in studies of calcium metabolism. Japanese quail have other characteristics that give them an advantage over the chicken. The quail has a short life cycle. The egg incubation time is 17 days, and by six weeks of age the female begins laying eggs. By eight weeks of age the female is laying at a maximal rate. Immature birds are readily sexed by the appearance of their breast feathers; the feathers are speckled in the female. The adult bird weighs approximately 150 grams. The latter characteristic permits storage of hundreds of birds in a small space. The food consumption of the adult Japanese quail is obviously much less than that of the chicken. The size of the bird, although not ideal for certain specific uses, is adequate for most purposes. The birds easily can be injected intravenously using the wing vein, and blood samples, adequate for most analyses, can be obtained by cardiac puncture under halothane anesthesia. They are particularly suited for studies requiring large numbers of adult birds.

AVIAN REPRODUCTION AND THE VITAMIN D ENDOCRINE SYSTEM

An important part of our research efforts during the past decade has been devoted to a study of the regulation of the renal vitamin D endocrine system in the Japanese quail. These studies emanated from our initial observation that reproductive activity in female Japanese quail leads to increased production of 1,25-$(OH)_2D_3$, the hormonal form of vitamin D_3 (Kenny, 1975, 1976; Kenny *et al.*, 1974). Similar observations were made in the chicken (Losty *et al.*, 1974). In these series of experiments, homogenates of kidneys removed from reproductively active female Japanese quail were incubated with tritiated 25-(OH)D$_3$ and the metabolites were extracted and identified by chromagraphic methods. Kidneys removed from birds with and without an egg in the oviduct revealed that ovulation resulted in enhanced production of 1,25-(OH)$_2$D$_3$ (Fig. 6). Further examination of this phenomenon in relation to the ovulatory cycle revealed that 1,25-(OH)$_2$D$_3$ production is enhanced throughout the 24 hours following ovulation (Fig. 7). Particularly important is the finding that its synthesis is already enhanced during the first six hours after ovulation, at a time before any calcification of the eggshell begins. If, following oviposition, no ovulation occurs, then 1,25-(OH)$_2$D$_3$ production rapidly decreases within the first hours following oviposition. This study revealed for the first time a physiological state, namely the reproductive period in the female bird, in which endogenous control over 1,25-(OH)$_2$D$_3$ production is exhibited without any previous manipulation (dietary, surgical, pharmacological, and the like) of the animals.

In reporting this study (Kenny, 1976) we posed the following questions. In the case of the reproductively active female bird, do the gonadal hormones serve directly or indirectly as messengers involved in the control of the renal 25-hydroxyvitamin D$_3$-1-hydroxylase system? Might a certain qualitative and quantitative combination of gonadal hormones stimulate the synthesis of this renal enzyme? We suggested that this possibility was worthy of study and that the egg-laying Japanese quail represents a convenient model for studying the physiological and biochemical control of the renal vitamin D endocrine system. In 1973, experiments were initiated in my laboratory at the University of Missouri with the aim of answering these questions. Japanese quail were injected with various combinations of estradiol and progesterone, either singly or in combination, with the hope of demonstrating pharmacological effects on the renal vitamin D endocrine system. The move of my laboratory to The University of Texas Medical Branch in Galveston delayed progress until 1975 when Dr. Samarendra N. Baksi joined my laboratory and was able to continue this line of research. This led to a series of experiments in which the gonadal hormones, particularly estradiol, were injected into immature Japanese quail, both male and female, under chronic (4 weeks), sub-acute (3 days), and acute (single dose) injection conditions. The studies have resulted in a series of reports (Baksi and Kenny, 1977*a*, 1977*b*, 1977*c*, 1978*a*, 1978*b*). In the sub-acute study (Baksi and Kenny, 1977*a*, 1977*b*) immature males and females were injected with either estradiol benzoate (3 mg/kg) or progesterone (5 mg/kg) daily for three days. The birds were killed on the fourth day and *in vitro* production by kidney homogenates of 1,25-(OH)$_2$D$_3$ and 24,25-(OH)$_2$D$_3$ was determined. The data for the immature male Japanese quail are presented in Fig. 8. Estradiol resulted in a marked increase in 1,25-(OH)$_2$D$_3$ production coupled with a suppression of 24,25-(OH)$_2$D$_3$ production. The effects of the hormone treatments on other parameters are presented in Table 2. In the acute injection studies (Baksi and

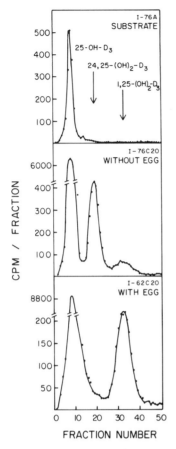

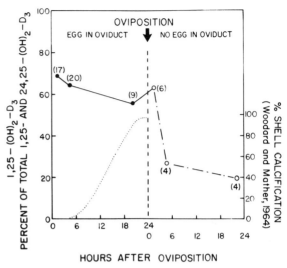

FRACTION NUMBER

FIG. 6 (left).—Vitamin D metabolism in the adult female
bird with and without an egg in the oviduct. Kidney homogenates were prepared from the birds and incubated with
the substrate, 25-hydroxyvitamin D_3, and the metabolites
produced were separated by Sephadex LH-20 chromatography. Upper panel: substrate alone. Middle panel: chromatographic profile of the metabolites from a bird without
an egg reveals that the major metabolite is the relatively
inactive form of vitamin D, 24,25-dihydroxyvitamin D_3.
Lower panel: chromatographic profile of the metabolites
from a bird with an egg in the oviduct indicating that the
major metabolite is the active hormonal form, 1,25-
dihydroxyvitamin D_3. Taken from Kenny (1976).

FIG. 7 (right).—*In vitro* renal production 1,25-$(OH)_2$-D_3, expressed as a percent 1,25- and 24,25-
$(OH)_2$-D_3 biosynthesis, at different periods (1-3, 3-6, and 18-24 hours) after oviposition in Japanese quail with (closed circles connected by an unbroken line) and without (open circles connected by a broken line) an egg in the oviduct. Ovulation, when it occurs, is within 30 minutes
after oviposition (Woodard and Mather, 1964). The production of 1,25-$(OH)_2$-D_3 is already
enhanced (69 and 54%) at 1-3 and 3-6 hours after oviposition/ovulation (left-side of figure) at a
time before calcification of the eggshell has begun, and remains enhanced throughout the 24-
hour ovulatory cycle. If oviposition is not followed immediately by ovulation (right side of figure) then 1,25-$(OH)_2$-D_3 production, although still enhanced 1-3 hours after oviposition, falls
rapidly and significantly ($P<0.05$) to 27 and 20% at 3-6 and 18-24 hours after oviposition, respectively. Data, taken from Woodard and Mather (1964) and indicating eggshell calcification rate,
are plotted on left side of figure (dotted line). Number in parentheses represents number of birds
contributing data to mean value. Taken from Kenny (1976).

Kenny, 1978*b*), immature female Japanese quail received a single intramuscular injection of estradiol benzoate. The renal 1-and 24-hydroxylase activities were studied at one, four, 12, and 24 hours after the injection. Three
doses of estradiol benzoate were used, namely 0.01, 0.1, and 1.0 mg/kg. The
data are presented in Fig. 9. Doses as low as 0.01 mg/kg were able to stimulate the 1-hydroxylase activity and suppress the 24-hydroxylase activity at 12

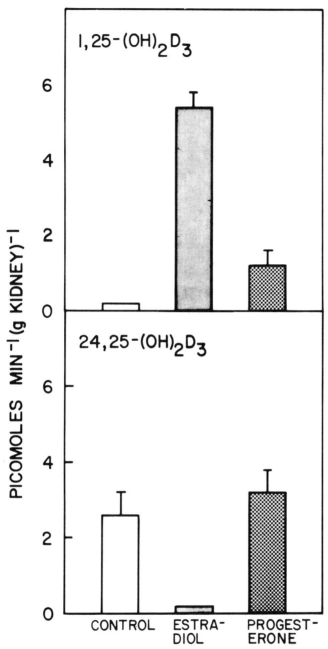

FIG. 8.—Vitamin D metabolism in immature male Japanese quail injected with estradiol or progesterone for 3 days. Upper panel: estradiol stimulates the production of 1,25-dihydroxyvitamin D_3 whereas progesterone gave a much smaller response. Lower panel: estradiol depressed the synthesis of the less active metabolite, 24,25-dihydroxyvitamin D_3. Taken from Baksi and Kenny (1977a).

TABLE 2.—*Effect of estradiol and progesterone injections on oviduct, testis, kidney, and femur weights and plasma calcium and inorganic phosphate in immature 4-week-old female and male Japanese quail. Values given are means ± SE. Data from Baksi and Kenny (1977a, b).*

Treatment	No. of birds	Body weight (g)	Oviduct or testis weight (g/100g)	Kidney weight (g/100g)	Plasma Ca (mg/100 ml)	Plasma P (mg/100 ml)	Femur weight (mg/cm)
			Females				
Oil control	10	76±3	0.13±0.003	0.82±0.07	10.3±0.5	6.2±0.8	30.1±1.6
Estradiol[a]	5	75±3	1.29±0.06	1.40±0.06*	53.7±3.2*	30.1±1.8	33.0±1.2
Progesterone[b]	5	71±2	0.14±0.003	1.42±0.04*	11.5±0.4	8.2±0.9	33.7±1.1
			Males				
Oil control	10	65±3	0.12±0.05	1.24±0.06	10.2±0.4	7.7±0.7	25.9±0.8
Estradiol[a]	5	66±3	0.15±0.004	1.21±0.06	46.7±2.8*	30.7±5.7*	30.1±1.1*
Progesterone[b]	5	63±2	0.24±0.04	1.35±0.07	9.6±0.9	10.1±1.2	30.2±1.4*

[a]Estradiol benzoate, 3 mg/kg intramuscular injection daily for 3 days.
[b]Progesterone, 5 mg/kg intramuscular injection daily for 3 days.
*$P<0.01$ vs. appropriate controls.

and 24 hours. Stimulation of the 1-hydroxylase activity occurred as quickly as four hours after injection with the 1.0 mg/kg dose. A subsequent study has shown that the lowest dose (0.01 mg/kg) is within the physiological range; plasma estradiol levels following this dose do not rise above levels encountered physiologically (Kenny, 1979). The response of the renal vitamin D endocrine system to estradiol injection can be blocked by prior administration of an antiestrogen, tamoxifen (Baski and Kenny, 1979c). The response of the renal vitamin D endocrine system to PTH, which is a well-established regulator of the system (Baksi and Kenny, 1979b), is not blocked by tamoxifen (Baksi and Kenny, 1979a).

These observations in the Japanese quail led us to conclude that reproductive activity in the female or administration of exogenous estradiol lead to increased production of 1,25-$(OH)_2D_3$, observations which have been supported and confirmed by many other laboratories (for detailed review, see Kenny, 1981). Do these findings have any clinical relevance? Gallagher *et al.* (1979) have reported that, in a study of postmenopausal women, there are differences in both plasma concentrations of 1,25-$(OH)_2D$ and in fractional calcium absorption between such women with osteoporosis and those without osteoporosis (Table 3). Postmenopausal women with osteoporosis had lower levels of both calcium absorption and plasma 1,25-$(OH)_2D$. In addition, Gallagher *et al.* (1978) have reported that administration of estrogen (conjugated estrogens, 1,25 and 2.5 mg/day cyclically for six months) to postmenopausal osteoporotic women resulted in a significant ($P<0.005$) rise in the concentration of 1,25-$(OH)_2D$ in the plasma from a pre-treatment mean value of 22.6 to 36.3 pg/ml, a value which is indistinguishable from the value of 35.2 pg/ml found in age-matched postmenopausal women without osteoporosis (Table 4). Is it possible that estrogen deficiency results in deficient 1,25-$(OH)_2D$ production and that the latter contributes to the development of osteoporosis in postmenopausal women? There may be a

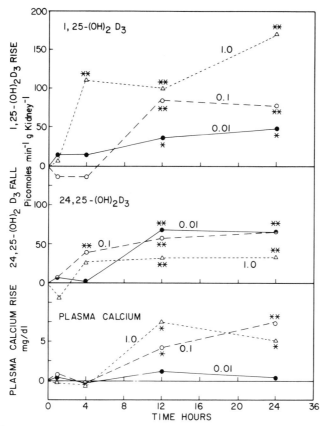

Fig. 9.—Effect of estradiol benzoate injection on *in vitro* renal metabolism of tritiated 25-(OH)D₃ and on plasma calcium in normal immature (4-week-old) female Japanese quail. Estradiol benzoate was injected at 0.01 (●---●), 1.0 (O---O), or 1.0 (△---△) mg/kg intramuscularly. The kidneys were removed 1, 4, 12 or 24 hours later, homogenized and incubated with 10^{-7} M [³H]-25 (OH)D₃ for 20 minutes. The rise in 1,25-(OH)₂D₃ or fall in 24,25-(OH)₂D₃ production is expressed in pmoles minute^{-1} (g kidney)$^{-1}$ over control birds receiving vehicle alone. Asterisks indicate significant responses. (* $P<0.05$; ** $P<0.01$). Taken from Baksi and Kenny (1978*b*).

pathophysiological link between estrogen and vitamin D status in the postmenopausal condition. On the other hand, in a recent study by Baran *et al.* (1980) in young normal women, plasma estradiol and 1,25-(OH)₂D concentrations were measured at two points in the menstrual cycle (day three and day 13); there was no correlation between the estradiol and 1,25-(OH)₂D concentrations (Table 5). Thus, there appears to be little relationship between these two endocrine factors in the young healthy woman; this does not preclude the existence of a relationship in the postmenopausal woman.

RADIORECEPTOR ASSAY FOR 1,25-DIHYDROXYVITAMIN D

The active form of vitamin D₃, namely 1,25-(OH)₂D₃, in common with other steroid hormones, binds to cytosolic receptors present in target cells.

TABLE 3.—*Calcium absorption and plasma vitamin D metabolite concentrations in postmeno-pausal osteoporotic patients and age-matched postmenopausal control women. Data from Gallagher et al. (1979). Values are means ± SE.*

| Subjects | | Ca absorption (fraction/6h) | Plasma concentrations | |
Type	Number		25-(OH)D (ng/ml)	1,25-(OH)$_2$D (pg/ml)
Control	20	0.61±0.02	15.9±1.3	33.2±2.3
Osteoporotic	27	0.49±0.02**	19.5±1.2	25.9±1.5*

*$P < 0.005$
**$P < 0.001$

The steroid-receptor complex moves, by a temperature-dependent process, to the nucleus where it binds to chromatin material and stimulates the transcription of mRNA's involved in the translational synthesis of a protein or proteins necessary for calcium transport (DeLuca, 1979; Norman, 1979; Bikle et al., 1979). In the past, cytosolic receptors for 1,25-(OH)$_2$D$_3$ have been prepared mostly from intestinal mucosa isolated from immature chickens that were fed a vitamin D-deficient diet for several weeks (Brumbaugh et al., 1974a, 1974b; Eisman et al., 1976; Lambert et al., 1978; Lund and Sorensen, 1979; Mason et al., 1980).

We have developed a sensitive and specific radioreceptor assay for 1,25-(OH)$_2$D$_3$ incorporating several important innovations (Abel de la Cruz et al., 1980a, 1980b; Pollard et al., 1980). First, Japanese quail (rather than chicken) intestinal mucosa was used as the source of receptor. Secondly, egg-laying females were used instead of immature birds. Thirdly, the birds were maintained on a normal diet rather than subjecting them to the time-consuming, and, in the case of immature Japanese quail, often lethal treatment of several weeks on a vitamin D-deficient diet. A description of the kinetic characteristics of the Japanese quail gut cytosolic receptor and of selected applications of the radioreceptor assay to the determination of plasma 1,25(OH)$_2$D concentrations follows.

Kinetic Properties of Japanese Quail Gut Cytosolic Receptor

The receptor-chromatin complex technique developed by Haussler and his associates (Brumbaugh et al., 1974a, 1974b) for the separation of bound from free 1,25-(OH)$_2$D$_3$ was used. The duodenum, freed of pancreas and mesentery, was removed from reproductively active female Japanese quail. The duodenum was emptied of its contents and rinsed. The preparation, as it was purified in subsequent steps, was maintained at 4°C or as close to that temperature as possible. The mucosa was scraped from the serosa and rinsed in buffer solution. The minced mucosa was homogenized with a Potter-Elvehjem homogenizer provided with a Teflon pestle and centrifuged to yield a supernatant (crude cytosol) and a pellet (crude nuclei) fraction. Further centrifugations, culminating in an ultracentrifugation step, yielded a purified cytosolic preparation which was stored until it was required for

TABLE 4.—*Effects of estrogen and calcitriol on calcium absorption and plasma 1,25-(OH)₂D concentrations in postmenopausal osteoporotic women. Data from Gallagher et al. (1978). Values are means ± SE.*

Subject	Treatment	Time[a]	No. of patients	Ca absorption (fraction/6h)	Plasma 1,25-(OH)₂D (pg/ml)
Osteoporotic	Placebo	Before	9	0.47±0.03	
		After	9	0.49±0.02	
Osteoporotic	Calcitriol[b]	Before	10	0.50±0.03	
		After	10	0.66±0.03**	
Osteoporotic	Estrogen[c]	Before	11	0.52±0.08	22.6±3.0
		After	11	0.64±0.13**	36.3±4.5*
Normal[d]	None		20	0.61±0.02	35.2±2.0

[a]Treatment period was 6 months.
[b]Calcitriol [1,25-(OH)₂D₃]: 0.4 µg/day.
[c]Estrogen: conjugated estrogens, 1.25 to 2.5 mg/day cyclically.
[d]Normal: age-matched postmenopausal women *without* osteoporosis.
*$P<0.005$, **$P<0.001$

mixing with the chromatin preparation. The 1,200 × G pellet obtained in an earlier centrifugation step was resuspended in buffer solution and subjected to successive centrifugations, resuspensions, and homogenizations to yield a clear final chromatin pellet following centrifugation at 65,000 × g. The cytosolic receptor fraction was thawed and half of its volume was used to resuspend the chromatin pellet. The cytosolic receptor-chromatin suspension was homogenized, forced through a 22 gauge needle, separated into small aliquots, frozen immediately in liquid nitrogen, and stored in a freezer at −20°C.

Radioreceptor Assay Method

Tritiated and unlabelled, 1,25(OH)₂D₃ was dissolved in 10 µl of 95 per cent ethanol at appropriate concentrations and was added to the assay tubes. Normally, 41 pg (0.1 pmole) of tritiated 1,25-(OH)₂D₃ with a specific activity of 82 Ci/mmole were added to each tube yielding a total count of approximately 18,000 dpm. To each assay tube, 100 µl of the cytosolic receptor-chromatin preparation was added. The tubes were incubated for 25 minutes at 25°C. At the end of the incubation period, ice-cold buffer was added to each tube which was immediately placed in ice. The contents of each tube were applied uniformly to a glass fiber filter supported in a 12-place filter manifold. The glass fiber filter had been previously washed with two milliliters of buffer solution under a gentle vacuum. An additional two milliliters of buffer solution were added to each assay tube, mixed, and applied to each filter. The side of each filter cup was washed with two milliliters of buffer solution and the vacuum was increased until the filter appeared dry. The glass fiber filter was removed from the manifold, dried at 105°C for one hour or at room temperature overnight, and placed in 5.0 milliliters of scintillation fluid and counted.

TABLE 5.—*Plasma estradiol and 1,25-dihydroxyvitamin D concentrations on days 3 and 13 of the menstrual cycle. Data from Baran et al. (1980). Values are means ± SE.*

Menstrual cycle time	No. of subjects	Plasma estradiol (pg/ml)	Plasma 1,25-(OH)$_2$D (pg/ml)
Day 3	12	63±9	47±3
Day 13	12	229±65*	51±3

*$P < 0.025$

Optimal Conditions.—The optimal incubation time varied according to the temperature (Fig. 10). At 25°C, the per cent specific binding had reached a maximum of 93 per cent or more between 15 and 30 minutes. An incubation time of 25 minutes at 25°C was selected for routine use as the most convenient conditions yielding greater than 90 per cent binding.

The effect of pH on the binding of 1,25-(OH)$_2$D$_3$ to the receptor-chromatin preparation was examined over a pH range from 6.5 to 7.6. Two maxima for specific binding were revealed, one at pH 6.6 and the other at pH 7.4. A pH of 7.4 was selected for routine use as the specific binding was less variable at the higher pH value.

Standard Curve.—A typical curve using the Japanese quail intestinal receptor-chromatin preparation is presented in Fig. 11. The limit of detection is 5 pg/tube and the working range is from 5 to 80 pg/tube.

Specificity Study.—The data obtained from a study of the specificity of the radioreceptor assay are presented in Fig. 12. The affinity of the receptor for 1,25-(OH)$_2$D$_3$ was 1,000 times greater than that for 25-hydroxyvitamin D$_3$ and 24,25-dihydroxyvitamin D$_3$ and 1.0×10^6 times greater than that for vitamin D$_3$.

Kinetic Analysis of Binding.—The relationship of specific binding to hormone concentration was studied over a range of 0.001 to 8.0 nM of 1,25-(OH)$_2$D$_3$. The specific binding sites were saturable at concentrations of 2 nM and above. A Scratchard plot of these data reveals a straight line indicating a single binding site and an equilibrium dissociation constant (K_D), calculated from the slope of the line, of 0.73 nM.

Application of Radioreceptor Assay to the Determination of Plasma 1,25-Dihydroxyvitamin D Concentration

The radioreceptor assay for 1,25-(OH)$_2$D developed in our laboratory using egg-laying Japanese quail gut cytosolic receptor has been applied to the determination of 1,25-(OH)$_2$D concentrations in plasma obtained from human subjects, rats, and Japanese quail (Pollard *et al.*, 1980; Abel de la Cruz *et al.*, 1981) as well as in rainbow trout (Kenny, Uchiyama, and Pang, unpubl. observ.). A 2.0 ml plasma sample is extracted with methanol:methylene chloride (2:1) according to the method of Shephard *et al.* (1979). After extraction, the sample is chromatographed on a column (0.7 cm by 18 cm) of Sephadex LH-20 using a chloroform:hexane (65:35) solvent system. The sample is further purified by high pressure liquid chromotography on a

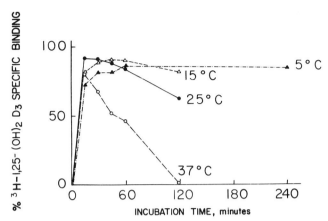

FIG. 10.—Specific binding of [23,24,-³H]-1,25-(OH)₂D₃ with egg-laying Japanese quail intestinal receptor-chromatin preparation expressed as percent of total binding. Specific binding is defined as the difference between the binding in the absence and presence of a 100-fold excess of added unlabelled 1,25-(OH)₂D₃. A total of 41 pg (18,000 dpm) of [23,24-₃H]-1,25-(OH)₂D₃, dissolved in 10 μl of 95% ethanol, was added to each tube together with 100 μl of the intestinal receptor-chromatin preparation. Each point represents the mean of duplicate tubes which were incubated at pH 7.4 for periods varying from 15 to 240 minutes and at temperatures of 5 (closed triangles), 15 (open triangles), 25 (closed circles), and 37°C (open circles). These data led to the selection of 25°C for 25 minutes as the incubation conditions for routine use; these conditions result in 93% specific binding. Taken from Abel de la Cruz and Kenny (unpubl. observ.).

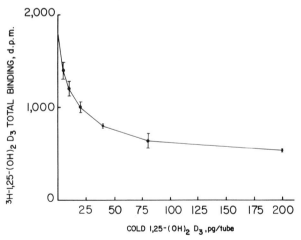

FIG. 11.—Standard curve, relating mean ± SE total binding of tracer [23,24-³H]-1,25-(OH)₂D₃ to increasing amounts of unlabelled 1,25-(OH)₂D₃ per tube, has a working range of 5 to 80 pg of 1,25-(OH)₂D₃ per tube. Incubation conditions were 25°C for 25 minutes at pH 7.4. Taken from Abel de la Cruz and Kenny (unpubl. observ.).

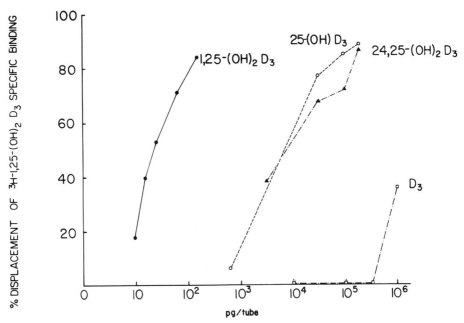

FIG. 12.—Specificity of 1,25-(OH)₂D₃ binding with the egg-laying quail intestinal receptor-chromatin preparation. To each tube containing 41 pg of [23,24-³H]-1,25-(OH)₂D₃, were added varying amounts of unlabelled 1,25-(OH)₂D₃, 25-(OH)D₃, 24,25-(OH)₂D₃, or vitamin D₃. The per cent displacement of the labelled 1,25-(OH)₂D₃ is plotted on the vertical axis. The affinity of 1,25-(OH)₂D₃ (closed circles) for the receptor-chromatin preparation was 1,000 times greater than that for either 25-(OH)D₃ (open circles) or 24,25-(OH)₂D₃ (closed triangles) and 1.0×10^6 greater than that for vitamin D₃ (open squares). Incubation conditions were 25°C for 25 minutes at pH 7.4. Taken from Abel de la Cruz and Kenny (unpubl. observ.).

Zorbax Sil column using propan-2-ol:hexane (1:10) at a flow rate of 1.5 ml/min. The data are presented in Table 6. Recoveries of added tracer 1,25-(OH)₂D₃ were all above 60 per cent; the plasma concentrations reported have been corrected for loss. Normal human values range between 20 and 56 pg/ml. These are close to the values reported by others in the literature (Kenny, 1981). In rats, normal values were found to range between 16 and 44 pg/ml. Application of a low calcium diet for 14 days significantly increased the mean value from 30 to 62 pg/ml. Normal immature male Japanese quail had a mean concentration of 18 pg/ml (only two samples). Again, imposition of a low calcium diet increased this mean value significantly to 64 pg/ml. The most striking finding was the extremely high concentrations of plasma 1,25-(OH)₂D found in rainbow trout. The mean value obtained from four pooled plasma samples was 242 ± 114.5 (SD) pg/ml, the individual values being 100, 209, 287, and 369 pg/ml respectively. Examination of kidney homogenates prepared from the same rainbow trout revealed that the renal 1-hydroxylase activity was undetectable. The high plasma 1,25-(OH)₂D concentrations in the face of a low production rate suggest that the elimina-

TABLE 6.—*Plasma 1,25-dihydroxyvitamin D concentrations in various species. Data from Abel de la Cruz* et al. *(1981) and Kenny* et al. *(1981).*

Species	Status	Diet	No. of samples	Recovery [%(range)]	Plasma 1,25-(OH)$_2$D [pg/ml(range)]
Human	Adult males	Normal Ca	3	84(75-100)	24(20-28)
	Adult females	Normal Ca	2	98(94-102)	38(20-56)
Rat	Adult males	Normal Ca	3	85(65-94)	30(16-44)
	Adult males	Low Ca (14 days)	4	78(66-90)	62(29-97)*
Jap. quail	4-wk males	Normal Ca	2	86(80-93)	18(18-18)
	4-wk males	Low Ca (10 days)	3	78(59-91)	64(32-88)**
Rainbow trout	Mixed sex (130 g)		4	69(62-75)	242(100-369)

*$P<0.05$, **$P<0.001$

tion rate of 1,25-(OH)$_2$D is very slow in such species, a phenomenon which is not unexpected in poikilotherms.

SUMMARY

The bird represents an excellent model for studying the endocrine aspects of calcium metabolism. The Japanese quail, in particular, is suited for such studies on the basis that it: 1) is capable of laying an egg every 24 hours over an extended period and thus is subjected to a daily loss amounting to 10 per cent of its body stores of calcium; 2) has a short life cycle involving an incubation period of 17 days and reaches maturity between six and eight weeks of age; 3) permits, because of its relatively small size (150 g at maturity), the storage of large numbers of birds in a relatively small space; 4) has a food consumption in the adult stages which is much less than that of the chicken and therefore is less expensive to maintain; and 5) can be readily sexed in the immature stage by the characteristic speckled feathers on the breast of the female.

Two major uses of the Japanese quail as an avian model for studies of calcium metabolism were described. Firstly, the discovery that ovulation leads to increased production of the hormonal form of vitamin D$_3$ led to further work which uncovered the observation that injection of an estrogen, such as estradiol, also leads to increased production of 1,25-(OH)$_2$D$_3$ by the kidney. These findings have some relevance to the problem of postmenopausal osteoporosis in women. Others have subsequently shown that postmenopausal women with osteoporosis have lower plasma 1,25-(OH)$_2$D concentrations than similar women without the bone disease, and that administration of estrogen to osteoporotic postmenopausal women will indeed increase the plasma 1,25-(OH)$_2$D concentrations to levels indistinguishable from those in postmenopausal women without osteoporosis. Secondly, the egg-laying Japanese quail was used as a source for preparing the gut cytosolic receptor for 1,25-(OH)$_2$D$_3$. The kinetic characteristics of this steroid-receptor interaction were determined and found to be similar to those of the chicken gut cytosolic receptor. The radioreceptor assay for 1,25-(OH)$_2$D, using the Japanese quail gut cytosolic receptor preparation, was applied to the determination of 1,25-(OH)$_2$D concentrations in the plasmas of human subjects, rats, Japanese quail, and rainbow trout. Imposition of a low calcium diet to both rats and Japanese quail increased plasma 1,25-(OH)$_2$D concentrations. The most interesting finding was that rainbow trout have extremely high levels of 1,25-(OH)$_2$D (242 pg/ml) associated with an undetectable rate of renal 1,25-(OH)$_2$D$_3$ production.

ACKNOWLEDGMENTS

The author is indebted to the collaboration of several colleagues whose contributions are duely acknowledged in the publications cited. In addition, the encouragement and helpful advice freely proffered by Dr. Mark R. Haussler of the Department of Biochemistry, University of Arizona College of Medicine, in the initial phase of the radioreceptor work, is greatly appreciated. These investigations were supported in part by the National Institutes of Health, the Dalton Research Center of the University of Missouri at Columbia, and the Instituto Nacional de Higiene Rafael Rangel, Venezuela.

LITERATURE CITED

ABEL DE LA CRUZ, L. A., C. G. DACKE, A. D. KENNY, AND S. K. POLLARD. 1980a. Modification of radioreceptor assay for 1,25-dihydroxyvitamin D using gut cytosolic receptor from adult male or immature Japanese quail. Brit. J. Pharm., 70:P179-P180.

ABEL DE LA CRUZ, L. A., J.-C. SHIEH, S. K. POLLARD, AND A. D. KENNY. 1980b. New radioreceptor assay for 1,25-dihydroxyvitamin D$_3$ using Japanese quail gut cytosolic receptor. Fed Proc., 35:559.

ABEL DE LA CRUZ, L. A., S. K. POLLARD, AND A. D. KENNY. 1981. Plasma 1,25-dihydroxyvitamin D levels determined by new radioreceptor assay using egg-laying Japanese quail gut cytosolic receptor. Pp. 366, in Hormonal Control of calcium metabolism (D. V. Cohn, R. V. Talmage, and J. L. Matthews, eds.), Exerpta Medica, Amsterdam, xviii+506 pp.

BAKSI, S. N., AND A. D. KENNY. 1977a. Vitamin D$_3$ metabolism in immature Japanese quail: effects of ovarian hormones. Endocrinology, 101:1216-1220.

———. 1977b. Influence of reproductive status on in-vitro renal production of 1,24,25-trihydroxyvitamin D$_3$ in Japanese quail. J. Reprod. Fert., 50:175-177.

———. 1977c. The effects of administration of antiestrogen (tamoxifen) in vitro on the metabolism of 25-hydroxyvitamin D$_3$ in vitro in the Japanese quail. Biochem. Pharm., 26:2439-2443.

———. 1978a. Vitamin D metabolism in Japanese quail: gonadal hormones and dietary calcium effects. Amer. J. Physiol., 234:E622-E628.

———. 1978b. Acute effects of estradiol on the renal vitamin D hydroxylases in Japanese quail. Biochem. Pharm., 27:2765-2768.

———. 1978c. Effect of lead ingestion on vitamin D$_3$ metabolism in Japanese quail. Res. Commun. Chem. Path. Pharmacol., 21:375-378.

———. 1979a. Parathyroid hormone stimulation of 1,25-dihydroxyvitamin D$_3$ production in antiestrogen-treated Japanese quail. Mol. Pharmacol., 16:932-940.

———. 1979b. Acute effects of parathyroid extract on renal vitamin D hydroxylases in Japanese quail. Pharmacology, 18:169-174.

———. 1979c. Vitamin D metabolism in Japanese quail: effects of lead exposure and dietary calcium. Toxicol. Appl. Pharmacol., 51:489-495.

———. 1979d. Vitamin D metabolism in aged Japanese quail: dietary calcium and estrogen effects. Calc. Tiss. Int., 28:172.

———. 1980. Estradiol-induced stimulation of 25-hydroxyvitamin D_3-1-hydroxylase in vitamin D-deficient Japanese quail. Pharmacology, 20:298-303.

BAKSI, S. N., A. D. KENNY, S. M. GALLI-GALLARDO, AND P. K. T. PANG. 1978. Vitamin D metabolism in bullfrogs and Japanese quail: effects of estradiol and prolactin. Gen. Comp. Endocrinol., 35:258-262.

BARAN, D. T., M. P. WHYTE, M. HAUSSLER, L. J. DEFTOS, E. SLATOPOLSKY, AND L. V. AVIOLI. 1980. Effect of the menstrual cycle on calcium-regulating hormones in the normal young woman. J. Clin. Endocrinol. Metab., 50:377-379.

BIKLE, D. D., R. L. MORRISSEY, AND D. T. ZOLOCK. 1979. The mechanism of action of vitamin D in the intestine. Amer. J. Clin. Nutr., 32:2322-2338.

BOELKINS, J. N., AND A. D. KENNY. 1973. Plasma calcitonin levels in Japanese quail. Endocrinology, 92:1754-1760.

BRUMBAUGH, P. F., D. H. HAUSSLER, R. BRESSLER, AND M. R. HAUSSLER. 1974a. Radioreceptor assay for 1α,25-dihydroxy vitamin D_3. Science, 183:1089-1091.

BRUMBAUGH, P. F., D. H. HAUSSLER, K. M. BURSAE, AND M. R. HAUSSLER. 1974b. Filter assay for 1α,25-dihydroxyvitamin D_3. Utilization of the hormone's target tissue chromatin receptor. Biochemistry, 13:4091-4097.

DACKE, C. G. 1979. Calcium regulation in sub-mammalian vertebrates. Academic Press, London, xv+222 pp.

DACKE, C. G., AND A. D. KENNY. 1973. Avian bioassay method for parathyroid hormone. Endocrinology, 92:463-470.

DACKE, C. G., J. N. BOELKINS, W. K. SMITH, AND A. D. KENNY. 1972. Plasma calcitonin levels in birds during the ovulation cycle J. Endocrinol., 54:369-370.

DACKE, C. G., X. J. MUSACCHIA, W. A. VOLKERT, AND A. D. KENNY. 1973. Cyclical fluctuations in the levels of blood calcium, pH and pCO_2 in Japanese quail. Comp. Biochem. Physiol., 44A:1267-1275.

DACKE, C. G., B. J. A. FURR, J. N. BOELKINS, AND A. D. KENNY. 1976. Sexually related changes in plasma calcitonin levels in Japanese quail. Comp. Biochem. Physiol., 55A:341-344.

DeLUCA, H. F. 1979. Recent advances in our understanding of the vitamin D endocrine system. J. Steroid Biochem., 11:35-52.

EISMAN, J. A., A. J. HAMSTRA, B. E. KREAM, AND H. F. DeLUCA. 1976. A sensitive, precise, and convenient method for determination of 1,25-dihydroxyvitamin D in human plasma. Arch. Biochem. Biophys., 176:235-243.

FEINBLATT, J. D., L. G. RAISZ, AND A. D. KENNY. 1973. Secretion of avian ultimobranchial calcitonin in organ culture. Endocrinology, 93:277-284.

GALLAGHER, J. C., B. L. RIGGS, A. HAMSTRA, AND H. F. DeLUCA. 1978. Effect of estrogen therapy on calcium absorption and vitamin D metabolism in postmenopausal osteoporosis. Clin. Res., 25:415A.

GALLAGHER, J. C., B. L. RIGGS, J. EISMAN, A. HAMSTRA, S. B. ARNAUD, AND H. F. DeLUCA. 1979. Intestinal calcium absorption and serum vitamin D metabolites in normal subjects and osteoporotic patients: effect of age and dietary calcium. J. Clin. Invest., 64:729-736.

KENNY, A. D. 1971. Determination of calcitonin in plasma by bioassay. Endocrinology, 89:1005-1013.

———. 1972. Introductory remarks. Pp. 9-11, in Calcium, parathyroid hormone and the calcitonins (R. V. Talmage and P. L. Munson, eds.), Exerpta Medica, Amsterdam, x+560 pp.

———. 1975. Regulation of vitamin D metabolism in egg-laying birds. Pp. 408-410, in Calcium-regulating hormones (R. V. Talmage, M. Owen, and J. A. Parsons, eds.), Exerpta Medica, Amsterdam, xiii+486 pp.

———. 1976. Vitamin D metabolism: physiological regulation in egg-laying Japanese quail. Amer. J. Physiol., 230:1609-1615.

————. 1979. Plasma estradiol levels in Japanese quail: relationship to renal vitamin D metabolism. Calc. Tiss. Int., 28:175.

————. 1981. Intestinal calcium absorption and its regulation. CRC Press, Boca Raton.

KENNY, A. D., AND S. N. BAKSI. 1976. Time of activation of renal 25-hydroxyvitamin D_3-1-hydroxylase in ovulating Japanese quail. Fed. Proc., 35:662.

————. 1978. Stimulation of renal-vitamin D endocrine system by physiological doses of estradiol in Japanese quail. Fed. Proc., 37:926.

KENNY, A. D., AND C. G. DACKE. 1974. The hypercalcaemic response to parathyroid hormone in Japanese quail. J. Endocrinol., 62:51-53.

KENNY, A. D., AND P. K. T. PANG. 1980. Failure of bPTH (1-34) to be inactivated by oxidation with hydrogen perioxide. Calc. Tiss. Int., 31:73.

KENNY, A. D., J. N. BOELKINS, C. G. DACKE, W. R. FLEMING, AND R. C. HANSON. 1972. Plasma calcitonin levels in birds and fish. Pp. 39-47, *in* Endocrinology 1971 (S. Taylor, ed.), William Heinemann Medical Books, London, xiv+509 pp.

KENNY, A. D., J. LAMB, N. R. DAVID, AND T. A. LOSTY. 1974. Regulation of vitamin D metabolism in egg-laying birds. Fed. Prod., 33:679.

KENNY, A. D., D. J. AHEARN, AND J. F. MAHER. 1976. Improved method for determining parathyroid hormone in biological material. Biochem. Med., 16:201-210.

LAMBERT, D. W., D. O. POTT, S. F. HODGSON, E. A. LINDMARK, B. J. WITRAK, AND B. A. ROOS. 1978. An improved method for measurement of 1,25-$(OH)_2D_3$ in human plasma. Endocrinol. Res. Comm., 5:293-310.

LOSTY, T. A., H. V. BIELLIER, AND A. D. KENNY. 1974. Regulation of vitamin D metabolism in egg-laying chickens. Program 56th Meeting Endocrine Soc., Atlanta, 1974, p. A-268.

LUND, B., AND O. H. SORENSEN. 1979. Measurement of circulating 1,25-dihydroxyvitamin D in man. Changes in serum concentration during treatment with 1α-hydroxycholecalciferol. Acta Endocrinol., 91:338-350.

MASON, R. S., D. LISSNER, H. S. GRUNSTEIN, AND S. POSEN. 1980. A simplified assay for dihydroxylated vitamin D metabolites in human serum: application to hyper- and hypovitaminosis D. Clin. Chem., 26:444-450.

NORMAN, A. W. 1979. Vitamin D: the calcium homeostatic hormone. Academic Press, New York, xvii+490 pp.

PANG, P. K. T., C. M. YANG, AND A. D. KENNY. 1980. The distinction between hypotensive and hypercalcemic actions of bovine parathyroid hormone. Calc. Tiss. Int., 31:74.

POLLARD, S. K., L. A. ABEL DE LA CRUZ, J.-C. SHIEH, AND A. D. KENNY. 1980. Application of a new radioreceptor assay for 1,25-dihydroxyvitamin D_3. Fed. Proc., 35:559.

SHEPARD, R. M., R. L. HORST, A. J. HAMSTRA, AND H. F. DeLUCA. 1979. Determination of vitamin D and its metabolites in plasma from normal and anephric man. Biochem. J., 182:55-69.

WOODARD, A. E., AND F. B. MATHER. 1964. The timing of ovulation, movement of the ovum through the oviduct, pigmentation and shell deposition in Japanese quail. Poultry Sci., 43:1427-1432.

Reprinted from
ASPECTS OF AVIAN ENDOCRINOLOGY:
PRACTICAL AND THEORETICAL IMPLICATIONS (C. G. Scanes *et al.*, eds.)
Grad. Studies, Texas Tech Univ., 1982, 26:1-411.

PROSTAGLANDINS: ARE THEY INVOLVED IN AVIAN CALCIUM HOMEOSTASIS?

CHRISTOPHER G. DACKE[1] AND ALEXANDER D. KENNY[2]

[1]*Department of Physiology, Marischal College, University of Aberdeen, Aberdeen, Scotland;* and [2]*Department of Pharmacology and Therapeutics, Texas Tech University Health Sciences Center, Lubbock, Texas 79430 USA*

The interaction of prostaglandins with mammalian calcium metabolism has been recognized for about a decade since the discovery that Prostaglandin E_2 (PGE_2) can stimulate resorption of ^{45}Ca from mouse bone *in vitro* (Klein and Raisz, 1970). Shortly after this discovery, it became apparent that hypercalcemia associated with many solid neoplastic diseases is not necessarily a result of excessive ectopic parathyroid hormone (PTH) secretion. Thus Seyberth *et al.* (1975) demonstrated that elevated PGE levels in blood plasma from patients suffering from solid neoplastic diseases were associated with raised plasma calcium levels. Furthermore, this hypercalcemia could be controlled with aspirin or indomethacin, both of which drugs are inhibitors of the enzyme prostaglandin synthetase (Flower, 1974).

Another classical paper concerning the hyercalcemic role of PGE_2 in mammals was that of Franklin and Tashjian (1975). They infused the PG into rats for four hours and observed the plasma calcium levels. Modest (10%) but significant increases in the plasma calcium level were obtained with PGE_2 at doses of 130 ng/min.

A probable mechanism for the hypercalcemic response to PGE_2 is increased bone resorption as observed in early studies of bone cultured *in vitro* by Klein and Raisz (1970) and also by Dietrich *et al.* (1975). These studies were carried out using bones of mammalian origin. Unfortunately, there do not appear to be any reports in the literature concerning the interaction between prostaglandins and avian bone.

Apart from a serving role as a bone resorptive factor, prostaglandins can also stimulate secretion of other calciotropic hormones or *vice versa*, although evidence for such effects is as yet flimsy. Thus Wark *et al.* (1979) reported a stimulatory *in vitro* effect of indomethacin (14 μM) and of cyclic AMP (1 mM) on production of 1,25-dihydroxyvitamin D_3 [1,25-$(OH)_2D_3$] from 25-hydroxyvitamin D_3 in chick kidney homogeonates. They also reported a stimulatory effect of PGE (10 μg/ml) on cyclic AMP production in this preparation, thus suggesting a tenuous link between prostaglandins and vitamin D_3 metabolism.

Meanwhile, in another report, Hammond and Ringer (1978) found that when indomethacin (50 mg/kg) was injected intravenously into egg-laying chickens (four to six hours before oviposition of the last egg of a sequence and one hour after intravenous injection of bovine LH), oviposition was delayed by about 15 hours and the egg had a significantly thickened shell. At the same time, plasma calcium levels, measured 11, 24, and 32 hours after indomethacin, were significantly lowered (by about 30%). The subsequent ovulation induced by the LH injection was not blocked by the indomethacin.

In a recent series of papers, Hertelendy and his colleagues have been measuring prostaglandin levels in the plasma of birds using specific radioimmunoassay procedures. Hertelendy and Biellier (1978) found PGE levels (the assay does not distinguish between PGE_1 and PGE_2) to be higher in the plasma of egg-laying chickens on days when an egg was in the oviduct than on days when the birds did not ovulate. Furthermore, within one hour of oviposition, there was a marked fall in the PGE concentrations. Plasma PGE levels, however, increased drastically in the last 30 minutes prior to oviposition (Wechsung *et al.*, 1978).

Similarly, Hertelendy and Biellier (1978) measured PGE and Prostaglandin F (PGF) levels in the plasma of Japanese quail and found both to be significantly higher in laying than in non-laying hens.

More recently Hammond *et al.* (1980) reported in more detail changes in plasma PGE and PGF levels during the chicken's ovulatory cycle. In particular, they found a sustained rise in levels of both prostaglandins between six and 16 hours after ovulation, that is, during the period of eggshell calcification. This had led Hertelendy (pers. comm.) to suggest that these changes may reflect the physiological mechanism responsible for initiation of mobilization of skeletal calcium.

Thus, we have good evidence that PGE might function in abnormal and possibly normal calcium metabolism in mammals. The evidence linking prostaglandins with calcium metabolism in birds is rather tenuous. However, the hypocalcemic response of egg-laying hens to the PG synthetase inhibitor, indomethacin, linked with current knowledge of plasma prostaglandin changes in egg-laying hens, suggests this as a fertile area for further research.

The following sections describe further investigations of the possible role of prostaglandins in avian calcium metabolism. Experiments have been conducted dealing with: 1) plasma calcium responses to prostaglandins or indomethacin in immature birds; 2) the effect of prostaglandins or indomethacin on vitamin D_3 metabolism in immature birds; and 3) the effect of indomethacin on plasma calcium levels in hens.

MATERIALS AND METHODS

Japanese quail for the vitamin D experiments were selected from a closed colony maintained at the Texas Tech University Health Sciences Center. They were fed Purina Game Bird Startena diet (Ralston Purina, St. Louis, Missouri) until the age of two weeks at which point half were transferred to the low calcium diet for two weeks (Baksi and Kenny, 1978), and the other half remained on the original diet.

The quail used for the other experiments examining blood calcium responses were obtained from a colony maintained in the Department of Physiology, University of Aberdeen. These birds were fed turkey starter crumbs (Nitrovit Ltd., Dalton, Thirsk, Yorkshire). Egg-laying hens received a turkey laying ration from the same source.

Day-old male chickens (White Leghorns), obtained from a hatchery near Aberdeen, were also fed the turkey starter diet.

All birds were maintained on 14-hour light/10-hour dark cycles. They were weighed at the start of the experiments and then caged individually.

Prostaglandins (PGE_1, PGE_2, and $PGF_{2\alpha}$), and indomethacin were obtained from the Sigma Chemical Company (St. Louis, Missouri). Each was dissolved in ethanol as a stock solution and then diluted 10-fold in 0.9% saline for injection. For the vitamin D experiments, prostaglandins were injected intravenously; subcutaneous injections were used for all other experiments. Estradiol benzoate (estradiol-3-benzoate), obtained from General Biochemicals (Chagrin Falls, Ohio), was dissolved in corn oil prior to intramuscular injection.

In vitro kidney incubation and separation of vitamin D metabolites procedures were carried out following the methods of Baksi and Kenny (1978). Estimates of $1,25-(OH)_2D_3$ production were made. Plasma calcium determinations were carried out by atomic absorption spectrophotometry using a Perkin-Elmer Model 303 spectrometer. All data are presented as mean response ± standard error of mean (SE), and Students' t-test was used to test for significance.

RESULTS

Plasma Calcium Responses to Prostaglandins and Indomethacin

Data from two experiments involving subcutaneous injection of PGE_1, PGE_2, $PGF_{2\alpha}$ or indomethacin into either immature (one-week-old) chickens or four-week old Japanese quail are shown in Table 1. In no case was there any significant change in plasma calcium level noted when compared with control birds at four hours after the injection. In the quail experiment, however, some ($P<0.05$) difference was seen between $PGF_{2\alpha}$- and PGE_2-injected groups, but this was not seen in the chicken experiment.

When higher (triple) doses of prostaglandins were used, they generally were lethal within a few minutes.

Effect of Prostaglandins and Indomethacin on Vitamin D Metabolism in Japanese Quail

Four experiments were carried out. In the first, PGE_2 (20 μg/bird), $PGF_{2\alpha}$ (20 μg/bird), or indomethacin (4 mg/bird), was injected into Japanese quail (weight range 68-96 grams) that had been fed a normal calcium (Purina Game Bird Startena) diet. A control group was injected with estradiol benzoate, a substance previously shown to enhance conversion of 25-hydroxyvitamin D_3 to $1,25-(OH)_2D_3$ in birds (Baksi and Kenny, 1977; Pike *et al.*, 1978).

TABLE 1.—*Plasma calcium responses of immature birds to prostaglandins or indomethacin.*

Species	Weight range (g)	Treatment	Dose/bird	No. of birds	Plasma Ca Mean±SE (mg/dl)
Chickens	80-85	Control		5	9.37±0.37
		PGE$_2$	40 μg	5	9.80±0.12
		PGF$_{2\alpha}$	40 μg	5	9.37±0.32
		Indomethacin	20 mg	5	10.21±0.37
Japanese quail	34-42	Control		5	9.82±0.24
		PGE$_2$	20 μg	5	10.05±0.21
		PGF$_{2\alpha}$	20 μg	5	8.90±0.37*
		Indomethacin	8 mg	5	9.49±0.25

All injections were subcutaneous (0.5 ml/bird); control injections were vehicle only (0.9% saline with 10% ethanol).

Plasma calcium levels were determined 4 hours after injection.

*Significantly different from PGE$_2$-injected group ($P<0.05$).

All drug doses are per bird; PG = prostaglandin; IND = indomethacin; E$_2$B = estradiol benzoate; LCD = low calcium diet.

E$_2$B was given 24 hours before kidney homogenate incubation; all other drugs and controls were injected 4 hours before kidney incubation.

1,25-(OH)$_2$D$_3$ production shown as mean ± SE; 24,25-(OH)$_2$D$_3$ production was not detectable in any experiment.

*Significantly different from control group ($P<0.05$).

The birds were anesthetized and the kidneys removed four hours after the injection (24 hours in the case of the estradiol benzoate-treated group) and incubated. In this experiment, however, production of the dihydroxylated metabolites, 1,25-(OH)$_2$D$_3$ and 24,25-(OH)$_2$D$_3$, proved to be minimal and undetectable for most of the birds. None of the treated groups showed any significant increase in production of either metabolite when compared with the control group.

Attention was therefore focused upon quail which had been fed the low calcium diet prior to incubation of the kidney homogenates, in an effort to enhance production of 1,25-(OH)$_2$D$_3$. Three experiments were carried out under these conditions.

Several birds injected with prostaglandins (20 μg/bird) died in the first experiment (A). Therefore, in experiment B, in the remaining birds receiving one third of this dose, 1,25-(OH)$_2$D$_3$ was detectable (Table 2). This metabolite was perhaps slightly increased by PGE$_2$, while two surviving PGF$_{2\alpha}$-treated birds showed lowered production of the metabolite. Indomethacin had no apparent effect in this experiment. Estradiol benzoate had the expected stimulatory effect on 1,25-(OH)$_2$D$_3$ production.

In two further experiments using the lower (non-lethal) doses of PG (6.7 μg/bird) we further examined these responses, but results from these two experiments (Table 2) again were not conclusive. Thus, whereas in experiment B all treatments had a variable effect on vitamin D metabolism, in experiment C all three prostaglandins showed a tendency to enhance 1,25-(OH)$_2$D$_3$ production, although none were significant. As shown in Table 2, the protocol between these two experiments differed slightly with respect to age of birds and time on the low calcium diet. Similarly, estradiol benzoate, which as mentioned previously is known to stimulate 1,25-(OH)$_2$D$_3$ produc-

TABLE 2.—*Effect of prostaglandins or indomethacin on vitamin D_3 metabolism in calcium-deprived Japanese quail.*

Experiment	Weight range (g)	Days on LCD	Treatment group	Dose	Route	No. of birds	1,25-$(OH)_2D_3$ production Mean ± SE (pmole $min^{-1}g^{-1}$)
A	66-104	11	Control			3	15.28± 1.50
			E_2B	0.1 mg	I.M.	4	27.93± 2.07*
			PGE_2	6.7 µg	I.V.	3	24.93± 4.10
			$PGF_{2\alpha}$	6.7 µg	I.V.	2	9.20± 1.09*
			IND	4.0 mg	I.P.	4	18.72± 5.72
B	98-123	8	Control			6	36.92± 5.18
			PGE_1	6.7 µg	I.V.	4	36.72± 7.57
			PGE_2	6.7 µg	I.V.	4	28.89± 6.43
			$PGF_{2\alpha}$	6.7 µg	I.V.	5	37.63± 3.30
C	83-116	12	Control			5	55.02± 5.45
			PGE_1	6.7 µg	I.V.	5	61.24± 5.23
			PGE_2	6.7 µg	I.V.	2	73.79±29.94
			E_2B	0.1 mg	I.M.	4	73.41±11.74

tion in Japanese quail, showed only a slight stimulatory effect in the third experiment, although the percentage increase was of a similar magnitude to that found in Experiment A.

Effect of Indomethacin on Plasma Calcium Levels of Egg-Laying Birds

The hypocalcemic response to indomethacin (8 mg/bird I.P.) at four hours in egg-laying Japanese quail hens, is shown in Fig. 1. Hens used in this experiment were palpated first to determine the presence or absence of an egg in the oviduct. Only hens that previously had ovulated were used and the indomethacin was injected approximately 12 hours after ovulation.

DISCUSSION

Immature chickens and Japanese quail do not appear to show calcemic responses, at least within a four-hour period, to injections of PGE_2, PGE_1, or $PGF_{2\alpha}$. However, it is not known if PGE_2 produces a hypercalcemic response when infused slowly into these birds as it does in rats (Franklin and Tashjian, 1975). The lack of response to prostaglandins by immature birds could indicate that tissue receptors are already responding maximally to high endogenous levels of prostaglandins. However, this hypothesis is not supported by the bird's lack of response to indomethacin, which could normally be expected to reduce any endogenous prostaglandin activity by virtue of its inhibitory effect on the enzyme prostaglandin synthetase. It will be of interest in the future to study the calcemic effects not only of prostaglandin infusion but also those of long-life PGE_2 analogues in immature birds. One might expect a hypercalcemic response to occur in birds where bone is rapidly remodelling, as it is in young growing rats, and where the principles of bone turnover are similar (Dacke, 1979).

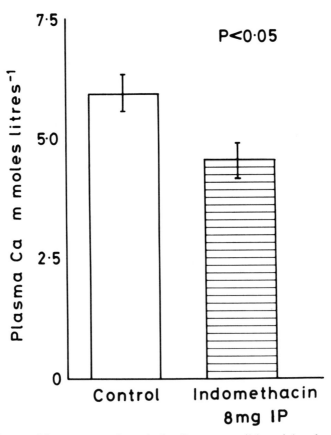

Fig. 1.—Plasma calcium responses in egglaying Japanese quail hens injected with indomethacin. Data are shown as mean plasma calcium level±SE. The indomethacin-injected group gave significantly ($P<0.05$) lower plasma calcium levels 4 hours after injection of indomethacin than did the control (vehicle-injected) group.

The experiments involving vitamin D_3 metabolism are even less conclusive. PGE_2, as well as PGE_1 or $PGF_{2\alpha}$, does not appear to have any consistent effect on metabolism of the vitamin in immature Japanese quail kidneys, although the results in two out of three experiments (shown on Table 2) do show a tendency for PGE_2 to stimulate conversion of 25-(OH)D_3 to 1,25-(OH)$_2D_3$. In the same experiment, estradiol benzoate showed a similar trend which reached significant proportions in one experiment. Elsewhere (Baksi and Kenny, 1977; Pike et al., 1978) estradiol benzoate has been shown to stimulate the production of the 1,25-(OH)$_2D_3$ metabolite; results obtained in this study might have reflected some problem with the experimental procedure or condition of the birds.

Similarly, in vivo injection of the prostaglandin synthetase inhibitor indomethacin does not have any significant effect, either stimulatory or inhibitory, at a high dosage, as has been claimed for the chick kidney in *vitro* 25-hydroxyvitamin D$_3$-1-hydroxylase system (Wark *et al.*, 1980). With indomethacin, problems of dosage, timing and route of injection are less likely to have been a problem than with prostaglandins; the negative result would therefore have more validity.

The most interesting result has been that concerning the hypocalcemic action of indomethacin in egg-laying hens. From the work of Hertelendy and co-workers (see Introduction) we know that the levels of PGE are raised in the plasma of chickens and quail when a calcifying egg is present in the oviduct. Hence if these substances play any role in the calcium metabolism of the egg-laying hen, inhibition of their synthesis with indomethacin might be expected to induce some change in that calcium metabolism, which in turn could be ultimately reflected by the plasma calcium levels. This appears to be the case since indomethacin has been demonstrated to have a hypocalcemic effect in chickens when an egg is in the oviduct (Hammond and Ringer, 1978), a finding which is confirmed in Japanese quail in the present study. These results indicate that the high circulating levels of PGE in the blood of egg-laying hens might be involved in some way in the translocation of calcium between gut, bone (including medullary bone), and eggshell. From mammalian studies (Klein and Raisz, 1970; Dietrich *et al.*, 1975) we know that bone resorption is implicated in the hypercalcemic responses to infused PGE$_2$. Perhaps indomethacin in the egg-laying hen acts by inhibiting the high prevailing rate of bone resorption which is presumably controlled partially by the PG. It is possible that the prostaglandins act in some way as a messenger for parathyroid hormone, or perhaps independently of this hormone. Future studies will resolve this problem. Medullary bone, which is an additional target organ to be found in the egg-laying hen, may present a potential target for prostaglandins. Another possible target organ for prostaglandins is the avian shell gland, which is capable of secreting large quantities of CaCO$_3$ into the eggshell (see Dacke, 1979). Indeed, the oviduct might respond to prostaglandins in a variety of ways to influence the production, passage and ultimate expulsion of the egg (Hertelendy, this vol.).

At the present time evidence to implicate prostaglandins in avian calcium metabolism should be considered preliminary. However, the fact that this evidence has been obtained in the egg-laying bird, where calcium metabolism presents one of its most interesting facets, must stimulate further work in this area.

ACKNOWLEDGMENTS

Part of this work concerned with vitamin D$_3$ metabolism was carried out during Dr. Dacke's sabbatical sojourn in the Department of Pharmacology and Therapeutics of the Texas Tech University Health Sciences Center in Lubbock. The senior author is grateful to the Welcome Trust for a travel grant during that period and to his host, Dr. Alexander D. Kenny. The technical assistance of Ingrid L. Greene and Jong-Chaur Shieh is also gratefully acknowledged. Supported in part by NIH Research Grant AM 19475 awarded to Dr. Kenny.

LITERATURE CITED

BAKSI, S. N., AND A. D. KENNY. 1977. Vitamin D_3 metabolism in immature Japanese quail: effect of ovarian hormones. Endocrinology, 101:1216-1220.

———. 1978. Vitamin D metabolism in Japanese quail: gonadal hormones and dietary calcium effects. Amer. J. Physiol., 234:E622-E628.

DACKE, C. G. 1979. Calcium regulation in sub-mammalian vertebrates. Academic Press, London, xv+222 pp.

DIETRICH, J. W., J. M. GOODSON, AND L. G. RAISZ. 1975. Stimulation of bone resorption by various prostaglandins in organ culture. Prostaglandins, 10:231-240.

FLOWER, R. J. 1974. Drugs which inhibit prostaglandin biosynthesis. Pharm. Rev., 26:33-67.

FRANKLIN, R. B., AND A. H. TASHJIAN, JR. 1975. Intravenous infusion of prostaglandin E_2 raises plasma calcium concentration in the rat. Endocrinology, 97:240-243.

HAMMOND, R. W., AND R. K. RINGER. 1978. Effect of indomethacin on the laying cycle, plasma calcium, and shell thickness in the laying hen. Poultry Sci., 57:1141.

HAMMOND, R. W., D. M. OLSON, R. B. FRENKEL, H. V. BIELLIER, AND F. HERTELENDY. 1980. Prostaglandins and steroid hormones in plasma and ovarian follicles during the ovulation cycle of the domestic hen (Gallus domesticus). Gen. Comp. Endocrinol., 42:195-202.

HERTELENDY, F., AND H. V. BIELLIER. 1978. Prostaglandin levels in avian blood and reproductive organs. Biol. Reprod., 18:204-211.

KLEIN, D. C., AND L. G. RAISZ. 1970. Prostaglandins: stimulation of bone resorption in tissue culture. Endocrinology, 86:1436-1440.

PIKE, J. W., E. SPANOS, K. W. COLSTON, I. MacINTYRE, AND M. R. HAUSSLER. 1978. Influence of estrogen on renal vitamin D hydroxylases and serum 1 alpha,25-$(OH)_2D_3$ in chicks. Amer. J. Physiol., 235:E338-E343.

SEYBERTH, H. W., G. V. SEGRE, J. L. MORGAN, B. J. SWEETMAN, J. T. POTTS, AND J. A. OATES. 1975. Prostaglandins as mediators of hypercalcemia associated with certain types of cancer. New England J. Med., 293:1278-1283.

WARK, J. D., R. G. LARKINS, J. A. EISMAN, AND T. J. MARTIN. 1979. Prostaglandins and 25-hydroxyvitamin-D-1α-hydroxylase. Pp. 563-566, in Vitamin D: basic research and its clinical application (A. W. Norman, K. Schaefer, D. U. Herrath, H.-G. Grigoleit, J. W. Coburn, H. F. DeLuca, E. B. Mawer, and T. Suda, eds.), Walter de Gruyter, Berlin, xxvi+1318 pp.

WECHSUNG, L., M. KORTEWEY, G. VERKONK, AND A. HOUVENAGHEL. 1978. Plasma levels of prostaglandin F related to oviposition in the domestic hen. Arch. Int. Pharmacodyn. Ther., 236:331-333.

Reprinted from
ASPECTS OF AVIAN ENDOCRINOLOGY:
PRACTICAL AND THEORETICAL IMPLICATIONS (C. G. Scanes *et al.*, eds.)
Grad. Studies, Texas Tech Univ., 1982, 26:1-411.

CALCIUM HOMEOSTASIS AND THE CONTROL OF OVULATION

MARTIN R. LUCK[1] AND COLIN G. SCANES[2]

[1]*Department of Physiology and Pharmacology, University of South-ampton, Bassett Crescent East, Southampton S09 3TU England; and [2]Department of Animal Sciences, Rutgers - The State University, New Brunswick, New Jersey 08903 USA*

CALCIUM HOMEOSTASIS AND THE CONTROL OF OVULATION

The extremely high dietary calcium requirement of the laying hen is well recognized, both as a matter of biological interest and as a commonplace necessity of poultry husbandry. It is estimated that the production of a single egg represents the loss of 10 per cent of the hen's total body calcium (Taylor, 1965). Thus, the maintenance of a production rate of six eggs per week requires a large and freely available calcium source, as well as a refined and well controlled system of calcium absorption and turnover.

Reducing the level of dietary calcium below about 2.5 per cent quickly leads to a drop in egg production as well as shell thinning. Lay virtually ceases when the calcium level reaches 0.3 per cent or less (Mehring and Titus, 1964; Roland *et al.*, 1973, 1974; Wakeling, 1977), and at this point a state of reproductive quiescence is established during which the reproductive organs regress, body weight is lost, and bone mineral is depleted. Gilbert's group (Gilbert, 1972, 1975; Blair and Gilbert, 1973; Gilbert and Blair, 1975; Gilbert *et al.*, 1978) have established, however, that a low level of calcium (0.05 per cent) is adequate for the maintenance of the reproductively quiescent bird for quite an extended period and has no deleterious effect on the subsequent performance when a high dietary calcium level is restored.

This observation, that egg production can be clearly distinguished in terms of calcium requirement from general body maintenance, indicates the importance to the bird of allowing calcium availability to be a regulating, and possibly limiting, factor in reproduction. It is the purpose of this article to review recent work that investigates some of the changes brought about by calcium deficiency and the mechanisms of the interaction between calcium homeostasis and reproduction.

Plasma Ionized Calcium

Of major importance in these studies has been the development of an assay for ionized calcium in avian plasma. It is well recognized that the non-ionized and largely protein bound fraction of the plasma calcium is unimportant in terms of calcium homeostatic mechanisms (Copp, 1969) and, except insofar as it equilibrates with the ionized fraction, can be considered as inactive. In the mammal the active ionized fraction contributes some 50 per cent to the total (Marshall, 1976), and changes in the ionized concentration are to a large extent reflected in measurements of total calcium. In the female fowl, the ionized fraction is only about a quarter of the total calcium because of the abundance of yolk proteins with a large calcium binding capacity. Hence, physiologically important changes are easily masked. Previous estimates of the active fraction have measured diffusible or ultrafilterable calcium as a good approximation. These, however, have involved some error in that they include in the estimate a small but potentially misleading quantity of nonprotein bound, nonionized calcium. A simple, accurate, and direct system of quantitating ionized calcium was required.

In the assay system devised (Luck and Scanes, 1979a), samples of minimally heparinized blood were drawn, centrifuged and assayed anaerobically, using an ion selective electrode against calcium chloride standards. Fig. 1 shows the concentrations of ionized calcium in samples of plasma taken from hens over non-clutch terminating cycles. The pattern clearly approaches a sigmoidal shape, with a peak (mean 1.60 mM) occurring between three and six hours after oviposition, followed by a fall during the period of shell calcification. The minimal level (mean 1.14 mM) at three to six hours before the next oviposition was significantly different from the peak level ($P<0.001$, by t-test). This difference was independent of the effects of repeated sampling (Luck and Scanes, 1979a). The pattern is consistent with changes in diffusible or ultrafilterable calcium during shell calcification (Taylor and Hertelendy, 1961; Dacke et al., 1973); there were no significant differences between the levels of total calcium in samples taken at the time of extreme ionized calcium concentration.

Effects of Exogenous Parathyroid Hormone and Calcitonin

The development of an assay for ionized calcium has enabled a more detailed study of the effects of parathyroid hormone (PTH) and calcitonin (CT) in the bird than was previously possible. In particular, it has been possible to study the responses of plasma calcium activity to the hormones in birds with and without a calcifying egg in the oviduct. In the absence of reliable information from hormone assays, the variations in response brought about by calcification give some indication of the relative importance of these hormones in avian calcium homeostasis.

These studies were performed using non-homologous preparations of PTH (bovine, MRC NIBSC) and CT (Salmon, Armour, Illinois) at doses (40 units each) whose physiological relevance is not known. Intravenous injections were given, with the blood samples taken at five, 15, 30, and 60 minutes, on two occasions: 1) during the supposed period of eggshell formation (15-16 hours after oviposition of a non-clutch terminating egg estimated from published information; see Luck and Scanes, 1979a); and 2) at a time when shell calcification was presumed not to be taking place (2-3 hours after oviposition).

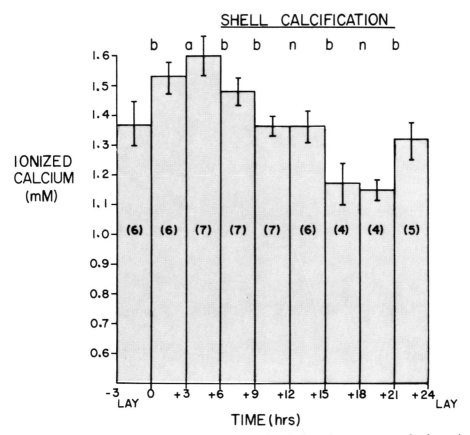

FIG. 1.—Concentration (mean ± SE) of ionized calcium in hen plasma over non-clutch terminating laying cycle. Observations (number of birds in parentheses) grouped into 3-hour periods
relative to time of lay of first egg. Significance of difference between adjacent groups are as follows: a, $P<0.001$; b, 0.01; n, not significant. Difference between first (-3 to 0 hour) and last
($+21$ to $+24$ hour) periods is not significant. Period of shell calcification estimated from published data. Figure based on Luck and Scanes (1979a).

The ionized calcium responses to PTH and CT are compared at different
times relative to oviposition (Fig. 2). The notable feature of the response to
PTH in non-egg-calcifying birds is the marked fall in concentration which
precedes the rise at 30 minutes. This response is similar to the transient
hypocalcemia previously observed following the injection of PTH into
mammals (Care et al., 1966; Parsons and Robinson, 1972) and birds (Mueller
et al., 1973; Boelkins et al., 1976). The effect has been attributed either to an
initial uptake of calcium by bone cells to facilitate triggering of subsequent
bone resorption (Parsons and Robinson, 1972) or to an alteration in skeletal
blood flow (Boelkins et al., 1976). The effect is not seen in egg-calcifying
birds, but there the increase in concentration is more prolonged. This is
taken to represent a simple enhancement of the already activated PTH-

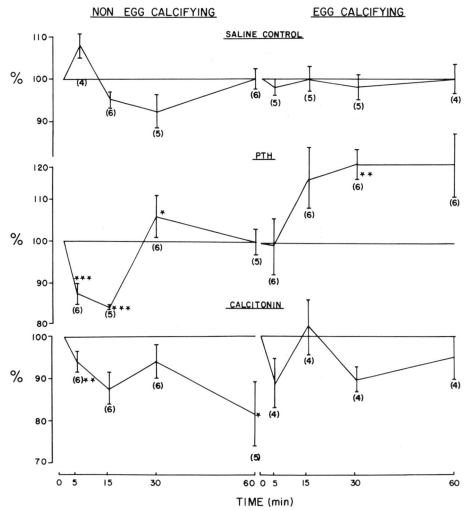

Fig. 2.—Effect of injections of saline (control), PTH, and CT on plasma ionized calcium concentration in non-egg-calcifying and egg-calcifying hens. Mean pre-injection concentrations in control birds were 1.56±0.02 mM and 1.25 ± 0.05 mM, respectively. Figure based on Luck et al. (1980).

induced bone resorption mechanism. Furthermore, it demonstrates that before injection, bone mobilization was not proceeding at its maximal rate and that the normal rate of shell calcification is not limited by the ability of the skeleton to release calcium in response to PTH.

The concentrations of total calcium and phosphorus were also estimated in these blood samples (for details see Luck et al., 1980). In non-egg-calcifying birds the plasma phosphorus concentration underwent a similar fall following PTH injection to that seen in ionized calcium. This probably resulted from a transient increase in phosphorus excretion rate (Levinsky

TABLE 1.—*Plasma LH concentrations (after 1 month) and organ weights postmortem (after 2 months) in hens fed normal and calcium-deficient diets. Mean ± followed by N in parentheses. Data from Luck and Scanes (1979b).*

Diet	LH (ng/ml)	Liver (g)	Ovary (g)	Oviduct (g)	Thyroid (2 lobes) (g)	Parathyroid (4 lobes) (g)
Normal	1.87±0.32(23)	33.5±2.5(8)	34.3±4.6(8)	49.9±2.0(8)	0.20±0.01(8)	0.029±0.003(8)
Calcium-deficient	0.91±0.05(16)	28.0±2.9(8)	17.4±5.4(8)	22.0±7.5(8)	0.21±0.02(8)	0.234±0.57(8)
P	<0.001	ns	<0.05	<0.01	ns	<0.001

and Davidson, 1957; Martindale, 1969). This was not seen in the egg-calcifying birds, presumably because the phosphorus excretion rate was already high and able to accommodate the additional bone phosphate released by the injection. With the exception of a small elevation at 30 minutes in non-egg-calcifying birds, none of the changes seen in ionized calcium were observed in the total calcium concentration, emphasizing the masking effect of the large protein bound fraction.

Calcitonin, given during the non-egg-calcifying period, was effective in reducing the ionized calcium both initially and in the longer term, but this effect was not seen in the hens injected during the egg-calcifying period. This suggests that the hormone was unable to overcome the enhanced bone resorption occurring during calcification. Although the insubstantial nature of the response to CT would be due to the non-homologous nature of the preparation, it more likely reflects the extreme sensitivity of the parathyroid glands to a slight fall in plasma calcium activity. Other workers have found a similar lack of response unless the birds are previously parathyroidectomized (Kraintz and Intscher, 1969; Lloyd et al., 1970; Sommerville, 1978). The secondary effect at 60 minutes suggests that the PTH damping action is a transient one, at least at this stage of the cycle when bone mineral is undergoing a net deposition.

Effects of a Low Calcium Diet

The effects of a low calcium diet on reproduction were studied in 24 laying hens (Light Sussex X Rhode Island Red or Hisex strains) transferred from a normal layers diet (3.0 per cent calcium) to a calcium deficient diet. Egg production was noticeably reduced in all birds within one week of calcium deprivation and in every case the final two or three eggs had thin or soft shells. Eight birds were observed to molt and there was an average loss of body weight of 21 per cent.

Table 1 shows some of the changes in organ weights which resulted from the removal of dietary calcium. Typically, the ovaries of calcium-deficient birds were devoid of large yolky follicles and failed to show the follicular size hierarchy seen in normal layers. After four weeks of treatment, the mean plasma LH level in 16 hens, measured by radioimmunoassay (Follett et al., 1972), was 0.91 ng/ml (±0.05 SE) compared with 1.87 ng/ml (± 0.32 SE, P<0.001) in calcium replete controls.

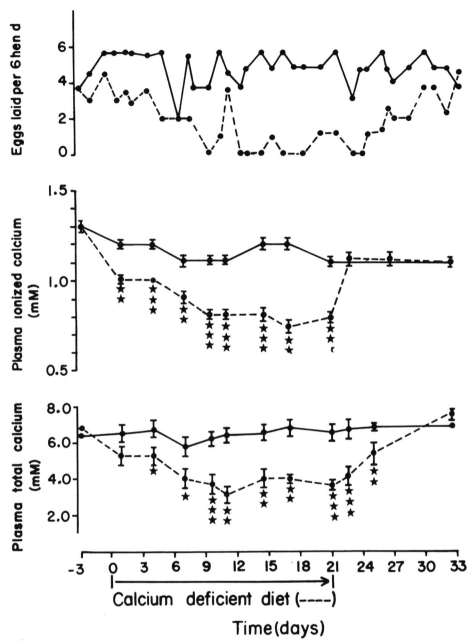

FIG. 3.—Plasma concentrations of calcium and rate of egg production in hens fed a normal or a calcium-deficient diet. The normal diet was reintroduced for the calcium-deficient hens after 21 days. Figure based on Luck and Scanes (1979b).

Effects on Blood Mineral

The changes in blood calcium and phosphorus which occur when the dietary calcium supply is withdrawn are shown in Fig. 3 (Luck and Scanes, 1979b). The sensitivity of the ionized calcium fraction to changes in the available calcium is clearly shown; by comparison, the total calcium concentration changed only gradually, again emphasizing the importance of estimating the active calcium fraction in studies of this kind. The pattern of loss of lay and fall in ionized calcium concentration are in agreement with the suggestion of Taylor (1965) that the ovulation-inducing activity of the pituitary-hypothalamic complex is dependent upon the maintenance of a certain threshold level of calcium activity in the blood. These observations suggest that such a level is represented by an ionized calcium concentration of about 1.0 mM, some 0.4 mM less than the average level in normally laying hens.

Microradiographic Studies.—At the point at which regular lay ceased (day 11), the mean total calcium in the blood reached its lowest level, due presumably to decreased circulating levels of yolk precursors and associated calcium. This suggests that the birds were able to maintain body calcium reserves quite well, despite negligible calcium intake, once the stress of egg production was ended. This lack of egg production is of obvious value, in terms of calcium conservation. This is also reflected in changes seen in the distribution of bone mineral. One of the major events preceding the onset of lay in the maturing hen is the deposition and establishment of a layer of medullary bone on the endosteal surface of a large proportion of the skeleton (Bloom *et al.*, 1941; Hurwitz, 1964; Taylor and Morris, 1964; Hurwitz and Bar, 1971). This layer, which forms in response to increasing levels of sex steroids, acts as a transient store for calcium, capable of rapid mobilization and ensuring a continuous supply of calcium for shell production.

Figs. 4 and 5 show microradiographs of mid-tibial cross-sections taken from normally laying hens. The outer cortical layer of dense structural bone is lined with a spongy layer of medullary bone, some of which is little more than lightly mineralized osteoid. The partial detachment of the medullary from the cortical layer is indicative of a site of active endosteal resorption (Simkiss, 1967). Fig. 6 shows, for comparison, a similar section of cockerel tibia. Note the complete absence of medullary bone and the difference between the slightly darker area of endosteal bone undergoing remodelling and the endosteal bone undergoing resorption in the hen sections.

Figs. 7 and 8 are microradiographs of sections from hens out of lay after receiving the calcium-deficient diet for nine weeks. The mineral of the medullary layer in these sections has been almost completely eroded, leaving only lightly mineralized osteoid. The width of the medullary layer in Fig. 8 appears to have expanded, with considerable signs of erosion on the endosteal surface of the cortical bone. This is in keeping with previous observations (Taylor and Moore, 1954, 1956; Zallone and Mueller, 1969) that during calcium deficiency medullary bone is maintained at the expense of cortical bone. However, the density of the medullary layer is clearly very low compared with that of the structural bone, and an additional compensatory layer of soft periosteal bone has formed, presumably as a response to mechanical stress, leading to an increase in overall bone diameter. The periosteal bone is of the fine cancellous type, unremodelled and much less dense than the well remodelled cortical bone, so that any strength it provides comes from its width rather than from its structural resilience.

It is well established (Pritchard, 1972; Gardner, 1972) that cortical bone growth ceases as an animal reaches maturity and that the osteoblastic and osteoclastic components of the periosteal surface tend to disappear. The existence of new periosteal bone provides support for the view (Pritchard, 1972) that, under abnormal conditions, osteoblastic activity can be resumed at this site. The absence of Haversian remodelling presumably shows that this bone lacks the osteoclastic activity found in cortical bone growth prior to maturity.

Fig. 9 shows a mid-tibial section from a hen maintained on the calcium-deficient diet but brought into lay (2 thin-shelled eggs) by the injection of a crude glycoprotein extract of chicken anterior pituitary glands (approximately equivalent to 0.5 pituitary per day). Active endosteal resorption, indicated by the detachment of a layer of bone at the inner cortical surface, can be seen, with an expanded, but well eroded, medullary layer. The severity of the calcium deficiency has also caused severe erosion of the periosteal layer and it was noted that this bird had great difficulty in standing and, on postmortem examination, had moderately pliable long bones.

It is interesting to observe that, in both the calcium-deficient non-laying hens and in the calcium-deficient hens induced to lay with pituitary extract, the bone deficiency was manifested as a loss of mineral and matrix. This was true of medullary bone, of medullary bone formed from cortical bone, and of periosteal bone. It suggests that secondary bone osteoid, once formed, remains capable of remineralization and defines the subsequent capacity of the skeleton to act as a buffer in times of dietary mineral insufficiency. This must be of particular importance at the onset of lay. According to Hurwitz and Griminger (1960) hens enter a state of negative calcium balance soon after the start of egg production, despite the previous formation of medullary bone, and Cos and Balloun (1970, 1971) suggested that some 30 eggs may be laid before the resulting skeletal depletion is overcome. Evidence obtained from bone densitometry indicates that this may be an overly optimistic assessment and that net replacement of mineral may continue for a much longer period.

Densitometric Studies—Measurements of relative bone mineral density were made using a non-invasive scanning technique which allowed a number of measurements to be made in the same bird over a period of experimental treatment. The apparatus consisted of a collimated ^{125}I gamma source and detector moving along a horizontal scan on opposite sides of a chamber. The bird to be scanned was placed in a Perspex® support above the chamber such that its right leg (previously plucked) was held vertically

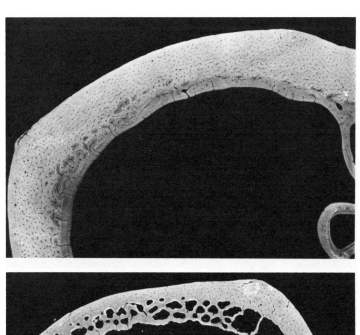

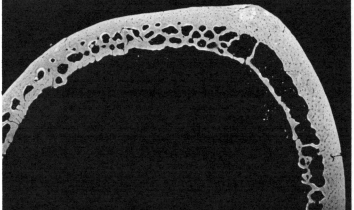

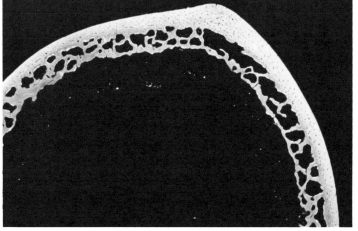

Figs. 4 (left), 5 (middle), 6 (right).—Microradiographs of chicken mid-tibial sections. See text for explanation.

in the path of the gamma beam, midway along the length of the tibia. The degree of attenuation caused to the beam by the leg at each point in the scan was monitored automatically to provide a density profile of bone tissue relative to the surrounding soft tissue (Fig. 10). From this profile an on-line computer calculated a scan integral of mass per unit length of scanned bone; this was an arbitrary unit of mineral density, directly related to bone ash content.

Table 2 shows two sets of tibia scan integral measurements separated by five weeks, in two groups of birds. The normal control group was laying regularly throughout; the treated group was given the calcium-deficient diet after the first scan and went out of lay during the first week. Prior to the first sets of scans all the hens had been in lay for approximately five months. Mean body weights in each group are given for comparison. The observation that bone mineral density increased substantially over the five-week period, without an accompanying increase in body weight, indicates that the increase in mineral deposition was not due to growth and suggests a structural rather than a dimensional change in the bone. Thus, net mineral deposition was continuing, despite each hen having previously laid some 100 eggs. It is the continuing remineralization which is absent from the calcium-deficient group rather than an actual reduction in bone mineral. Again, despite the redistribution of mineral shown in the microradiographs the loss of lay has protected the skeleton from wastage.

Localization of Sites of Mineral-Endocrine Interaction

As noted above, and as has been shown previously (Taylor et al., 1962; Taylor, 1965), it is possible to induce ovarian activity in anovulatory calcium-deficient hens by the injection of gonadotropins in the form of an anterior pituitary extract. It is clear from these studies that, once follicular development and ovulation are achieved, then limited calcium availability puts no further constraint on the passage of the egg, even if it is subsequently laid without a fully developed shell. Thus, the ovary remains potentially responsive to pituitary stimulation, suggesting that reproductive limitation is centrally mediated, either through the pituitary gland or the brain. A combination of in vivo and in vitro studies has been used to determine the site at which reproductive function is affected and to discover whether the effect is a direct result of plasma hypocalcemia, or an indirect one mediated by a secondary hormonal change.

In Vivo Studies

The ability of the pituitary gland to secrete gonadotropins during calcium deficiency was tested using intravenous injections of LHRH. Table 3 shows the effect of 25 μg synthetic ovine LHRH on plasma LH levels in normal and calcium-deficient hens. In normal birds there was a large but variable rise in LH concentration by 10 minutes after injection. By 60 minutes the concentration had fallen to less than its pre-injection level. In calcium-

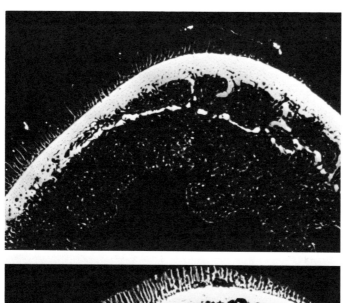
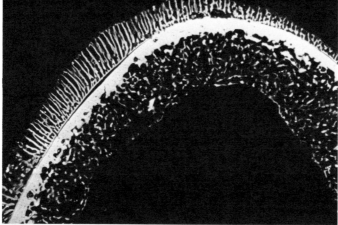
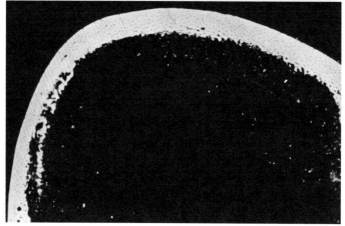

Figs. 7 (left), 8 (middle), 9 (right).—Microradiographs of chicken mid-tibial sections. See text for explanation.

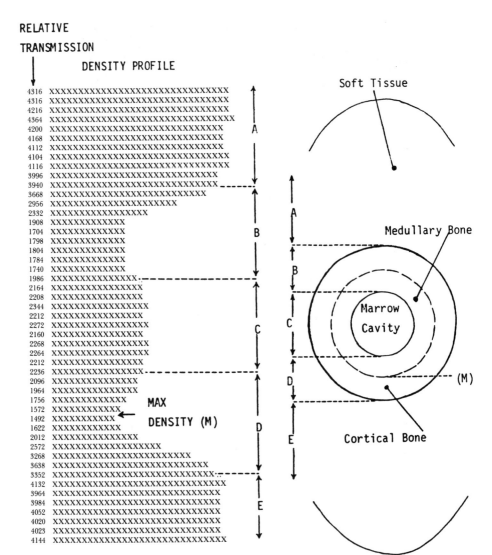

FIG. 10.—Example of densitometric bone scan and diagramatic representation of corresponding leg structure.

deficient birds the pattern was similar, although there was greater variability in the increase at five minutes and the levels were still elevated at 60 minutes.

This response demonstrates no apparent loss of pituitary sensitivity in the calcium-deficient bird. Indeed, the variability of the response of both groups perhaps masked a tendency towards greater sensitivity in the deficient group. This is taken to reflect the presence of a large quantity of readily releasable LH, stored in the pituitary as a result of absent or reduced endogenous release stimuli. It implies that gonadotropin synthesis continues in the non-laying bird if only at a reduced rate. Despite being reduced by nearly 40 per

TABLE 2.—*Effect of 5-week interval (A to B) on scan integral and body weight in hens fed normal and calcium-deficient diets. Mean ± SE followed by N in parentheses.*

Diet	Time	Scan integral	P	Body weight (kg)	P
Normal	A	25.5±0.2(5)	<0.001	2.12±0.08(5)	ns
	B	41.5±3.5(5)		2.18±0.11(5)	
Calcium deficient	A	25.0±0.2(6)		2.10±0.04(6)	ns
	B	26.5±0.7(6)		2.04±0.05(6)	

cent, the basal plasma LH level was still measurable, indicating that pituitary release activity was not completely suppressed even if the release associated with ovulation was abolished.

In order to test the ability of the hypothalamus to stimulate the pituitary gland during calcium deficiency, normal and calcium-deficient hens were injected subcutaneously with progesterone (1 mg) and the plasma response determined. Such progesterone stimulation is thought to mimic the ovarian positive feedback effect which brings about the preovulatory LH response in the normal cycle; a number of experiments have demonstrated that the steroid has no direct stimulatory effect on the pituitary (Ralph and Fraps, 1960; Fraser and Sharp, 1978; Luck and Scanes, 1980). The responses of the two groups of hens are different (Table 3) notably in that the calcium deficient group showed an initial fall in LH concentration without the sustained rise seen in the normal group. Thus, it is concluded that the hypothalamus has become refractory to progesterone and that it is here that the block to reproduction occurs.

In Vitro Studies

In order to investigate more specifically how direct and indirect effects of hypocalcemia affect pituitary function, a dispersed cell incubation technique was used. This technique has been described elsewhere (Bicknell and Follett, 1975; Godden *et al.*, 1977; Luck and Scanes, 1980) and involves the dispersal of anterior pituitary cells, and their incubation for two hours in Medium 199. The production of LH by the cells was determined by radioimmunoassay (Follett *et al.*, 1972) of the centrifuged incubate.

Dietary calcium deficiency leads to a reduction in the plasma ionized calcium concentration to 1.0 mM or less (Fig. 3). Furthermore, during the normal ovulatory cycle, a peak concentration of 1.6 mM occurs prior to the start of shell calcification (Fig. 1). This range of concentrations was reflected in the *in vitro* incubation system by using normal Medium 199 (1.8 mM calcium) and calcium-free Medium 199 with $CaCl_2$ added to a concentration of 0.9 mM. Table 4 shows the effects of these two media on LH production in the presence and absence of LHRH. The lower calcium concentration clearly depressed the basal LH secretion and prevented the stimulatory effect of LHRH.

This is clearly a more subtle effect of calcium concentration on the pituitary than the absolute dependence of LH release on the presence of calcium

TABLE 3.—*Effect of LHRH and progesterone on plasma LH concentration in hens fed normal and calcium-deficient diets. Based on Luck and Scanes (1979b).*

	LHRH (25 mg)			
Time after injection (min)	0 (ng/ml)	5 (%)	10 (%)	60 (%)
Normal diet	1.67±0.26 (10)	+14.0±10.8 (9)	+93.2±63.8 (10)	−26.2± 9.4 (10)
Calcium-deficient diet	1.03±0.10*(9)	+108.6±64.0 (8)	+89.2±71.3 (7)	+38.8±37.0*(6)

	PROGESTERONE (1 mg)			
Time after injection (min)	0 (ng ml⁻¹)	30 (%)	60 (%)	120 (%)
Normal diet	1.98±0.20(6)	+24.8±24.8(6)	+74.6±35.6(5)	+87.0±21.2(6)
Calcium-deficient diet	1.16±0.07*	−26.5± 5.5	+15.4± 9.6	−29.6± 6.4**

Values are mean percentage changes (± SE) from stated mean preinjection value (± SE) at time 0; number of hens in parentheses.

Significance of difference from respective control value: *$P<0.05$, **$P<0.01$.

described by Bonney and Cunningham (1977). *In vitro* experiments using calcium ionophores (Luck and Scanes, 1980) showed that an increase in the rate of calcium entry into pituitary cells could bring about an increase in the rate of LH release similar in magnitude to that achieved by a maximal concentration of LHRH. Thus, pituitary activity can be influenced directly by calcium concentration and we suggest that this influence is exerted, not only when calcium is unavailable, but perhaps also during normal reproduction. Indeed, it may well be that while shutting off of hypothalamic stimulation is the major anti-ovulatory event when dietary calcium is critically low, the pituitary effect may be important in modulating the ovulation rate according to less drastic changes in calcium availability.

Indirect Effects

One of the effects of calcium deficiency observed postmortem (Table 1) was a large increase in the size of the parathyroid glands. There is as yet no reliable assay for circulating avian PTH, but inference from mammalian systems and circumstantial evidence (Kenny and Dacke, 1975; Chan, 1976) make it probable that PTH secretion increases as hypocalcemia develops. The consequences of this would be the maintenance of plasma calcium activity by the release of bone mineral, increased renal calcium conservation, and increased gut absorption of calcium either directly or indirectly through modified cholecalciferol metabolism (Kenny and Dacke, 1974, 1975). It would be reasonable to expect that PTH secretion is also increased during the shell calcification phase of the egg cycle (as a result of the substantial fall in calcium activity at this time) facilitating the release of bone mineral for shell production.

Enhancement of 1,25-dihydroxyvitamin D_3 [1,25-$(OH)_2 D_3$] production during the avian egg-laying cycle has been demonstrated (Kenny, 1976; Baksi and Kenny, 1977; Sedrani and Taylor, 1977) and is consistent with the

TABLE 4.—*Effects of altered medium calcium concentration on LH secretion by pituitary cells in vitro (ng ml^{-1} 2 hours^{-1}). Mean ± SE followed by N in parentheses. Based on Luck and Scanes (1980).*

Calcium concentration (mM)	Unstimulated	Stimulated with LHRH (10 ng/ml)
1.8 (control)	100.0±7.2 (5) a	+232.1±49.6 (5) c
0.9	56.4±3.9 (10) b	78.1±11.5 (5) b

Results shown as percentage of the unstimulated control secretion.
Means with unshared letters are significantly different ($P<0.05$).

increased production of this metabolite in mammals during conditions of calcium demand. Recent studies by Abe *et al.* (1979) have shown distinct changes in 1,25-$(OH)_2D_3$ production during the ovulatory cycle. This rhythm appears to be the mirror image of the rhythm in plasma calcium activity (Fig. 1), with a minimum concentration at the time of maximal ionized calcium concentration immediately prior to the start of shell calcification. Similarly, the fall in plasma calcium activity as calcification proceeds is matched by a rise in 1,25-$(OH)_2D_3$ production. The two rhythms are clearly interrelated by the demand for calcium brought about by shell calcification.

To test the possibility that PTH and 1,25-$(OH)_2D_3$ also help to conserve calcium by some regulatory effect on ovulation, both hormones were used in the pituitary cell system (Table 5). PTH (bovine, MRC NIBSC) not only reduced the basal LH secretion, but also blocked the enhancing effect on LHRH. It should be noted, however, that a non-homologous preparation of the hormone was used and that, in the absence of information on endogenous levels, the dose used was somewhat arbitrary. High doses of 1,25-$(OH)_2D_3$ also prevented LHRH stimulation, but had no effect on basal LH secretion. Abe *et al.* (1979) found that the plasma concentration of 1,25-$(OH)_2D_3$ varied over the ovulation cycle from just under 0.1 ng/ml to over 0.3 ng/ml. Hence, the concentration employed was probably a good representation of the situation at late calcification.

TABLE 5.—*Effect of PTH and 1,25-$(OH)_2D_3$ on LH secretion by pituitary cells in vitro (ng ml^{-1} 2 hour^{-1}). Mean ± SE followed by N in parentheses. Based on Luck and Scanes (1980).*

Treatment (ng/ml)	Unstimulated	Stimulated with LHRH (10 ng/ml)
PTH		
0 (control)	100.0± 8.9 (5) a	232.0±20.3 (5) c
5.0	42.9± 4.6 (50) b	77.0± 8.7 (5) b
50	63.7±13.6 (6) b	52.7± 5.1 (5) b
1,25-$(OH)_2D_3$		
0 (control)	100.0±30.3 (10) a	231.6±49.6 (10) b
0.1	75.7±13.7 (10) a	245.6±53.0 (10) b
1.0	122.6± 7.0 (10) a	92.9± 9.5 (10) a
10	131.7±23.3 (10) a	102.4±12.1 (10) a

Results shown as percentage of the unstimulated control secretion.
Means with unshared letters are significantly different ($P<0.05$).

TABLE 6.—*Effects of altered medium phosphorus concentration on LH secretion by pituitary cells* in vitro *(ng ml^{-1} 2 hours^{-1}). Mean ± SE followed by N in parentheses. Based on Luck and Scanes (1980).*

Phosphorus concentration (mM)	Unstimulated	Stimulated with LHRH (10 ng/ml)
1.0 (control)	100.0±21.9 (5) a	156.8±10.2 (5) b
0.5	108.5±27.2 (5) a	151.3±35.3 (5) ab
0.25	197.8±25.5 (5) b	193.2±17.7 (5) b

Results shown as percentage of the unstimulated control secretion.
Means with unshared letters are significantly different (*P*<0.05).

Thus, it is possible that PTH contributes to the suppression of pituitary activity in the calcium-deficient bird, and that 1,25-$(OH)_2D_3$ has some regulatory influence associated with the pre-ovulatory LH surge. We proposed that this latter effect may contribute to the regulation of the clutch pattern by preventing a premature pre-ovulatory LH peak before calcification of the current egg is complete.

It will be seen from Fig. 1 that a major event associated with the restoration of calcium to the diet of deficient birds was a sharp fall (50 per cent) in the plasma phosphorus concentration. This was presumably the result of a sudden increase in the rate of bone mineral deposition. The effects of a reduced medium phosphorus concentration on pituitary activity are shown in Table 6. There is clearly a tendency for the reduced concentration to enhance both unstimulated and LHRH-stimulated pituitary activity; such a pituitary sensitizing effect may be of advantage in facilitating an early resumption of reproductive activity after a period of calcium deficiency.

GENERAL DISCUSSION

The interaction of calcium homeostasis with the control of reproduction has been investigated in these studies. Egg production is regulated with calcium availability, there being both an absolute dependence and a less drastic adjustment of ovulation frequency as calcium availability fluctuates. This interaction clearly operates on a number of levels: by direct interaction of blood calcium activity with the pituitary and hypothalamic complexes, and by indirect interaction mediated by alterations in vitamin D metabolism and the production of calcium regulating hormones.

The calcium-conserving result of these interactions appears to be closely related to the maintenance of skeletal integrity, and the studies show that the skeletal requirement has some priority over the reproductive requirement. This was most clearly demonstrated by the densitometric and microradiographic studies where it was seen that calcium restriction led to a loss of lay before the point at which structural bone would have been seriously depleted. It was the continuing remineralization of depleted medullary bone which ceased, thus maintaining an adequate level of blood calcium activity for the body's non-reproductive needs. This provides a probable explanation for the increase in egg productivity which can be achieved in the commercial hen by the use of a period of reproductive quiescence, induced by lighting or dietary means (Wakeling, 1977). It is likely that this period enables the skeleton to "recover" by giving priority to the process of remineralization without the additional burden of shell production.

Even where the quiescence is induced by dietary calcium restriction, only a small elevation of dietary calcium concentration above the minimal amount required for body maintenance would be needed to facilitate bone replenishment. The presence of replete skeletal calcium stores would then enable a maximal rate of egg production to be achieved when the high calcium diet was resumed. Presumably, the seasonality of the laying pattern in wild birds provides a similar beneficial effect.

The question arises as to whether the daily calcium cycle, demonstrated by ionized calcium measurements, plays any part in regulating the length of the ovulation cycle in the calcium-replete bird. It would be reasonable to expect that such a relatively time-consuming process as shell calcification influences the between-egg interval of which it is a major component. The existence of cycles of calcium activity and $1,25\text{-}(OH)_2D_3$ production, related to the calcium demand, provide direct and indirect mechanisms by which this influence could take place. The event that ultimately defines the length of the ovulatory cycle is the surge of LH release responsible for causing ovulation. This surge reaches a peak some four to six hours before ovulation (Furr et al., 1973), at a time when plasma calcium activity is rising and $1,25\text{-}(OH)_2D_3$ is falling to its minimal concentration (Abe et al., 1979). The LH surge is preceded by a steady fall that coincides with the fall in calcium activity and rise in $1,25\text{-}(OH)_2D_3$ output associated with shell calcification. Such close correspondence between these cycles makes the possibility of physiological interaction worthy of consideration. We suggest that future studies of the endocrine control of ovulation take this interaction into account. It may well be that the calcium cycle is of equal importance to the photoperiodic cycle in permitting or preventing LH release or entraining the cycle, but that this has been previously obscured.

ACKNOWLEDGMENTS

We are indebted to Professor A. D. Care for the facilities for the experimentation, to Dr. P. Atkinson, Department of Oral Biology, University of Leeds, and persons in his laboratory for the preparation of microradiographs and for assistance with their interpretation, and to Dr. A. Horsman, MRC Mineral Metabolism Unit, Leeds General Infirmary, for the use of bone densitometry equipment.

LITERATURE CITED

ABE, E., R. TANABE, T. SUDA, AND S. YOSHIKA. 1979. Circadian rhythm of 1,25-dihydroxyvitamin D₃ production in laying hens. Biochem. Biophys. Res. Commun., 88:500-507.

BAKSI, S. N., AND A. D. KENNY. 1977. Influence of reproductive status on *in vitro* renal production of 1,24,25-trihydroxyvitamin D₃ in Japanese quail. J. Reprod. Fert., 50:175-177.

BICKNELL, R. J., AND B. K. FOLLETT. 1975. A quantitative assay for luteinizing hormone releasing hormone (LHRH) using dispersed pituitary cells. Gen. Comp. Endocrinol., 26:141-152.

BLAIR, R., AND A. B. GILBERT. 1973. The influence of supplemental phosphorus in a low calcium diet designed to induce a resting phase in laying hens. Brit. Poult. Sci., 14:131-135.

BLOOM, W., M. A. BLOOM, AND F. C. McLEAN. 1941. Calcification and ossification. Medullary bone changes in the reproductive cycle of female pigeons. Anat. Rec., 81:433-466.

BOELKINS, J. N., M. MAZURKIEWICZ, P. E. MAZUR, AND W. J. MUELLER. 1976. Changes in blood flow to bones during the hypocalcemic and hypercalcemic phases of the response to parathyroid hormone. Endocrinology, 98:403-412.

BONNEY, R. C., AND F. J. CUNNINGHAM. 1977. Effect of ionic environment on the release of LH from chicken anterior pituitary cells. Mol. Cell. Endocrinol., 7:245-251.

CARE, A. D., W. M. KEYNES, AND T. DUNCAN. 1966. An investigation into the parathyroid origin of calcitonin. J. Endocrinol., 34:299-318.

CHAN, A. S. 1976. Effect of a low calcium diet on the ultrastructure of the parathyroid gland of chicks. Cell Tiss. Res., 173:71-76.

COPP, D. H. 1969. Review: endocrine control of calcium homeostasis. J. Endocrinol., 43:137-161.

COX, A. C., AND S. L. BALLOUN. 1970. Depletion of femur bone mineral at the onset of egg production in S.C.W. Leghorn pullets. Poultry Sci., 49:1463-1468.

————. 1971. Depletion of femur bone mineral after the onset of egg production in a commercial strain of leghorns and in broiler type pullets. Poultry Sci., 50:1429-1433.

DACKE, C. G., X. J. MUSACCHIA, W. A. VOLKERT, AND A. D. KENNY. 1973. Cyclical fluctuations in the levels of blood calcium, pH and pCO₂ in Japanese quail. Comp. Biochem. Physiol., 44A:1267-1275.

FOLLETT, B. K., C. G. SCANES, AND F. J. CUNNINGHAM. 1972. A radioimmunoassay for avian luteinizing hormone. J. Endocrinol., 52:359-378.

FRASER, H. M., AND P. J. SHARP. 1978. Prevention of positive feedback in the hen by antibodies to luteinizing hormone releasing hormone. J. Endocrinol., 76:181-182.

FURR, B. J. A., R. C. BONNEY, J. R. ENGLAND, AND F. J. CUNNINGHAM. 1973. Luteinizing hormone and progesterone in peripheral blood during the ovulatory cycle of the hen *Gallus domesticus*. J. Endocrinol., 57:159-169.

GARDNER, E. 1972. Osteogenesis in the human embryo and fetus. Pp. 77-188, *in* The biochemistry and physiology of bone, second ed., Vol. III (G. H. Bourne, ed.), Academic Press, London, xviii+584 pp.

GILBERT, A. B. 1972. The role of calcium in regulating reproductive activity in the domestic hen. J. Reprod. Fert., 29:150-151.

————. 1975. Low calcium diets for pausing layers. World's Poult. Sci. J., 30:309.

GILBERT, A. B., AND R. BLAIR. 1975. A comparison of the effects of two low calcium diets on egg production in the domestic fowl. Brit. Poult. Sci., 16:547-552.

GILBERT, A. B., J. PEDDIE, P. W. TEAGUE, AND C. G. MITCHELL. 1978. The effect of delaying the onset of laying in pullets with a low calcium diet on subsequent egg production. Brit. Poult. Sci., 19:21-34.

GODDEN, P. M. M., M. R. LUCK, AND C. G. SCANES. 1977. The effect of luteinizing hormone releasing hormone and steroids on the release of LH and FSH from incubated turkey pituitary cells. Acta Endocrinol., Copenhagen, 85:713-717.

HURWITZ, S. 1964. Calcium metabolism of pullets at the onset of egg production, as influenced by dietary calcium level. Poultry Sci., 53:1462-1472.

HURWITZ, S., AND A. BAR. 1971. The effect of prelaying mineral nutrition on the development, performance and mineral metabolism of pullets. Poultry Sci., 50:1044-1055.

HURWITZ, S., AND P. GRIMINGER. 1960. Observations on the calcium balance of laying hens. J. Agric. Sci., 54:373-377.

KENNY, A. D. 1976. Vitamin D metabolism: physiological regulation in egg-laying Japanese quail. Amer. J. Physiol., 230:1609-1615.

KENNY, A. D., AND C. G. DACKE. 1974. The hypercalcaemic response to parathyroid hormone in Japanese quail. J. Endocrinol., 62:15-23.

——. 1975. Parathyroid hormone and calcium metabolism. World Rev. Nutr. Diet., 20:231-298.

KRAINTZ, L., AND K. INTSCHER. 1969. Effect of calcitonin on the domestic fowl. Can. J. Physiol. Pharm., 47:313-315.

LEVINSKY, N. G., AND D. G. DAVIDSON. 1957. Renal action of parathyroid extract in the chicken. Amer. J. Physiol., 191:530-536.

LLOYD, J. W., R. A. PETERSON, AND W. E. COLLINS. 1970. Effects of an avian ultimobranchial extract in the domestic fowl. Poultry Sci., 49:1117-1121.

LUCK, M. R., AND C. G. SCANES. 1979a. Plasma levels of ionized calcium in the laying hen (Gallus domesticus). Comp. Biochem. Physiol., 63A:177-181.

——. 1979b. The relationship between reproductive activity and blood calcium in the calcium deficient hen. Brit. Poult. Sci., 20:559-564.

——. 1980. Ionic and endocrine factors influencing the secretion of luteinizing hormone by chicken anterior pituitary cells in vitro. Gen. Comp. Endocrinol., 41:260-265.

LUCK, M. R., B. A. SOMMERVILLE, AND C. G. SCANES. 1980. The effect of egg-shell calcification on the response of plasma calcium activity to parathyroid hormone and calcitonin in the domestic fowl (Gallus domesticus). Comp. Biochem. Physiol., 65A:151-154.

MARSHALL, R. W. 1976. Plasma fractions. Pp. 62-185, in Calcium, phosphate and magnesium metabolism (B.E.C. Nordin, ed.), Churchill Livingstone, London, x+683 pp.

MARTINDALE, L. 1969. Phosphate excretion in the laying hen. J. Physiol., London, 203:82P-83P.

MEHRING, A. L., AND H. W. TITUS. 1964. The effects of low levels of calcium in the diet of laying chickens. Poultry Sci., 43:1405-1414.

MUELLER, W. J., K. L. HALL, C. A. MAURER, JR., AND I. G. JOSHUA. 1973. Plasma calcium and inorganic phosphate response of laying hens to parathyroid hormone. Endocrinology, 92:853-856.

PARSONS, J. A., AND C. J. ROBINSON. 1972. The earliest effects of parathyroid hormone and calcitonin on blood-bone calcium distribution. Pp. 399-406, in Calcium, parathyroid hormone and the calcitonins (R. V. Talmage and P. L. Munson, eds.), Exerpta Medica, Amsterdam, ix+560 pp.

PRITCHARD, J. J. 1972. The osteoblast. Pp. 21-43, in The biochemistry and physiology of bone, Second ed., Vol. I (G. H. Bourne, ed.), Academic Press, London, xvi+376 pp.

RALPH, C. L., AND R. M. FRAPS. 1960. Induction of ovulation in the hen by injection of progesterone into the brain. Endocrinology, 66:269-272.

ROLAND, D. A., D. R. SLOAN, H. R. WILSON, AND R. H. HARMS. 1973. Influence of dietary calcium deficiency on yolk and serum calcium, yolk and organ weights and other selected production criteria of the pullet. Poultry Sci., 52:2220-2223.

——. 1974. Relationship of calcium to reproductive abnormalities in the laying hen (Gallus domesticus). J. Nutr., 104:1079-1085.

SEDRANI, S., AND T. G. TAYLOR. 1977. Metabolism of 25-hydroxycholecalciferol in Japanese quail in relation to reproduction. J. Endocrinol., 72:405-406.

SIMKISS, K. 1967. Calcium in reproductive physiology. Chapman and Hall, London, xiv+264 pp.

SOMMERVILLE, B. A. 1978. Effect of parathyroid hormone and calcitonin on calcium and phosphorus levels in the plasma and urine of the chicken. J. Endocrinol., 77:52P.

TAYLOR, T. G. 1965. Calcium-endocrine relationships in the laying hen. Proc. Nutr. Soc., 24:49-54.

TAYLOR, T. G., AND F. HERTELENDY. 1961. Changes in the blood calcium associated with egg shell calcification in the domestic fowl. 2. Changes in the diffusible calcium. Poultry Sci., 40:115-123.

TAYLOR, T. G., AND J. H. MOORE. 1954. Skeletal depletion in hens laying on a low calcium diet. Brit. J. Nutr., 8:112-124.

———. 1956. The effect of calcium depletion on the chemical composition of bone minerals in laying hens. Brit. J. Nutr., 10:250-263.

TAYLOR, T. G., AND T. R. MORRIS. 1964. The effects of early and late maturing on the skeletons of pullets. World's Poult. Sci. J., 20:294-297.

TAYLOR, T. G., T. R. MORRIS, AND F. HERTELENDY. 1962. The effect of pituitary hormones on ovulation in calcium deficient pullets. Vet. Rec., 74:123-125.

WAKELING, D. E. 1977. Induced moulting: a review of the literature, current practice and areas for further research. World's Poult. Sci. J., 33:12-20.

ZALLONE, A. Z., AND W. J. MUELLER. 1969. Medullary bone of laying hens during calcium depletion and repletion. Calc. Tiss. Res., 4:136-146.

Reprinted from
Aspects of Avian Endocrinology:
Practical and Theoretical Implications (C. G. Scanes *et al.*, eds.)
Grad. Studies, Texas Tech Univ., 1982, 26:1-411.

METABOLISM OF 25-HYDROXYVITAMIN D3
BY CULTURED AVIAN CELLS

Russell T. Turner[1,2], Pakawan Duvall[3], J. Edward Puzas[4],
and Guy A. Howard[3]

[1]*Veterans Administration Medical Center, Charleston, South
Carolina 29403 USA;* [2]*Department of Pharmacology, Medical
University of South Carolina, Charleston, South Carolina 29403
USA;* [3]*Veterans Administration Medical Center, Tacoma, Wash-
ington 98493 USA; and* [4]*Department of Orthopaedics, University
of Rochester School of Medicine, Rochester, New York 14642 USA*

Vitamin D functions as a prohormone. Subsequent to its synthesis in der-
mal cells by photolysis of 7-dehydrocholesterol, vitamin D undergoes a series
of hydroxylations, the first occurring at carbon 25 to produce 25-
hydroxyvitamin D_3 [25-$(OH)D_3$]. 25-$(OH)D_3$ is further hydroxylated to more
polar metabolites, including 1,25-dihydroxyvitamin D_3[1,25-$(OH)_2D_3$], the
most potent known vitamin D metabolite in enhancing intestinal mineral
transport and in mobilizing bone mineral, and 24,25-$(OH)_2D_3$, a major
vitamin D metabolite whose physiological significance is less clearly understood (Fraser and Kodicek, 1970; Holick and Clark, 1978; Pochon and DeL-
uca, 1969).

The metabolism of 25$(OH)D_3$ to more polar products is closely regulated.
Although some progress has been made toward understanding the control of
25-$(OH)D_3$ metabolism by *in vivo* studies (reviewed by Holick and Clark,
1978; Fraser, 1980), the many uncontrolled variables inherent in whole
animal experiments point out the need for *in vitro* models. We describe here
an avian cell culture model that has proven useful for studies of vitamin D
metabolism, particularly in regard to the metabolic fate of 25-$(OH)D_3$, tissue
distribution of 25-$(OH)D_3$ metabolizing enzymes, and regulation of those
enzymes.

METHODS

Cells were dispersed from selective tissues of 15-16 day chicken embryos
(chorioallantoic membrane, kidney, liver, heart, intestine, and skin), two-day
old chickens (calvaria and kidney), two-week old chickens (kidney), and
adult male Japanese quail (kidney, liver, heart, and skin) by methods that
have been described in detail for the respective tissues (Howard *et al.*, 1979;
Turner *et al.*, 1980*b*; and Puzas *et al.*, 1980). Unless noted otherwise, the
animals were maintained on control diets containing adequate vitamin D,
calcium, and phosphorus. The dispersed cells were either incubated with [3]H-
25-$(OH)D_3$ in suspension, as has been described for rat kidney cells by

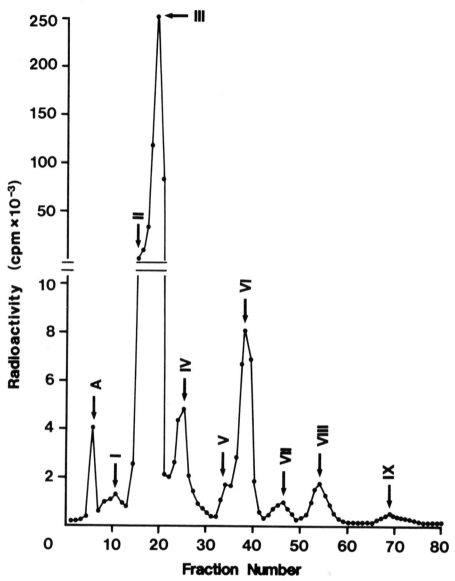

Fig. 1.—Sephadex LH-20 chromatography (1 × 56 cm column, 15 g Sephadex) of ^{3}H-25-(OH)D$_{3}$ metabolites from freshly isolated chick kidney cells. The cells were dispersed from birds fed adequate vitamin D and were incubated at 4 × 10^{6} cells/ml with 10 nM ^{3}H-25-(OH)D$_{3}$ for 1 hour at 37°C. The metabolites were extracted with dichloromethane, dried under N$_{2}$, and then separated using hexane:chloroform:methanol (9:1:1) as the solvent. The peak designated A is at the void volume. Peak III is ^{3}H-25-(OH)D$_{3}$ whereas peaks VI, VIII, and XI have been identified as ^{3}H-24,25-(OH)$_{2}$D$_{3}$, ^{3}H-1,25-(OH)^{2}D$_{3}$, and ^{3}H-1, 24,25(OH)$_{3}$D$_{3}$, respectively. The identities of peaks I, II, IV, V, and VII are not known.

TABLE 1.—*Production of selected vitamin D metabolites by several* in vitro *quail kidney models.*

Model system	Rate of production [fmol min^{-1} (10^6 cells)$^{-1}$]		
	Peak IV	24,25-(OH)$_2$D$_3$	1,25-(OH)$_2$D$_3$
Freshly isolated cells	390	1230	390
Homogenates of whole kidney	3	9	2
Mitochondria	37	66	15

Turner *et al.* (1980*a*), or plated and grown in 60 millimeter tissue culture dishes (Howard *et al.*, 1979; Turner *et al.*, 1980*b*; and Puzas *et al.*, 1980). Embryonic cells were grown in serum-free BJG$_b$ medium and other cells were grown in serum-free McCoys 5a medium. In one experiment, kidney cells from adult male Japanese quail kidney were grown in McCoys 5a medium supplemented with 10 per cent fetal calf serum. The metabolism of ^{3}H-25-(OH)D$_3$ by cultured avian cells was tested in confluent cultures as described by Howard *et al.* (1979). Vitamin D metabolites were extracted from the cultures with dichloromethane and separated by Sephadex LH-20 column chromatography (Howard *et al.*, 1979) followed by high performance liquid chromatography (HPLC) (Turner *et al.*, 1980*b*).

RESULTS

Metabolism of ^{3}H-25(OH)D$_3$ by Freshly Isolated Kidney Cells

Cells that were isolated from two-week-old chicks and suspended in serum-free McCoys 5a medium metabolized ^{3}H-25-(OH)D$_3$ to at least eight products. A representative Sephadex LH-20 profile of the distribution of these metabolites is shown in Fig. 1. Three of the metabolites have been identified with reasonable certainty as 1,25-(OH)$_2$D$_3$, 24,25-(OH)$_2$D$_3$, and 1,24,25-(OH)$_3$D$_3$ (Howard *et al.*, 1979). The structures of the other products remain unknown but are under investigation.

We have previously shown that the metabolism of 25-(OH)D$_3$ by freshly isolated rat kidney cells accurately reflects vitamin D metabolism *in vivo* (Turner *et al.*, 1980*a*). To verify that finding in birds we compared cells isolated from two-week old chicks that have been maintained from hatching on a rachitic diet with those fed a vitamin D normal diet. As expected (Figs. 1 and 2), renal 1-hydroxylase activity was elevated and 24-hydroxylase suppressed in the cells isolated from the vitamin D deficient chicks. The metabolism of 25-(OH)D$_3$ by freshly isolated cells is compared in Table 1 to two commonly used *in vitro* assays for vitamin D metabolism: whole kidney homogenates and isolated mitochondria. Although the qualitative distributions of the polar products in the three model systems were similar, isolated intact cells were much more efficient in metabolizing 25-(OH)D$_3$ than either homogenates (100-fold increase in specific activity) or mitochondria (10-fold increase in specific activity). As a result of the higher specific activities of the

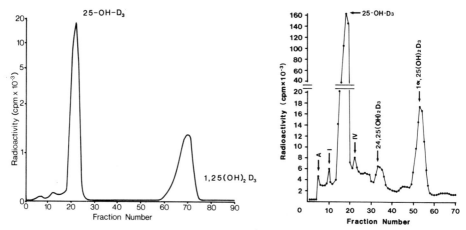

Fig. 2 (left).—Sephadex LH-20 chromatography of ^{3}H-25-(OH)D$_3$ metabolites from suspensions of freshly isolated chick kidney cells dispersed from birds fed a rachitic diet. The isolation and incubation conditions were the same as in Fig. 1. ^{3}H-25-(OH)D$_3$ was metabolized exclusively to ^{3}H-1,25-(OH)$_2$D$_3$ by kidney cells from the rachitic animals.

Fig. 3 (right).—Sephadex LH-20 chromatography of ^{3}H-25-(OH)D$_3$ metabolites from confluent monolayers of kidney cells dispersed from adult Japanese quail that had been fed a vitamin D normal diet. The cells (3 dishes containing a total of 3×10^6 cells) were incubated with 10 nM ^{3}H-25-(OH)D$_3$ for 2 hours. The metabolites were extracted and separated as described in Fig. 1. Notice the similarity of this profile with that produced by freshly isolated cells from rachitic chicks (Fig. 2).

1- and 24-hydroxylases, freshly isolated cells might be useful in assaying vitamin D metabolism when only small amounts of kidney are available.

Metabolism of 25-(OH)D$_3$ by Cultured Avian Kidney Cells

Cells that have been dispersed from adult male Japanese quail grew to confluence when plated in McCoy's 5a medium with and without serum. However, it is important to note that ^{3}H-25-(OH)D$_3$ (10 nM) was metabolized to more polar products only in serum-free cultures (Fig. 3). Henry (1979) has shown that the inhibitory effect of serum could be overcome by washing cells with serum-free medium followed by use of a much higher substrate concentration [>100 nM ^{3}H-25-(OH)D$_3$] during assay. It is interesting that cultured quail kidney cells metabolized ^{3}H-25-(OH)D$_3$ to only one major product, ^{3}H-1, 25-(OH)$_2$D$_3$, as if they were obtained from vitamin D-deficient animals instead of birds fed a vitamin D normal diet. A similar observation was made when cells were grown in Medium 199, a cell culture medium containing calciferol (Howard *et al.*, 1979). These findings suggest that 25-(OH)D$_3$ was depleted from kidney cells during culture and that they responded to this *in vitro* deficiency in an appropriate manner, that is, stimulation of 1-hydroxylase activity and inhibition of the 24-hydroxylase enzyme. To test the hypothesis that vitamin D metabolism is regulated by vitamin D metabolites, we added 10 nM of non-radioactive 25-(OH)D$_3$ to confluent kidney cell cultures and periodically assayed 1- and 24-hydroxylase

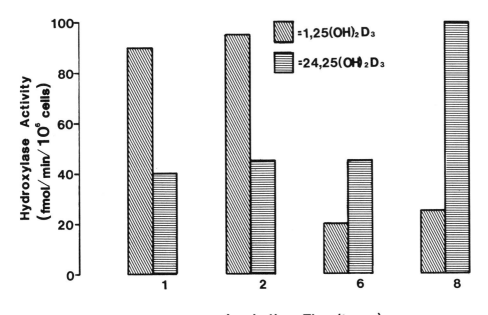

FIG. 4.—Metabolism of ³H-25-(OH)D₃ by monocultures of quail kidney cells as a function of incubation time. 1.3×10^6 cells were preincubated with 10 nM 1,25-(OH)₂D₃ for the times indicated on the abscissa, then assayed for 1- and 24-hydroxylase activites. Areas under the elution profiles as described in Fig. 1 were quantified and expressed as on the ordinate for 1,25-(OH)₂D₃ and 24,25-(OH)₂D₃.

activities. The results are shown in Fig. 4. As expected, preincubation with 25-(OH)D₃ resulted in a dramatic alteration of vitamin D metabolism by cultured cells to a profile resembling that of vitamin D normal animals at sacrifice (Fig. 1). Incubation with 1,25-(OH)₂D₃ had a similar effect on vitamin D metabolism *in vitro* (Howard *et al.*, 1979; Spanos *et al.*, 1978; Henry, 1979), suggesting that the effect of 25-(OH)D₃ was due to its further metabolism to 1,25-(OH)₂D₃. The regulation of vitamin D metabolism in cultured avian kidney cells by vitamin D metabolites as well as other regulatory factors such as PTH and sex hormones is of great interest and is under investigation in several laboratories (Henry, 1977; Howard *et al.*, 1979; Spanos *et al.*, 1978; Trechsel *et al.*, 1979).

Metabolism of 25-(OH)D₃ by Cultured Cells From Selected Embryonic Chicken Organs

The tissue distribution of the 1- and 24-hydroxylases was determined in 16-day chicken embryos by assaying for enzyme activity in cells isolated from selected organs. Cells from kidney, calvarium, and chorioallantoic membrane hydroxylated 25-(OH)D₃ to 1,25-(OH)₂D₃ and 24,25-(OH)₂D₃, whereas cells isolated from liver, skin, and heart did not appear to have that capacity (Tables 2 and 3). The specific activities of the 1-hydroxylase in embryonic

TABLE 2.—*Distribution of 1- and 24-hydroxylases in culture embryonic chicken cells.*

Cell origin	Specific activity [fmol min^{-1} (10^6 cells)$^{-1}$]	
	24-hydroxylase	1-hydroxylase
Kidney	0.2	7.2
Calvaria	3.8	1.2
Chorioallantoic membrane	ND[1]	3.4
Heart	ND	ND
Liver	ND	ND
Skin	ND	ND

[1]ND = below limit of detection [<0.1 fmol min^{-1} (10^6 cells)$^{-1}$].

chicken cells decreased in the following order: kidney > chorioallantoic membrane > calvarium, whereas 24-hydroxylase decreased as follows: calvarium > kidney > chorioallantoic membrane.

The Effect of Age on Metabolism of 25-(OH)D₃ by Cultured Cells

Relatively little is known concerning the effect of development on vitamin D metabolism in cultured cells, particularly with regard to extra-renal sites. Kidney cells isolated from embryonic chickens, two-day-old chicks, two-week-old chicks, and adult quail all grew to confluence following dispersal into culture and all synthesized 1,25-(OH)₂D₃ and 24,25-(OH)₂D₃ *in vitro.* Furthermore, as in the case of embryonic tissue, cells isolated from liver, heart, and skin of adult quail did not synthesize dihydroxyvitamin D metabolites. The specific activities of the 1- and 24-hydroxylases [0.9 and 2.2 fmol min^{-1} (10^{6}cells)$^{-1}$, respectively] in calvarial cells from 2-day-old chicks were similar to those in cells from 16-day embryos (Table 2), indicating that the extra-renal metabolism of 25-(OH)D₃ is not strictly limited to embryonic tissues. Attempts to isolate calvarial cells from older animals has proven difficult. However, intestinal cells from embryonic chickens appear to synthe-

TABLE 3.—*Metabolism of 25-(OH)D₃ by cultured embryonic chicken cells.*

Treatment	Specific activity [fmol min^{-1} (10^6 cells)$^{-1}$]	
	1-hydroxylase	24-hydroxylase
Calvarial Cells		
Control	0.81±0.09	3.14±0.02
Preincubation with 10 nM of 1,25-(OH)₂D₃	0.29±0.02	7.69±0.13
Chorioallantoic Membrane Cells		
Control	2.84±0.43	ND[1]
Preincubation with 10 nM of 1,25-(OH)₂D₃	0.92±0.09	1.02±0.09

[1]ND = below limit of detection [<0.1 fmol min^{-1} (10^6 cells)$^{-1}$].
Means ± SE.

size small amounts of 1,25-(OH)₂D₃ as well as 24,25-(OH)₂D₃. If confirmed, the intestine may prove to be a useful model for study of the extra-renal synthesis of vitamin D during development.

Regulation of Vitamin D Metabolism at Extra-renal Sites

Incubation with 1,25-(OH)₂D₃ altered the subsequent metabolism of ³H-25-(OH)D₃ by cultured cells from calvarium and chorioallantoic membrane (Table 3). Following an 8-hour preincubation with 10 nM 1,25-(OH)₂D₃, 1-hydroxylase specific activity was reduced to 35.8 per cent of the control value in calvarial cells and 32.4 per cent in cells from chorioallantoic membrane. There was a concurrent increase in 24-hydroxylase activity in both calvarial cells (245% of control value) and chorioallantoic membrane cells (>1,000% of the control). These results suggest that 1,25-(OH)₂D₃ regulates the metabolism of 25-(OH)D₃ at extra-renal sites as well as in kidney cells.

DISCUSSION

We have been able to explant viable cells from several avian tissues and grow then to confluence in serum-free, chemically defined medium. Elimination of serum as a necessary growth promoter constitutes a significant advance over previous cell culture models for vitamin D metabolism. Serum contains high-affinity, high-capacity binding sites for 25-(OH)D₃ that act to prevent *in vitro* expression of 1- and 24-hydroxylases by sequestering the substrate (Henry, 1979; Ghazarian *et al.*, 1978). Furthermore, serum contains variable amounts of substances such as steroid hormones that might have specific effects on vitamin D metabolism.

Confluent monolayers of avian kidney cells efficiently metabolized 25-(OH)₂D₃ to 24,25-(OH)₂D₃ and 1,24,25-(OH)₃D₃, as well as to several unidentified products (that is metabolites I, II, IV, V, and VII). The *in vitro* metabolism of 25-(OH)D₃ did not appear to be an artifact of cell isolation or culture. The synthesis of 1,25-(OH)₂D₃ and 24,25-(OH)₂D₃ did not occur in the absence of cells. Most of the metabolites produced by cultured kidney cells have been previously reported to be made by kidney homogenates (Turner *et al.*, 1979) and, in addition, have been identified in serum (Turner *et al.*, unpubl. data). Finally, the same metabolites were produced by freshly isolated kidney cells assayed in suspension, suspension culture being an *in vitro* model that faithfully reflects *in vivo* vitamin D metabolism at time of sacrifice (Turner *et al.*, 1980a).

We have investigated the tissue distribution of 1- and 24-hydroxylases in birds and have identified both enzymes in cultured cells isolated from kidney, calvarium, and chorioallantoic membrane. The presence of 1-hydrozylase activity in cells from the latter two tissues was unexpected because the kidney has long been regarded as the unique site of synthesis of 1,25-(OH)₂D₃ (Fraser and Kodicek, 1970). However, there is at least one precedence for extra-renal 1-hydroxylase activity: the placenta has recently been reported to produce 1,25-$(OH)_2D_3$ (Weisman *et al.*, 1979; Tanaka *et al.*, 1979). The presence of 24-hydroxylase activity in cells from calvarium and chorioallantoic membrane is less surprising because extra-renal 24,25-$(OH)_2D_3$ production has been reported in several tissues, including intestine and cartilage (Garabedian *et al.*, 1978).

The significance of extra-renal metabolism of 25-$(OH)D_3$ is unclear, but probably it is significant that the calvarium and chorioallantoic membrane are both vitamin D target organs (Narbaitz and Tolnai, 1978) concerned with mineral transport. Production of active vitamin D metabolites in vitamin D target organs may provide a means of local regulation of the response to vitamin D. This duplication of synthetic activity might provide a more versatile control of calcium and phosphorus metabolism. For example, under certain physiological conditions there could be some advantage for a differential response by gut and bone as mediated by different amounts or types of locally determined vitamin D metabolites.

The results of the present study indicate that extra-renal metabolism of 25-$(OH)D_3$ is altered *in vitro* by at least one important regulating factor [1,25-$(OH)_2D_3$] in a manner qualitatively similar to that in kidney cells. In all avian tissues in which 1-hydroxylase activity has been identified, 1,25-$(OH)_2D_3$ inhibits the activity of that enzyme while enhancing 24-hydroxylase activity. However, it is important to note that the absolute activities of 1- and 24-hydroxylases as well as the relative distribution of the two enzymes appear to be tissue specific.

In summary, we have described a primary cell culture model that we have shown to be applicable to studies of vitamin D metabolism in avian cells from several tissues (kidney, liver, skin, heart, intestine, calvarium, and chorioallantoic membrane), stages of development (embryonic, chick, and adult bird), and different species (chicken and quail). By use of this model we have shown that cells from kidney, chorioallantoic membrane, and calvarium metabolize 25-$(OH)D_3$ to 1,25-$(OH)_2D_3$. This finding constitutes evidence that the distribution of the 25-hydroxyvitamin D_3-1-hydroxylase enzyme is more widespread than previously believed. We expect that the *in vitro* model described here will be of use in further defining the metabolic pathways of vitamin D and their regulation in extra-renal sites as well as in the kidney.

ACKNOWLEDGMENTS

We thank Brian L. Bottemiller and Mario D. Forte for their expert technical assistance. This work was supported by the Research Service of the Veterans Administration.

LITERATURE CITED

FRASER, D. R. 1980. Regulation of metabolism of vitamin D. Physiol. Rev., 60:551-613.
FRASER, D. R., AND E. KODICEK. 1970. Unique biosynthesis by kidney of a biologically active vitamin D metabolite. Nature (London), 228:764-766.

GARABEDIAN, M., M. BAILLY-DUBOIS, M. T. CORVOL, E. PEZANT, AND S. BALSAN. 1978. Vitamin D and cartilage. I. *In vitro* metabolism of 25-hydroxycholecalciferol by cartilage. Endocrinology, 102:1262-1268.

GHAZARIAN, J. G., B. KREAM, K. M. BOTHAM, M. W. NICKELLS, AND H. F. DELUCA. 1978. Rat plasma 25-hydroxyvitamin D binding protein: an inhibitor of the 25-hydroxyvitamin D$_3$-1α-hydroxylase. Arch. Biochem. Biophys., 189:212-220.

HENRY, H. L. 1977. Metabolism of 25-hydroxyvitamin D$_3$ by primary cultures of chick kidney cells. Biochem. Biophys. Res. Commun., 74:768-774.

———. 1979. Regulation of the hydroxylation of 25-hydroxyvitamin D$_3$ *in vivo* and in primary cultures of chick kidney cells. J. Biol. Chem., 254:2722-2729.

HOLICK, M. F., AND M. B. CLARK. 1978. The photo-biogenesis and metabolism of vitamin D. Fed. Proc., 37:2567-2574.

HOWARD, G. A., R. T. TURNER, B. L. BOTTEMILLER, AND J. L. RADER. 1979. Serum-free culture of Japanese quail kidney cells: regulation of vitamin D metabolism. Biochem. Biophys. Acta, 587:495-506.

NARBAITZ, R., AND S. TOLNAI. 1978. Effects produced by the administration of high doses of 1,25-dihydroxycholecalciferol to the chick embryo. Calc. Tiss. Res., 26:221-226.

PONCHON, G., AND H. F. DELUCA. 1969. The role of the liver in the metabolism of vitamin D. J. Clin. Invest., 48:1273-1279.

PUZAS, J. E., R. T. TURNER, M. D. FORTE, A. D. KENNY, AND D. J. BAYLINK. 1980. Metabolism of 25(OH)D$_3$ to 1,25(OH)$_2$D$_3$ and 24,25(OH)$_2$D$_3$ by chick chorioallantoic cells in culture. Gen. Comp. Endocrinol., 42:116-122.

SPANOS, E., D. I. BARRETT, K. T. CHONG, AND I. MACINTYRE. 1978. Effect of oestrogen and 1,25-dihydroxycholecalciferol metabolism in primary chick kidney-cell cultures. Biochem. J., 174:231-236.

TANAKA, Y., B. HALLORAN, H. SCHNOES, AND H. F. DELUCA. 1979. *In vitro* production of 1,25-dihydroxyvitamin D$_3$ by rat placental tissue. Proc. Nat. Acad. Sci., U.S.A., 76:5033-5035.

TRECHSEL, U., J-P. BONJOUR, AND H. FLEISCH. 1979. Regulation of the metabolism of 25-hydroxyvitamin D$_3$ in primary cultures of chick kidney cells. J. Clin. Invest., 64:206-217.

TURNER, R. T., J. I. RADER, L. P. ELIEL, AND G. A. HOWARD. 1979. Metabolism of 25-hydroxyvitamin D$_3$ during photo-induced reproductive development in female Japanese quail. Gen. Comp. Endocrinol., 37:211-219.

TURNER, R. T., B. L. BOTTEMILLER, G. A. HOWARD, AND D. J. BAYLINK. 1980a. *In vitro* metabolism of 25-hydroxyvitamin D$_3$ by isolated rat kidney cells. Proc. Nat. Acad. Sci., U.S.A., 77:1537-1540.

TURNER, R. T., J. E. PUZAS, M. D. FORTE, G. E. LESTER, T. K. GRAY, G. A. HOWARD, AND D. J. BAYLINK. 1980b *In vitro* synthesis of 1α 25-dihydroxycholecalciferol and 24,25-dihydroxycholecalciferol by isolated calvarial cells. Proc. Nat. Acad. Sci., U.S.A., 77:5720-5724.

WEISMAN, Y., A. HARRELL, S. EDELSTEIN, M. DAVID, Z. SPIRER, AND A. GOLANDER. 1979. 1α, 25-dihydroxyvitamin D$_3$ *in vitro* synthesis by human decidua and placenta. Nature (London), 281:317-319.

SECTION V

AVIAN OSMOREGULATORY MECHANISMS IN PHYLOGENETIC PERSPECTIVE

INTRODUCTION

Colin G. Scanes

Department of Physiology, Rutgers-The State University, New Brunswick, New Jersey 08903 USA

Birds offer a number of interesting models for the study of osmoregulation and its hormonal control. Obvious examples include the salt gland for salt excretion and the production of nearly solid urine containing urate salts. Avian species are found in very disparate environments including tropical (both arid deserts and more moist habitats), temperate, tundra and polar, as well as marine and fresh water habitats. These environmental conditions have profoundly influenced the hormonal control mechanisms which effect water and electrolyte homeostasis in these animals.

This section considers only several of many aspects of osmoregulation in birds. Balment places the osmoregulatory control systems in birds into an evolutionary perspective. In particular, he considers the importance of both the metanephric kidney and the extrarenal organs (for example, salt gland and lower intestine) in the control of salt and water balance.

The role of prolactin in avian salt and water balance is given detailed consideration. Phillips and Harvey give a critical but balanced analysis of most of the available information on the role of prolactin in the control of avian osmoregulation. This account is strengthened by their inclusion of substantive unpublished data. Ensor presents a succinct case for the importance of prolactin in osmoregulatory homeostasis in birds and places this action of prolactin in a phylogenetic perspective.

The final paper in the section by Tanabe discusses the ontogeny of steroidogenesis (corticosterone and the sex steroids) by the avian adrenal gland and gonads. This topic impinges on osmoregulation in birds peripherally because of the influence of corticosterone on the salt gland. Tanabe also presents perhaps one of the most intriguing ideas found in this volume, when she proposes that the high activity of the female gonad is related to the female heterozygosity (ZW) (compared to the male heterozygosity XY in male mammals).

Reprinted from
ASPECTS OF AVIAN ENDOCRINOLOGY:
PRACTICAL AND THEORETICAL IMPLICATIONS (C. G. Scanes *et al.*, eds.)
Grad. Studies, Texas Tech Univ., 1982, 26:1-411.

EVOLUTIONARY CONSIDERATION OF
VERTEBRATE OSMOREGULATORY MECHANISMS

R. J. BALMENT

*Department of Zoology, University of Manchester, Manchester
M13 9PL, England*

Attempts to place evolutionary considerations on any aspect of vertebrate physiology are fraught with problems and, of necessity, must involve many assumptions and generalizations. The development of osmoregulatory mechanisms can only be deduced from extant species, but animals adapt and specialize to particular environments, thereby overshadowing the precise delineation of progressive steps in the lines of evolutionary change. This problem is exacerbated by the scanty information available for nonmammalian species. Nonetheless, such phylogenetic analysis is not only rewarding in our understanding of endocrinology, but can also reveal some broad trends that may help in the comprehension of wider issues.

Maintenance of body fluid by careful balance of water and electrolyte uptake and excretion is an important vertebrate characteristic. The body fluid composition of modern adult vertebrates is very similar, although the pattern is somewhat marred by the elasmobranchs and myxinoids. Mechanisms are directed toward the maintenance of a constant internal environment, and deviations in one direction bring opposite reactions, according to Bernard's principle of negative feedback for homeostasis (Greene, 1957; Langley, 1965). Thus individual ion concentration, fluid volume, and osmotic concentration are sustained in diverse osmotic environments. The greatest necessity is the management of sodium and chloride ions coupled with potassium ions. Thence follows regulation of osmotic pressure of blood and blood volume.

If we are to gain insight into the possible pattern of evolutionary development of the endocrines involved in the osmoregulatory process, mere demonstration of hormone presence and activity is insufficient; it is essential to evaluate the contribution of such responses to the overall water and electrolyte management problems of each group. An initial evaluation of the evolutionary development of vertebrate osmoregulatory organs therefore would be helpful in our present task.

In all vertebrate groups, the kidney plays a major role in fluid volume regulation, which may reflect the primitive function of the ancestral glomerular kidney. By contrast, in lower vertebrates, management of sodium, potassium, and chloride content is largely performed by extra-renal organs (gills, rectal gland, skin, bladder, and 'salt' gland). Only in the Amniota (reptiles, birds, and mammals) does the kidney emerge as an important site of electrolyte management. Following presumptive lines of vertebrate evolu-

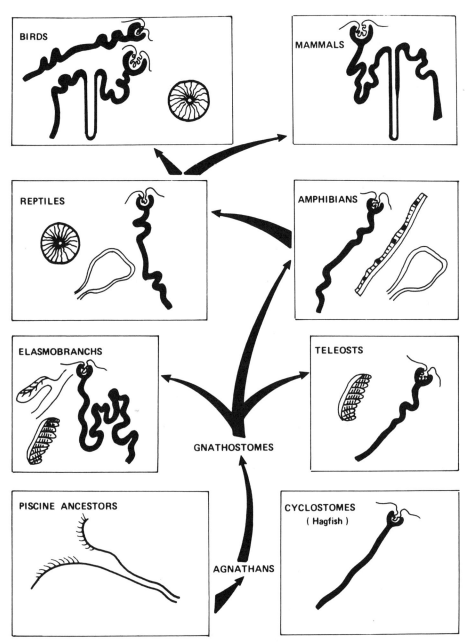

FIG. 1.—Osmoregulatory organs of the vertebrates. The extra-renal organs (gill, rectal gland, bladder, skin, and salt gland) of lower vertebrates are gradually replaced by the increasingly elaborate renal apparatus.

tion, it is possible to trace a gradual elaboration of renal function and a reciprocal diminution of reliance upon extra-renal devices (see Fig. 1). One might suggest that a linking thread, in our present consideration of the evolution of vertebrate osmoregulatory mechanisms, is the development of the renal apparatus.

The basic renal unit, the nephron or renal tubule, has a glomerular component for filtering blood and tubular components for both reabsorption from, and secretion into, the forming urine. The hypothetical ancestral form of nephron (Goodrich, 1958) consisted of a coelomoduct, into which later was inserted the glomerular tuft. Such a structure is similar to that found as the nephron of larval hagfish. In adult myxinoids, the vertebrate type nephron is seen with glomerular tuft and simple tubule leading to the cloaca (Fig. 1). Later vertebrate groups have segments added to this basic tubular structure. These components, joined together to form the kidney, are so ideally suited for the excretion of large volumes of water that this has been taken as evidence supporting a freshwater origin for vertebrates (Marshall and Smith, 1930; Smith, 1932).

Fossil evidence, however, fails to support this concept, and, avoiding designation of the hagfish as a secondary and degenerate form, it can be proposed that this marine group is representative of an interphase animal between the invertebrate and vertebrate groups. Moreover, the "kidney" of the decapod crustaceans functions in much the same way as the vertebrate kidney, that is, as a filtration-reabsorption-secretion system (Ramsay, 1968). In this case, it must have arisen from a marine ancestry and yet, in crustaceans, which have since invaded the freshwater environment, an ion absorption segment has been added. The picture presented is that of a crustacean kidney developed primitively as an organ of ionic balance and that has secondarily assumed the function of fluid excretion in those animals that entered estuarian and freshwater niches. Similar considerations might suggest, contrary to the embedded Homer Smith concept, that the glomerulous of the vertebrate kidney appeared first in organisms in the seawater environment as a device for the control of ionic balance. This be later proved to be a useful preadaptation to freshwater habitats (Robertson, 1957). In the conquest of the new hypotonic environment, the function of volume regulation was added to the kidney in association with the development of an extra segment, the proximal tubule, for the major obligatory reabsorption of ions. The subsequent evolution of the kidney involved the addition of extra tubule segments that thereby further modified the functional capacity of the basic renal unit. Thus, urine volume regulated by the rate of glomerular filtration and the various other established functions such as divalent ion secretion and urinary dilution, are retained in one form or another by all other vertebrates.

The movement to land is associated with the addition of further distal nephron segments. Selective, specific reabsorption and secretion of sodium, potassium, and water is thus possible. The importance of these segments increases from Amphibia to reptiles and birds. With the emergence of mammals, particularly the dominant Eutheria, a further distinctive segment, the loop of Henle, is inserted in the nephron (Fig. 1). This loop is capable of generating a concentration gradient within the renal interstitium and augments water reabsorption from the urine via the collecting duct. Loops of Henle are also found in some birds where they are interspersed with simpler reptilian nephron types (Braun and Dantzler, 1972).

Birds and mammals are distinguished in the vertebrate series in having the capacity to produce a urine that is hypertonic to plasma. This has important implications not only in osmotic regulation but also in solute excretion. Solutes can be excreted in hypertonic urine without excessive depletion of body water. Thus, compared with the other vertebrate groups, mammals, and to a lesser extent birds, have an increased capacity for renal salt excretion. The highly developed renal concentrating mechanism of the Mammalia permits salt excretion to be managed entirely by the kidney; no extrarenal salt glands are required. The more modest renal concentrating capacity of birds (2 or 3 times plasma at most) might be supplemented by cloacal reabsorption but many marine or estuarine species, faced with a high salt intake, rely on auxiliary extra-renal salt glands for the excretion of highly hypertonic saline solutions. However, the limited concentrating ability of the avian kidney and the absence of any concentrating ability of the reptilian kidney do not necessarily reflect the true capacity of these kidneys for the excretion of inorganic cations. Complexing of cations with uric acid in these groups allows considerable solute excretion with little contribution to urine osmolarity. As much as 75 per cent of the sodium and 34 per cent of potassium in the ureteral urine of the domestic chicken is associated with the urate precipitates (McNabb et al., 1973). In the remaining lower vertebrate groups, where the kidney is capable of producing hypotonic or at best isotonic urine, the important monovalent ions sodium, potassium, and chloride are largely under extra-renal (gill, rectal gland, skin, bladder) management.

The gradual emergence of the kidney in the vertebrate series as the major osmoregulatory organ is reflected in the activities of the two principle endocrine centers (neurohypophysis and adrenal cortex) involved in osmomineral management. The interplay of neurohypophysial and adrenocortical hormone actions centered on the kidney is clearly seen in the eutherian mammal. In other vertebrates, either or both of these sets of hormones act upon extra-renal, osmoregulatory organs.

The neurohypophysis of jawed vertebrates contains two octapeptide hormones. They epitomize, as do steroids, the conversion of basic chemical compounds universally found in living matter into recognizable hormones. Arginine vasotocin (AVT), the basic neurohypophysial peptide, is found in all nonmammalian groups, from cyclostomes to birds, and is only replaced in mammals by arginine and lysine vasopressin. Thus, AVT has survived 400 million years in several distinct evolutionary lines (Sawyer, 1967) as the primary neurohypophysial peptide involved in osmomineral management.

In jawed vertebrates, the basic peptide is present with a second neutral octa-peptide that shows considerably less conservative phyletic variation, the properties of which are largely attributed to reproductive physiology. Birds share the same peptide, mesotocin, with lungfishes, amphibians, and rep-tiles. Recent work in our laboratory questions this dismissal of a hydroso-motic role for the neutral peptide, even in the Mammalia (Balment et al., 1981).

More than 50 different steroids have been isolated from and attributed to the adrenal cortex, yet only a small number appear in the venous blood car-rying adrenocortical secretions and, of these, few have relevant biological activity. Functional adrenal steroids influence carbohydrate, lipid, and pro-tein metabolism, and are associated with anti-inflammatory effects: cortisol and corticosterone, and the less dominant cortisone, 11-dehydrocortisone, 11-deoxycortisol, and 21-deoxycortisol. The second functional category of adre-nocortical steroids regulates electrolyte relationships, especially those of sodium and potassium, and here aldosterone and deoxycorticosterone are especially important. The distinction between these glucocorticoid and mineralocorticoid activities is one of convenience but frequently not of fact and, once we leave the eutherian mammals, their relative importance varies from group to group. Representatives of the Tetrapoda so far examined pos-sess aldosterone as do lungfishes. However, its production in gnathostomat-ous fish is in doubt. Birds, reptiles, and elasmobranchs also have 11-hydroxycorticosterone in significant amounts.

In teleost fish, glomerular filtration and urine flow rates can be controlled by neurohypophysial hormones without tubular components (Fig. 2). This might involve alteration in both the number of filtering nephrons (glomeru-lar intermittency) and the rate of filtration in individual glomeruli (Brown et al., 1978). However, tubular responsiveness to AVT and isotocin with respect to osmotically important ions might still be significant. It might be suggested that regulation of renal excretion by neurohypophysial peptides through changes in glomerular filtration and specific tubular electrolyte handling is an early primitive feature of gnathostomes. The water and elec-trolyte contents of teleosts are greatly influenced by neurohypophysial pep-tides acting at both renal and extra-renal sites (Table 1). It might be sup-posed that these actions are short term ones, giving quick responses to environmental changes while the corticosteroids, having in many cases parallel influences, provide a longer term adaptive control. Ionic control is primarily displayed by the gills, and both uptake and excretion of sodium, depending on the tonicity of the external fluid, are affected by neurohypo-physial and adrenocortical hormones.

Not enough is known about the Dipnoi, or even the elasmobranchs, to fit them with any certainty into this type of discussion. Amphibians, like fish, have a mesonephros which does not show dominant reactions to the two sets of hormones under discussion. It is again the extra-renal sites which attract the major attention—especially the skin and bladder (Table 1). This is, of

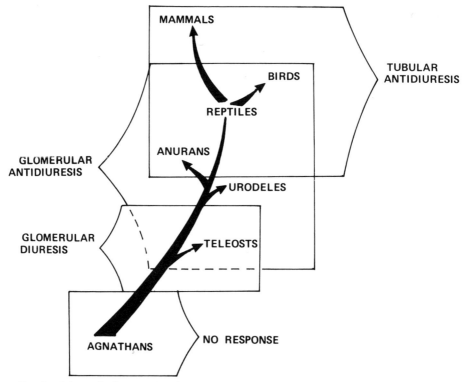

Fig. 2.—Schematic diagram showing the proposed phyletic distribution of renal responses to the neurohypophysial hormones (modified from Sawyer, 1972).

course, a question of degree because it is clear that the kidney is an essential organ; there is a sharing of responsibility for the homeostasis of the organism between various structures. In the Amphibia, neurohypophysial hormones can influence urine production through altered glomerular filtration rate and, to a lesser extent in anurans, distal tubular water reabsorption (see Fig. 2). Peptide hormones also might affect handling of renal tubular sodium. The major site of sodium management, at the skin and bladder, comes under the influence of both hormone centers (Table 1). Aldosterone is present in the Amphibia, but no clear role has yet been assigned to this hormone.

The emergence of the Amniota is associated with a trend toward specialization of hormone action. Thus the neurohypophysial peptides become largely involved with management of water content, and adrenocorticosteroids are concerned with renal and extra-renal salt handling (see Table 1). The amniote metanephric kidney is also seen to take a more dominant osmoregulatory role. The tubular component of the neurohypophysial peptide-induced antidiuresis becomes more prominent, whereas the glomerular component is perhaps of importance in more severe dehydration (see Fig. 2), and might have no physiological significance in mammals. It is of interest here that in

TABLE 1.—*Summary of neurohypophysial and adrenocortical hormone actions on vertebrate sodium metabolism.*

	Neurohypophysial Peptide	Adrenal Steroid
	Extra-Renal	
Mammals		
Birds		Loss by salt gland
Reptiles		Loss by salt gland
		Uptake across bladder
Amphibians	Uptake across skin & bladder	Uptake across skin & bladder
Teleosts	Transport across gills	Transport across gills
Elasmobranchs	?	?
	Renal Tubule	
Mammals	(Fractional excretion)	Fractional excretion
Birds	Fractional excretion	Fractional excretion
Reptiles	Fractional excretion	Fractional excretion
Amphibians	Fractional excretion	?
Teleosts		?
Elasmobranchs	?	?

the heterogenous avian kidney, the AVT-induced reduction in glomerular filtration rate is largely a result of a reduction in the number of filtering reptilian, rather than mammalian, type nephrons (Braun and Dantzler, 1974). However, a direct effect of AVT on distal tubule water permeability has yet to be demonstrated in avian species and might only be present in some ophidian and chelonian reptiles (see Dantzler, 1978). Indeed, AVT does not apparently stimulate renal cyclic AMP production, suggesting that the avian distal nephron lacks AVT receptors. The corticosteroids have a profound influence on total osmotic concentration of body fluids through their actions upon the movements of sodium, potassium, and chloride ions. Inevitably, this also allows influence on total fluid volume, thus underlying, in the mammals at least, the conjugation of actions of vasopressin and aldosterone. A similar intertwining of influences may be claimed for the birds and reptiles, though with less force.

It is of interest to pursue the analysis of kidney and adrenal interrelationships. In addition to the established influence of adrenal steroids on renal function, the kidney plays a clear role in the determination of adrenal morphology and secretory activity. The close anatomical association of the two tissues belies a common embryological origin. The Gnathostomata present a recognizable morphological evolution of the adrenal gland that is bound up with the kidney, gonads, and neural crest (Chester Jones, 1976). This is exemplified during embryological development in that the eventual fate of the adrenocortical homologue is closely related to the fate of the mesonephric blastema. Amphibia and elasmobranch fishes have a persistent mesonephros in the adult, and the adrenocortical homologue is accordingly

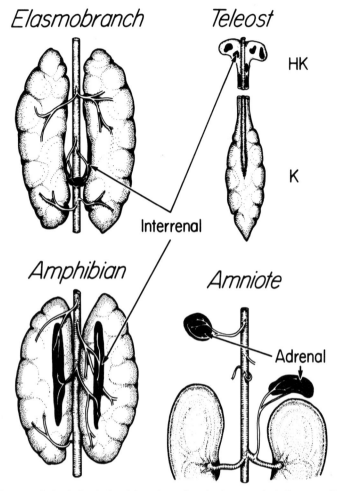

Fig. 3.—The morphological relationships of renal and adrenocortical (inter-renal) tissues in the major vertebrate groups. HK, head kidney; K, kidney (renal tissue is stippled).

found associated with the renal tissue (Fig. 3). Teleost fish are unusual in possessing a functional pronephros in early larval stages, at which time the adrenocortical tissue is laid down. The adult teleost mesonephros develops later, posterior to the pronephros, leaving behind the anterior adrenocortical tissue. The remaining amniote classes (Mammalia, Aves, and Reptilia) have discrete encapsulated adrenal glands (Holmes and Phillips, 1976; Idelman, 1978; Lofts, 1978). This is an inevitable concomitant of the appearance of the metanephros, leaving the adrenocortical tissue separated with the meso-nephros of the embryo which has only vestigial remnants in the adult.

The regulation of 17-hydroxycorticosteroids, as well as aldosterone in most non-mammalian groups examined thus far, is under negative feedback control based upon the hypothalamic/pituitary CRF-ACTH system. The Mammalia differ in that aldosterone synthesis shows considerable independence of pituitary control and, instead, the renal renin-angiotensin-system (RAS) is a major influence (see Davis, 1975). Although aldosterone titers in fish can be slight and spasmodic, it might be the major component of steroid secretion in some tetrapod vertebrates. In amphibians, reptiles, and birds, aldosterone has a full complement of glucocorticoid activity and, because of the large amounts secreted, it might have a physiological role in carbohydrate metabolism in these groups. Although aldosterone still retains its glucocorticoid activity in mammals, it has no physiological role in carbohydrate metabolism because of the low levels produced. Indeed, it is perhaps then not surprising to find that the mode of aldosterone secretion has changed and shows some independence of the pituitary. The basic elements of the renal RAS are present in all other jawed vertebrates except possibly elasmobranchs (Sokabe and Ogawa, 1974), but its physiological influence upon adrenal steroidogenesis remains equivocal in non-mammalian vertebrates.

The phylogeny of these differences in mode of aldosterone regulation is unclear. The reliance of the so-called mineralocorticoid release upon a renal hormone, renin, might indeed, be an early acquisition and central nervous control of ACTH release, and be an attribute of those species that begin delicately to control less obvious features of their metabolism such as liver glycogen. In mammals and birds, renin is released in response to sodium depletion. This contrasts with representative teleosts and amphibians in which elevated environmental salinities might not change or indeed increase plasma renin activity (PRA). Phylogenetically, therefore, there seems to be a dichotomy between homoiothermic and poikilothermic vertebrates; in the former, the renin release results in a stimulation of aldosterone production and thence renal sodium retention. Teleost fishes present further problems in that sodium loading provokes an increase in PRA and an increase in adrenocortical activity. This contrasts markedly with the tetrapod responses. It might be that elevated renin activities in marine environments are related not so much to renal sodium excretion, but to the relative polydipsia and reduced rates of glomerular filtration (Hirano et al., 1978).

The dipsogenic activity of angiotensin, at least in mammals and birds, and even perhaps teleost fishes, reflects an interplay between the kidney, the adrenocortical steroids, and neuropophysial antidiuretic peptides. Angiotensin and antidiuretic hormone in mammals are in negative feedback with one another, and this can satisfactorily explain general aspects of water balance. In the mammalian, and probably also avian, kidney, urinary dilution in the ascending limb of the loop of Henle and distal tubular convolutions is dependent upon adrenal steroids, whereas subsequent water reabsorption (the extent of which depends upon the corticomedullary gradient established by the diluting process) is governed by antidiuretic hormone. The amount of water available at these various sites is of course governed by the rate of glomerular filtration. Angiotensin thus might be viewed phylogenetically as a renal product that changes glomerular filtration by vasoconstriction, and

influences electrolyte and water balance by acting on both adrenal steroid
output and neurohypophysial hormone release. Whether these functions
were acquired simultaneously is not known. An attractive hypothesis is that
the proximity of the sites of renin release, and renin's intrarenal role of controlling the renal vasculature and interrenal steroid production, were functionally related. With gradual separation of renal and adrenocortical tissues,
angiotensin took a more systemic role to influence not only antidiuretic
hormone release but also to regulate systemic blood pressure. Such a broadening of renal-adrenocortical interactions allows a more comprehensive and
precise integration of the mechanisms of body fluid management.

A theme to which comparative endocrinology constantly returns is one
that considers the evolution of hormonal pattern in relation to tissue
response. The latter might indeed not be a central question in contemplat-
ing the appearance of complex endocrine interrelations. Thus it has been
shown that mammalian ACTH acts in cyclostomes to raise titers of corticosteroids (Idler *et al.*, 1971), but the presence of these steroids as normally circulating materials is still in doubt. Tissue reactivity does not have to evolve,
but the production of a familiar, widespread chemical type does. However,
the evolutionary trend toward refinement and specialization of target tissues,
including the osmoregulatory organs, is paralleled by changes in the actions
of the controlling hormones. Thus, on one hand, the development of separate and specific ionic and water regulatory mechanisms within the kidney, to
the exclusion of extra-renal devices, is matched on the other by the separation of hormone activities into largely ionic regulatory functions for steroid
hormones and water management roles for the neurohypophysial peptides.
In the eutherian mammal, the peak of specialization is perhaps reached with
the emergence of new specific antidiuretic hormones (vasopressins) and a
potent, largely pituitary-independent, mineralocorticoid-aldosterone. The
evolution of hormones has also seen refinement and perhaps multiplication
of factors that control their secretion. Thus the mammalian adrenal cortex
comes under the influence of many factors including ACTH, the renin-
angiotensin-system, potassium, and serotonin. At what stage in evolution
these influences gained predominance cannot be judged readily, and there
remains a great deal to be done before tentative generalizations can be made.

LITERATURE CITED

BALMENT, R. J., M. J. BRIMBLE, AND M. L. FORSLING. 1981. Release of oxytocin induced by
    salt loading and its influence on renal excretion in the male rat. J. Physiol. (London),
    308:439-449.
BRAUN, E. J., AND W. H. DANTZLER. 1972. Function of mammalian-type and reptilian-type
    nephrons in kidney of desert quail. Amer. J. Physiol., 222:617-629.
———. 1974. Effects of ADH on single-nephron glomerular filtration rates in the avian kid-
    ney. Amer. J. Physiol., 226:1-8.
BROWN, J. A, B. A. JACKSON, J. OLIVER, AND I. W. HENDERSON. 1978. Single nephron filtra-
    tion rates (SNGFR) in the trout, *Salmo gairdneri*. Pflugers Archiv European J. Phy-
    siol., 377:101-108.

CHESTER JONES, I. 1976. Evolutionary aspects of the adrenal cortex and its homologues. J. Endocrinol., 71:1P-31P.

DANTZLER, W. H. 1978. Some renal glomerular and tubular mechanisms involved in osmotic and volume regulation in reptiles and birds. Pp. 187-201, *in* Osmotic and volume regulation (C. Barker Jorgensen and E. Skadhauge, eds.), XI Alfred Benzon Symposium, Munksgaard, Copenhagen, 512 pp.

DAVIS, J. O. 1975. Regulation of aldosterone secretion. Pp. 77-106, *in* Handbook of physiology, Section 7, Endocrinology, Vol. VI (H. Blaschko, G. Sayer, and A. D. Smith, eds.), Amer. Physiol. Soc., Washington, D.C., xii+742 pp.

GOODRICH, E. S. 1958. Studies on the structure and development of vertebrates. Vol. 2, Dover Publ., Inc., New York, and Constable and Co., Ltd., London, lxix+485 pp.

GREENE, H. C. G. 1957. An introduction to the study of experimental medicine. Translation of Claude Bernard's work of 1855, 1857. Dover Publ., Inc., New York, 154 pp.

HIRANO, T., Y. YAKEI, AND H. KOBAYASHI. 1978. Angiotension and drinking in the eel and the frog. Pp. 123-128, *in* Osmotic and volume regulation (C. Barker Jorgensen and E. Skadhauge, eds.), XI Alfred Benzon Symposium, Munksgaard, Copenhagen, 512 pp.

HOLMES, W. N., AND J. G. PHILLIPS. 1976. The adrenal cortex of birds. Pp. 293-420, *in* General, comparative and clinical endocrinology of the adrenal cortex, Vol. 1 (I. Chester Jones and I. W. Henderson, eds.), Academic Press, London, New York, and San Francisco, xv+461 pp.

IDELMAN, I. 1978. The structure of the mammalian adrenal cortex. Pp. 1-200, *in* General, comparative and clinical endocrinology of the adrenal cortex, Vol. 2 (I. Chester Jones and I. W. Henderson, eds.), Academic Press, London, New York, and San Francisco, xv+461 pp.

IDLER, D. R., G. B. SANGALANG, AND M. WEISBART. 1971. Are corticoids present in the blood of all fish? Pp. 983-989, *in* Hormonal steroids (V. H. T. James and L. Martini, eds.), Excerpta Medica, I.C.S. 219, Amsterdam, xvi+1063 pp.

LANGLEY, L. L. 1965. Homeostasis. Reinhold Book Corp., New York, Amsterdam, and London, xi+362 pp.

LOFTS, B. 1978. The adrenal gland in Reptilia. Pp. 292-369, *in* General, comparative and clinical endocrinology of the adrenal cortex, Vol. 2 (I. Chester Jones and I. W. Henderson, eds.), Academic Press, London, New York, and San Francisco, xv+461 pp.

MARSHALL, E. K., AND H. W. SMITH. 1930. The glomerular development of the vertebrate kidney in relation to habitat. Biol. Bull. Mar. Biol. Lab., Woods Hole, 59:135-153.

McNABB, R. A., F. M. A. McNABB, AND A. P. HINTON. 1973. The excretion of urate and cationic electrolytes by the kidney of the male domestic fowl (*Gallus domesticus*). J. Comp. Physiol., 82;47-57.

RAMSAY, J. A. 1968. A physiological approach to the lower animals. Second Edition, Cambridge Univ. Press, Cambridge, 148 pp.

ROBERTSON, J. D. 1957. The habitat of early vertebrates. Biol. Rev., 32:156-187.

SAWYER, W. H. 1967. Evolution of antidiuretic hormones. J. Med., 42:678-686.

———. 1972. Neurohypophysial hormones and water and sodium excretion in African lungfish. Gen. Comp. Endocrinol., Suppl., 3:345-349.

SMITH, H. W. 1932. Water regulation and its evolution in fishes. Quart. Rev. Biol., 7:1-26.

SOKABE, H., AND M. OGAWA. 1974. Comparative studies of the juxtaglomerular apparatus. Int. Rev. Cytol., 37:271-327.

Reprinted from
ASPECTS OF AVIAN ENDOCRINOLOGY:
PRACTICAL AND THEORETICAL IMPLICATIONS (C. G. Scanes *et al.*, eds.)
Grad. Studies, Texas Tech Univ., 1982, 26:1-411.

A REAPPRAISAL OF THE ROLE OF PROLACTIN
IN OSMOREGULATION

J. G. PHILLIPS AND S. HARVEY

The Wolfson Institute, University of Hull, Hull HU6 7 RX
England

Prolactin has many physiological actions (see Bern and Nicoll, 1968; Chadwick, 1969; Nicoll and Bern, 1972; Nicoll, 1974, 1978, for reviews) but in mammals and other vertebrates it is mainly involved in reproduction and the production of milk and milk-like secretions for the young (Bern, 1975*a*, 1975*b*; Chadwick, 1977; De Vlaming, 1980). As this action is a significant factor in the salt and water metabolism of the parent, it is not surprising that prolactin is also intimately involved in osmoregulation, an activity that may have been its primary role; osmoregulatory effects of prolactin have been observed in species (including most birds) that do not produce milk-like secretions (Chadwick, 1977). The role of prolactin in osmoregulation has been considered in detail by Bern (1975*a*) and De Vlaming (1980), although reviews on the hormonal control of salt and water balance in birds (for example, Wright *et al.*, 1967; Holmes and Wright, 1968; Holmes, 1978; Holmes and Pierce, 1978; Skadhauge, 1978) have rarely considered the importance of prolactin (Phillips and Ensor, 1972; Ensor, 1975, 1978; Thomas and Phillips, 1978).

Recognition of the importance of prolactin in osmoregulation stems from the now classical work of Pickford and Phillips (1959) who showed that this hormone had the unique capacity to enable hypophysectomised fish (*Fundulus heteroclitus*) to survive in freshwater. The ability of prolactin to maintain electrolyte homeostasis in these euryhaline teleosts pointed to the possibility that this hormone had as its locus of action the extrarenal site of ion transport contained within the gill epithelium. This observation encouraged subsequent studies in birds. These studies explored the possibility that prolactin might effect extrarenal excretion in marine species having nasal salt glands (Peaker and Linzell, 1975).

The first studies to assess the possible role of prolactin in avian osmoregulation were by Peaker and Phillips (1969) and Peaker *et al.* (1970). These authors showed that the administration of exogenous ovine prolactin (10 i.u./kg) to mallard ducks in which the salt glands had been minimally stimulated by intravenous saline infusion (MacFarland, 1964) resulted in a significant enhancement of the secretion rate (Table 1). Furthermore, the onset was considerably earlier than that induced by treatment with adrenocorticotropin (ACTH). The rapidity of this effect could not be attributable to changes in blood composition and suggested that prolactin had a direct effect on salt gland function.

TABLE 1. —*Effect of prolactin on nasal secretion by the minimally stimulated nasal gland of the duck.*

Min after appearance of first drop of nasal fluid	Mean nasal fluid output[a,b] (drops/min)			
	Control treatment (4)	Prolactin treatment (4)	SE (diff.)	P
0-5	1.90	2.75	±0.015	<0.001
5-10	2.25	3.00	±0.015	<0.001
10-15	1.67	2.25	±0.534	NS
15-20	1.17	2.10	±0.653	NS
20-25	0.80	1.35	±0.820	NS

[a]Four ducks were infused intravenously with 10% NaCl at 1 ml/min until the first drop of nasal fluid appeared when 20 i.u. ovine prolactin were administered i.v. Number of observations in parentheses; NS = not significant.

[b]Because of marked biological variation between individuals, differences between means were compared by the pared t-test, each animal acting as its own control. From Peaker, *et al.* (1970).

A causal relationship between prolactin and nasal gland activity was further suggested by Phillips and Ensor (1972) and Ensor *et al.* (1973) who found that the inhibition of salt gland function in ducks following hypophysectomy (Wright *et al.*, 1966, 1967; Holmes *et al.*, 1972) could be partly restored by replacement therapy with ovine prolactin or a crude preparation of duck prolactin (Fig. 1). These results also suggest that prolactin might have a direct effect on the salt gland, possibly by a synergism with adrenal corticosteroids (Chan *et al.*, 1970; Holmes *et al.*, 1972).

Further evidence for a role of prolactin in the control of salt gland function has also been derived from the demonstration that extrarenal excretion in response to a standard salt load is enhanced in birds subjected to severe dehydration following cold stress (4 hours at 10°C). This increase in activity was accompanied by a decrease in pituitary prolactin content, and was presumed to indicate an increase in circulating prolactin levels (Ensor *et al.*, 1976). Dehydration is also thought to be responsible for the daily fall in pituitary prolactin level in ducks following early morning feeding (Ensor and Phillips, 1970a). A relationship between prolactin secretion and electrolyte balance has also been attributed to the fact that the level of pituitary prolactin in ducks (Fig. 2) and herring gulls falls when exposed for five days to hypertonic saline drinking water (Ensor and Phillips, 1970a). This fall in prolactin content occurs at times when plasma sodium and potassium concentrations are increased, and it accompanies a rise in plasma osmolality and

FIG. 1 (top).—The effect of adenohypophysectomy in the duck and replacement therapy with ovine and avian (duck) prolactin on nasal gland secretion rate (5 birds/group). Birds were hypophysectomized and after 2 weeks infused along with the sham-operated group with 10% NaCl. Birds were then injected with prolactin for 3 days and reinfused with 10% NaCl (from Phillips and Ensor, 1972).

FIG. 2 (bottom).—Effect of chronic administration of hypertonic saline (as drinking water) on pituitary prolactin levels. Sixteen-week-old domestic ducks (*Anas platyrhynchos*) were maintained on 0.3 M NaCl for up to 5 days (from Ensor and Phillips, 1970a).

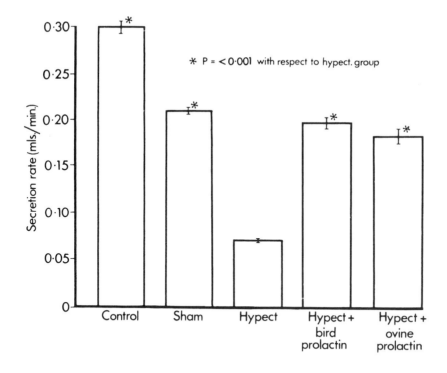

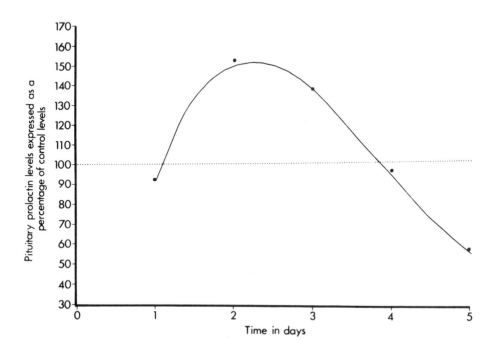

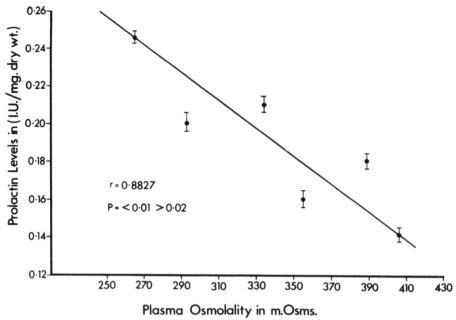

F<small>IG</small>. 3.—Relationship between plasma osmolality and pituitary prolactin concentration (r=0.883; P<0.02). Treatments from left to right are: freshwater controls; 40% seawater; heat stress; 80% seawater; water deprivation; 800 mM NaCl. Points represent mean ± s<small>E</small>, N=6 (from Ensor, 1975).

hypertrophy of the salt glands. Moreover, dehydration resulting from water deprivation, heat stress, and hyperosmotic saline has also been found to reduce pituitary prolactin concentrations in ducks (Table 2); the stimulus for this fall in pituitary content was a rise in plasma osmolality (Fig. 3). The effect of low pituitary prolactin levels is thought to be increased prolactin secretion which might be expected to affect salt and water balance by the activation of the salt glands or other extrarenal pathways of excretion (Ensor, 1975).

T<small>ABLE</small> 2.—*Pituitary prolactin levels in dehydrated birds. Results expressed as concentration of prolactin in i.u./mg with 95% confidence limits, and as a percentage of the control levels. From Ensor and Phillips (1972).*

Treatment	Prolactin level	% Control levels
Freshwater (control)	0.23 (0.207-0.253)	
40% seawater	0.20 (0.186-0.214)	86.9
80% seawater	0.16 (0.148-0.172)	69.5
800 mM NaCl	0.14 (0.127-0.153)	60.8
Freshwater (control)	0.26 (0.248-0.272)	
Water deprivation	0.18 (0.169-0.191)	69.2
Heat stress	0.21 (0.198-0.222)	80.7

In the natural environment, this effect of prolactin on nasal gland function was considered possibly to be related to the change in habitat of the bird at the end of its breeding season; the bird anticipating subsequent periods of hypersalinity. For instance, Mallards (from which domestic ducks are descended) rear their young on freshwater and then tend to migrate to estuaries for the winter. Consequently, if prolactin is also associated with migration in the duck, as it is in other birds (Meier *et al.*, 1965; Meier, 1975), it might predispose the bird to the sudden change from freshwater to saltwater conditions by enabling the previously inactive salt glands to start secreting at a high rate as soon as the bird drinks saline water. In this way prolactin would maintain the bird in a temporary ionic equilibrium and might stimulate the subsequent hypertrophy and hyperplasia of the salt glands (Peaker and Linzell, 1975). However, this hypothesis, as Ensor and Phillips (1970*a*) state, would not seem to cover species such as gulls which tend to migrate inland to winter freshwater situations after breeding.

If increased prolactin secretion does enhance extrarenal excretion, it could, by elimination of excess sodium, offset in a temporal way the deleterious effects of dehydration known to occur when birds face extremes of environmental temperatures or are denied access to drinking water. Such conditions might be expected to pose an osmoregulatory threat to some birds during their breeding cycles. In some species the adults immobilize themselves, often in great heat, on the eggs and often deprive themselves of food or water for several hours or days (Cade and Greenwald, 1966; Nelson, 1968; Ensor and Phillips, 1970*a*, 1972); the loss of respiratory water under such circumstances compounds the stress of dehydration. Under such circumstances the pituitary prolactin content of breeding and brooding birds is, however, high (for example, Eisner, 1960; Gourdji and Tixier-Vidal, 1966; Ensor, 1978). The possibility that prolactin might stimulate extrarenal secretion under such circumstances has not been assessed, but increased nasal gland function and saltwater adaptation has usually been associated with low pituitary prolactin concentrations (Ensor and Phillips, 1970*a*, 1972; Ensor *et al.*, 1976), although a diurnal pattern of pituitary prolactin content has been observed (Ensor and Phillips, 1970*b*) which paralleled changes in nasal gland activity.

The fact that there appears to be no simple relationship between nasal gland activity and pituitary prolactin concentration provides some evidence that prolactin secretion might not be causally related to salt gland function. This fact is emphasized by the observation that dehydration induced by water deprivation, heat stress, or exposure to hypertonic saline reduces pituitary prolactin levels (Ensor and Phillips, 1970*a*, 1972; Phillips and Ensor, 1972), although it is well established that dehydration greatly inhibits salt gland activity (Douglas and Neely, 1969; Ensor and Phillips, 1972; Ensor, 1975, 1978; Butler, 1980; Phillips and Harvey, 1980).

The difficulty in the interpretation of this information is partly because it is difficult to relate changes in pituitary hormone content with changes in its rate of release into the circulation and ultimately with its level at its

TABLE 3.—*Plasma prolactin levels (ng/ml) in domestic ducks faced with environmental challenge. Means ± SE (N = 8).*

Pretreatment → Treatment	Days of exposure				
	0	1	3	5	7
Freshwater → Freshwater	12.1±	11.1±0.6	10.9±0.7	10.5±0.9	10.8±0.7
Freshwater → 0.2 M NaCl	12.6±0.3	15.0±0.6	12.0±0.6	12.8±0.8	12.8±0.8
0.2M NaCl → 0.2 M NaCl	9.5±0.9	12.9±0.6	10.6±0.9	9.8±0.7	10.1±0.5
0.2M NaCl → Freshwater	11.0±1.2	13.3±0.9	12.7±0.6	12.7±0.7	11.1±1.1

target sites. The cautious interpretation of these results has, understandably therefore, been stressed several times in previous reviews (Phillips and Ensor, 1972; Ensor, 1975, 1978). This caveat is especially important when it is considered that increased and decreased pituitary prolactin content has been cited as evidence of enhanced prolactin release and that non-specific changes in the pituitary prolactin level can occur under a variety of experimental situations (Nicoll, 1972). Additionally, there is apparently very little correlation between pituitary and plasma prolactin levels in ducks (Harvey and Phillips, unpubl. observ.). For instance, environmental stresses such as food or water deprivation and exposure to hypertonic saline, which reduce the pituitary prolactin content (Ensor, 1975, 1978), have been found to have no effect on peripheral plasma prolactin concentrations (Tables 3 and 4) determined by a heterologous radioimmunoassay (McNeilly *et al.*, 1978). Furthermore, dehydration induced by saline-loading, exercise, or hemorrhage also has no effect on the concentration of immunoreactive plasma prolactin (Tables 5 and 6). These results suggest, therefore, that prolactin might have less importance in avian osmoregulation than that previously concluded from data derived from pituitary prolactin content. Although this might be the case, it must also be remembered that interpretation of data based solely on peripheral plasma prolactin levels must also be done with some degree of caution because its concentration in the circulation can be held constant by classical feedback and homeostatic mechanisms and hence might not reflect changes in the rate of its secretion or metabolism. A final cautionary note should also be sounded when considering these results, as the prolactin radioimmunoassay employed is a heterologous one which might not (Nicoll, 1975) be fully validated for the measurement of prolactin in ducks. However,

TABLE 4.—*Prolactin level (ng/ml plasma) following food or water deprivation in freshwater and saltwalter (0.2 M NaCl) maintained ducks (Anas platyrynchos). Means ± SE (N = 10).*

Hours	Freshwater maintained			Saline maintained		
	Controls	Fasted	Dehydrated	Controls	Fasted	Dehydrated
0	19.0±4.8	25.1±4.3	16.2±1.2	19.82±6.4	19.6±1.6	24.0±2.8
12	18.2±2.5	21.4±1.5	17.0±1.4	24.2 ±3.7	27.4±2.2	20.8±4.0
24	23.4±5.1	21.0±1.8	16.8±1.7	23.8 ±2.6	24.0±2.3	20.6±1.4
36	24.2±5.8	23.0±5.3	17.2±1.7	17.6 ±1.7	20.0±3.0	23.0±2.0

TABLE 5.—*Effect of salt loading (18 ml/kg of 500 mM NaCl) on plasma prolactin levels (ng/ml) in normal and dehydrated[a] domestic ducks* (Anas platyrhynchos). *Means* ± SE (N = 8).

Time after saline loading (minutes)	Controls	Dehydrated
0	11.2±0.6	10.8±1.0
10	10.2±1.1	11.4±0.7
20	10.4±0.9	11.2±1.4
30	8.4±1.2	11.8±1.0
40	8.2±1.4	10.6±1.3
50	9.6±0.9	9.8±1.8
60	9.0±1.2	10.4±0.8
70	10.4±1.0	12.2±0.8
80	10.6±0.8	8.2±1.2
90	10.8±0.6	11.0±0.8
100	11.2±1.2	12.4±0.7
110	9.6±0.8	12.4±0.8

[a]Dehydration was induced by the removal of approximately 30% of the bird's blood volume after removal of the pretreatment sample.

it must be pointed out that 1) in this assay duck pituitary extracts and plasmas produce dose-response inhibition curves parallel to the ovine prolactin standard, and 2) using this assay, the circulating prolactin levels measured in ducks during periods of egg laying, broodiness, incubation of eggs, and rearing of the young are physiologically meaningful (Goldsmith and Williams, 1980). Nevertheless, in view of the potential significance of this data, the possible role of prolactin in extrarenal (salt gland) excretion has been reassessed. This possibility has been tested by the intravenous administration of bovine prolactin (10 i.u./kg) to ducks in which the salt glands had been maximally stimulated by a standard intravenous salt load (18 ml/kg of 500 mM NaCl) or minimally stimulated by the infusion of the least amount of hypertonic (300 mM NaCl) saline required to induce secretion. In addition, this possibility was also assessed by comparing the extrarenal response of normal and bromocryptine-treated birds (in which prolactin secretion would be expected to be low) to a standard salt load. The results of these experiments are as follows.

Bromocryptine treatment (intramuscular injection of 1 mg CB154/day for 10 days) did not delay the onset or reduce the volume of the extrarenal

TABLE 6.—*Effect of tread-mill exercise[a] on plasma prolactin levels (ng/ml) in domestic ducks* (Anas platyrhynchos) *Means* ± SE (N = 8).

Treatment	Period of Excercise (min)						
	0	15	30	45	60	75	90
Nonexercised (control)	10.8±1.7	10.5±1.5	10.9±1.5	8.4±1.4	9.0±1.6	8.6±1.7	10.1±1.2
Exercised	12.1±1.8	14.1±1.8	10.5±1.3	10.4±1.3	8.6±1.3	10.3±1.1	11.0±1.5

[a]The exercised birds walked on a tread mill inclined at 3°, at a speed of 1.1 km/h.

TABLE 7.—*Effect of bovine prolactin (10 i.u./kg) on the rate of extrarenal excretion by minimally[a] stimulated salt glands of domestic ducks* (Anas platyrhynchos).

Bird no.	Treatment	Pretreatment volume (μl) of salt gland excretion. Means ± SE. of the volume collected during 5 consecutive 2-minute intervals	Post-treatment volume (μl) of salt gland excretion. Means ± SE. of the volume collected during 5 consecutive 2-minute intervals	
1	Control	0.244 ±0.30	0.301±0.021	NS[b]
2	Control	0.352 ±0.012	0.366±0.008	NS
3	Control	0.136 ±0.023	0.088±0.010	NS
4	Control	0.612 ±0.013	0.670±0.027	NS
5	Prolactin	0.057 ±0.009	0.082±0.017	NS
6	Prolactin	0.294 ±0.015	0.297±0.010	NS
7	Prolactin	0.1654±0.014	0.156±0.009	NS
8	Prolactin	0.222 ±0.024	0.250±0.015	NS

[a]Induced by the infusion of the least amount of 300 mM required to start secretion.
[b]No significant difference between pre and post-treatment response.

response to salt loading, and the administration of bovine prolactin failed to increase this response in normal or CB154 treated birds (Fig. 4). This was not due to prolactin being unable to exert an effect in maximally stimulated birds as it also failed to increase salt gland excretion in minimally stimulated birds (Fig. 5). The fact that no effect of prolactin on the bird's extrarenal response could be discerned when volumes of nasal gland secretion were collected at two-minute intervals (Table 7) also suggests that the failure of prolactin to affect secretion rate was not due to a transitory effect. This effect was not detected in Figs. 4 and 5, as might have been surmised from the results of Peaker *et al.* (1970).

These results are, therefore, at variance with those of Peaker *et al.* (1970), and no satisfactory explanation for this discrepancy can be given. However it must be pointed out that the effect noted by Peaker *et al.* (1970) was quite variable and of small magnitude. Additionally, it was observed only in four birds and might have been of little physiological significance. This conclusion is strengthened by the fact that the increase in nasal gland secretion in cold stressed ducks (Ensor *et al.*, 1976) was probably not due to the presumed increase in prolactin secretion (assumed from low pituitary prolactin levels) but to the increase in heart rate of the cold stressed birds and the increase in blood flow through the salt gland. The necessity for a reappraisal of the role of prolactin in salt gland function stems from the realization that the partial

→

FIG. 4 (top).—The effect of intravenous salt-loading (18 ml/kg of 500 mM NaCl) on the rate of nasal gland secretion in normal birds and those pretreated (1 mg/bird/day for 10 days) with bromocryptine (CB154). The controls were injected with the vehicle. Each bird was intravenously injected with bovine prolactin (10 i.u./kg) 150 minutes after the onset of secretion. Means ± SE of 5 control birds and 8 CB154-treated birds.

FIG. 5 (bottom).—The effect of intravenous bovine prolactin administration (10 i.u./kg) on the rate of nasal gland secretion in ducks minimally stimulated by the infusion of the least amount of 300 mM NaCl required to induce secretion, eight birds per group.

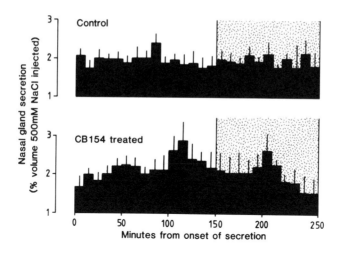

After prolactin injection (10 i.u./kg)

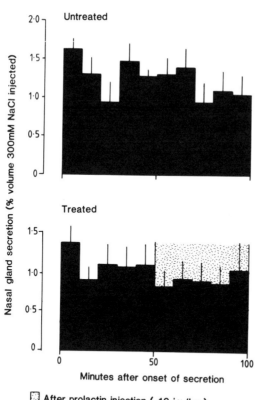

After prolactin injection (10 i.u./kg)

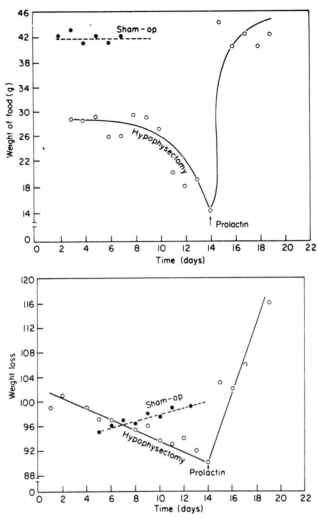

Fig. 6.—*Above*, the effect of hypophysectomy and replacement therapy on food intake in the duck *Anas platyrhynchos*. Each point represents the mean of 12 birds (from Ensor, 1975). *Below*, the effect of hypophysectomy and replacement therapy on body weight in the duck *Anas platyrhynchos*. Weight loss is expressed as a percentage of the original body weight. Each point represents the mean of 12 birds (from Ensor, 1975).

restoration of the extrarenal capacity of hypophysectomised ducks by prolactin replacement therapy (Phillips and Ensor, 1972; Ensor *et al.*, 1973) was probably not due to a direct effect on the salt gland but to a behavioral effect of prolactin on food and water intake. During the 14-day period between the hypophysectomy of these birds and initial salt loading, food and water intake as well as body weight decreased markedly (Ensor, 1975). Food intake fell nearly to zero and the birds lost eight to 12 per cent of their preoperative body weight. A single injection of prolactin on day 15 produced a remarkable change in food and water intake with a resulting increase in body weight (Fig. 6); this improvement was maintained for the three days that injections were given and by the third day the birds were fully rehydrated. A similar effect of prolactin has also been observed in hypophysectomised pigeons (Bates *et al.*, 1962) and intact ducks (Ensor, 1975, 1978). Consequently, as food and water deprivation and dehydration invariably result in reduced rates of nasal gland secretion (Phillips and Harvey, 1980), the stimulatory effect of prolactin on salt gland activity is probably due to the ability of prolactin to stimulate food and water intake.

This reappraisal of the role of prolactin in avian osmoregulation in no way vitiates its importance, as its behavioral effect on appetite would endow the bird with a considerable ecological advantage. In addition, there is some evidence that prolactin might also affect electrolyte balance by effects on other renal or extrarenal pathways of excretion (Ensor, 1975). For instance, it has been suggested that prolactin might have an indirect effect on salt gland activity by an action on the kidney to increase sodium retention, a role which has been established for prolactin in mammals (Lockett and Nail, 1965). Evidence for such an action in birds is, however, wanting, as changes in pituitary prolactin levels occur at times when there are no significant changes in plasma electrolyte levels (Ensor and Phillips, 1970a).

Prolactin may, however, affect avian osmoregulation by altering cloacal output. Birds exposed to or infused with high concentrations of saline have reduced rates of urine flow, and low pituitary prolactin concentrations are found in such birds (Ensor and Phillips, 1972; Ensor, 1975). Further evidence for such an antidiuretic effect of prolactin has been presented by Ensor (1975); the administration of ovine prolactin to ducks results in a dose-related fall in urine flow rate (Fig. 7). The observation that the pituitary prolactin content of ducks treated with a diuretic (Amiloride) was also inversely related to urine volume was also suggested to demonstrate the antidiuretic effect of the hormone. However, in this latter case, low prolactin levels were associated with high rates of urine flow (Fig. 7), whereas under experimentally induced dehydration conditions, low prolactin levels were accompanied by decreased cloacal output (Ensor and Phillips, 1972), making interpretations of the results difficult. The possibility that prolactin might affect osmoregulation by inducing renal function or by increasing cloacal or rectal water reabsorption has, however, been further investigated by Ensor (1978). Ensor demonstrated that injections of prolactin do not reduce glomerular filtration rate (GFR) in normal or salt-loaded ducks (Table 8). This suggests that prolactin might reduce urine flow by an extrarenal site of action, it being unlikely that the reduction in flow rate could be achieved by tubular reabsorption. Evidence for an extrarenal site of action might be derived from the fact that prolactin administration increases the uptake of tritiated water from the intact rectal-cloacal complex and isolated rectum of the ducks.

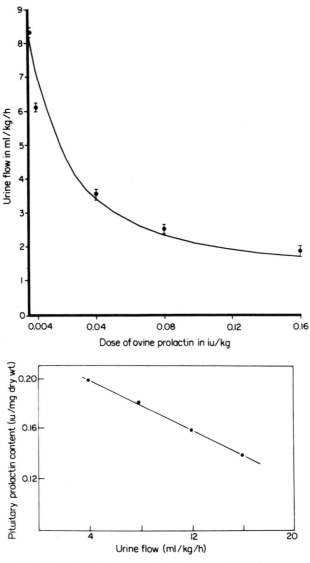

Fig. 7.—*Above*, the effect of prolactin on urine flow rate, six birds per group (from Ensor, 1975). *Below*, the effect of urine flow on pituitary prolactin levels in the duck *Anas platyrhynchos*, six birds per group (from Ensor, 1975).

The close involvement of prolactin in the salt and water balance of birds is further suggested by the interdependence between the circulating concentrations of prolactin with sodium and potassium levels; the circulatory levels of both sodium and potassium are directly correlated with the plasma prolactin level in domestic fowl subjected to a variety of experimental treatments (Bolton *et al.*, 1976; Chadwick *et al.*, 1977). Moreover, elevated potassium levels in *in vitro* incubations stimulate the release of prolactin (as

TABLE 8.—*Effect of injections of ovine prolactin (0.12 i.u./kg) on glomerular filtration rate (GFR) in control and saline-loaded (0.7 ml/min 10% NaCl) ducks,* Anas platyrhynchos. *From Ensor (1978).*

Treatment	GFR (ml/kg/min)
Control	1.47±0.12
Control and prolactin	1.52±0.16
Saline	0.82±0.10
Saline and Prolactin	0.96±0.11

measured by bioassay and densitometry following electrophoresis) from incubated chicken and duck pituitaries (Tixier-Vidal and Gourdji, 1972; Chadwick *et al.,* 1978). These findings could suggest that prolactin in the chicken induces potassium retention, possibly as a result of increased sodium excretion, possibly by activation of Na^+/K^+ - ATPase exchange mechanisms.

In the domestic fowl (in contrast to the duck) the increase in plasma sodium level as a result of intravenous or oral loading with NaCl does dramatically increase plasma prolactin levels (Bolton *et al.,* 1975; Scanes *et al.,* 1976; Tai and Chadwick, 1976; Chadwick, 1980; Morley *et al.,* 1980a), as shown in Table 9. Similar increases in plasma prolactin concentration also occur, as shown in Table 9, when domestic fowl are deprived of drinking water for as little as six hours or if fed 20 per cent NaCl in the diet for 14 days (Harvey *et al.,* 1979; Chadwick, 1980). Moreover, dehydration-induced increases in the circulating prolactin level have been correlated with accompanying changes in the cytology of the pituitary lactotrophs (Fig. 8). Tai and Chadwick (1976) demonstrated that the intravenous injection of 2 g of NaCl in a 30 per cent aqueous solution enhanced prolactin secretion within 10 minutes of injection and increased the number of polymorphic granules of the prolactin cells (Fig. 8). In addition, small and presumably newly synthesised granules are visible in the golgi apparatus of the lactotrophs following salt loading and exocytosis of granules, and merocrine secretion of the vessels formed from extended endoplasmic reticulum is evident (Fig. 8).

These results suggest that the elevated concentration of plasma sodium resulting from dehydratory salt loading stimulates the synthesis and release

TABLE 9.—*Effect of dehydration on plasma prolactin concentrations in domestic fowl. Means ± SE (N).*

Treatment	Prolactin (ng/ml)		Reference
	Untreated	Treated	
Water deprivation (6 h)	205± 6 (8)	310±25 (8)**	Harvey *et al.,* 1979
Water deprivation (12 h)	209± 7 (8)	440±41 (8)***	,,
Water deprivation (18 h)	204± 8 (8)	334±20 (8)***	,,
Water deprivation (24 h)	184± 7 (8)	360±24 (8)***	,,
Oral salt (2 g) loading	92±10 (4)	150±14 (4)*	Scanes *et al.,* 1976
Intravenous salt (2 g) loading	85±14 (5)	140±14 (5)*	,,

*Significantly different from corresponding control, $P< 0.05$, **$P<0.001$, ***$P<0.001$

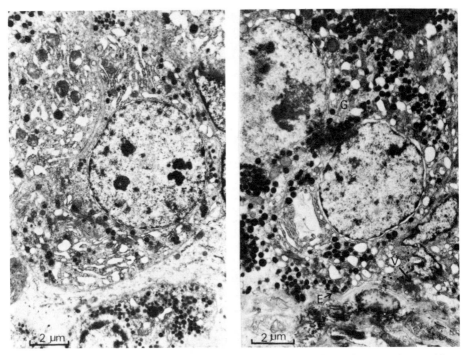

FIG. 8.—*Left*, prolactin secreting cell from cephalic lobe of pituitary of untreated control laying hen. *Right*, prolactin cell from pituitary of salt-loaded laying hen, killed 10 minutes after sodium chloride injection. G, Golgi apparatus; E, exocytosis; V, vesicle (from Tai and Chadwick, 1976).

of pituitary prolactin, which acts to restore osmotic balance by eliminating the excess sodium. Strong evidence in support of this hypothesis can be derived from the observation that the administration of ovine prolactin to domestic fowl lowers the concentration of plasma sodium (Morley *et al.*, 1980). In the dehydrated or salt loaded chicken, this effect of prolactin might be achieved by a depression in the net transport of sodium across the intestine or colon; prolactin administration reduces the *in vitro* uptake of sodium and water by both the jejunum and rectum (Table 10) and reduces the *in vivo* absorption of sodium by the jejunum (Morley *et al.*, 1981). Prolactin also has a similar effect on sodium transport across the intestine of the eel (Hirano and Utida, 1968; Utida and Hirano, 1971). The increase in plasma prolactin concentration in the chicken following dehydration and sodium loading (Harvey *et al.*, 1979; Morley *et al.*, 1980) therefore probably is responsible in part for that decrease in net sodium transport across the large intestine of the salt loaded bird (Thomas *et al.*, 1975, 1979; Choshniak *et al.*, 1977; Thomas and Skadhauge, 1979). The increase in plasma prolactin concentration that occurs in domestic fowl injected with aldosterone (Morley, therefore presumably is an homeostatis response to reduce the stimulatory

TABLE 10.—*Effect of prolactin on water uptake and net sodium uptake by everted gut sacs of 9-day-old domestic fowl* (Gallus domesticus). *Means ± SE (N). From Morley et al. (1980).*

Region of gut	Water Uptake g H_2O/g wet wt/hr		Net Sodium Uptake cpm/g wet wt.	
	Control	Prolactin treated[a]	Control	Prolactin treated[a]
Duodenum	0.173±0.020 (9)	0.171±0.031 (9)	2075± 352 (9)	1809± 613 (8)
Jejunum	0.294±0.028 (9)	0.177±0.022 (8)*	4102± 497 (9)	1209± 660 (6)*
Ileum	0.786±0.093 (9)	0.834±0.093 (10)	12439±2242 (6)	12979±1733 (8)
Rectum	0.492±0.048 (9)	0.277±0.052 (9)*	7267±1792 (5)	2694±1849 (5)*

*Significantly different from the corresponding control value, $P<0.05$.
[a]Pretreated by 7 daily subcutaneous injections of 100 µg ovine prolactin.

effect of aldosterone on cloacal sodium reabsorption (Thomas and Skadhauge, 1979) in an effort to maintain electrolyte balance.

Such an inhibitory effect of prolactin on intestinal sodium uptake might be essential for the acquisition of osmotically free water in species such as the domestic fowl which lack salt glands and cannot excrete excess salt at high concentrations. Thus the lack of a plasma prolactin response to saline loading or adaptation in the duck (Tables 3 and 4) could be due to the ability of these marine species of birds to extrarenally excrete sodium via the salt glands, which are probably not directly influenced by changes in prolactin secretion. The fact that there is a fundamental difference between such species of birds is demonstrated by the observation that both net water and net sodium transport are increased in salt water loaded or adapted ducks (Crocker and Holmes, 1971; Holmes, 1978). This increase in the mucosal transfer of salt and water is in fact essential for the sustained function of the nasal gland in birds that are continuously exposed to hypertonic drinking water.

In conclusion, the role of prolactin in the osmoregulation of birds without salt glands appears to be different from its role in birds with the capacity for extrarenal excretion. However, this is perhaps not surprising in view of the fact that differences in the hormonal control of osmoregulation probably exist even between different species of granivorous birds as a result of differences in their natural environment and the relative importance of renal or extrarenal (cloacal) pathways in the control of their salt and water balance (Skadhauge, 1978).

SUMMARY

The influence of prolactin on the salt and water balance of birds has been reassessed. In birds with the capacity for extrarenal excretion, prolactin probably has no direct effect on salt gland activity, but might have an indirect effect via alterations in renal, intestinal, or cloacal function. In birds with and without salt glands, prolactin respectively increases and decreases the uptake of water from the cloaca and intestine, resulting in decreased urine flow in birds which possess salt glands. The intestinal absorption of sodium is inhibited by prolactin in birds lacking salt glands and is probably unaffected in birds with the capacity for extrarenal salt excretion. In both groups of birds, prolactin prevents hydration by stimulating food and water intake, and in both groups is involved in the excretion of excess sodium. Osmoregulatory differences between birds with and without salt glands are reflected by differences in prolactin secretion.

ACKNOWLEDGMENTS

The authors would like to express their thanks to Dr. A. S. McNeilly (MRC Reproductive Endocrinology Unit, Edinburgh) for the generous gift of materials used in the heterologous prolactin radioimmunoassay. This work was supported by grants from the Science Research Council (GR/A29869 and GR/B 47782).

LITERATURE CITED

BATES, R. W., R. A. MILLER, AND M. M. GARRISON. 1962. Evidence in the hypophysectomised pigeon of synergism among prolactin, growth hormone, thyroxine and prednisone upon weight of the body, digestive tract, kidney and bodystores. Endocrinology, 71:345-360.

BERN, H. A. 1975a. Prolactin and osmoregulation. Amer. Zool., 15:937-949.

———. 1975b. On two possible primary activities of prolactins: osmoregulatory and developmental. Vech. Dtsch. Zool. Ges., 1975:40-46.

BERN, H. A., AND C. S. NICOLL. 1968. The comparative endocrinology of prolactin. Recent Prog. Horm. Res., 24:681-720.

BOLTON, N. J., C. G. SCANES, AND A. CHADWICK. 1975. A radioimmunoassay for prolactin in the circulation of birds. J. Endocrinol., 67:51P.

———. 1976. Plasma electrolytes and prolactin in the domestic fowl. IRCS Med. Sci., 4:517.

BUTLER, D. G. 1980. Functional nasal glands in adrenalectomised domestic ducks. Gen. Comp. Endocrinol., 40:15-26.

CADE, T. J., AND C. GREENWALD. 1966. Nasal salt gland secretion in falconiform birds. Condor, 68:338-350.

CHADWICK, A. 1969. Effects of prolactin in homiothermic vertebrates. Gen. Comp. Endocrinol., Suppl., 2:63-68.

———. 1977. Comparison of milk-like secretions found in non-mammals. Symp. Zool. Soc. London, 41:341-358.

———. 1980. Comparative aspects of the action of the hormone prolactin with particular reference to the domestic fowl. Gen. Comp. Endocrinol., 40:317-318.

CHADWICK, A., T. R. HALL, AND N. J. BOLTON. 1978. The effect of hypothalamic extract and K on prolactin secretion by the pituitary gland of the domestic fowl. IRCS Med. Sci., 6:238.

CHAN, D. K. O., I. P. CALLARD, AND I. CHESTER JONES. 1970. Observations on the water and electrolyte composition of the iguanid lizard *Dipsosaurus dorsalis dorsalis* (Baird and Baird) with special reference to the control by the pituitary gland and adrenal cortex. Gen. Comp. Endocrinol., 15:374-387.

CHOSHNIAK, I., B. G. MUNCK, AND E. SKADHAUGE. 1977. Sodium chloride transport across the chicken coprodaeum. Basic characteristics and dependence on sodium chloride intake. J. Physiol. (London), 271:489-504.

CROCKER, A. D., AND W. N. HOLMES. 1971. Intestinal absorption in ducklings (*Anas platyrhynchos*) maintained on freshwater and hypertonic saline. Comp. Biochem. Physiol., 40A:203-211.

De Vlaming, V. L. 1980. Actions of prolactin among the vertebrates. Pp. 561-642, in Hormones and evolution (E. J. W. Barrington, ed.), Academic Press, New York, xxi+989 pp.

Douglas, D. S., and S. M. Neely. 1969. The effect of dehydration on salt gland performance. Amer. Zool., 9:1095.

Eisner, E. 1960. The relationship of hormones to the reproductive behaviour of birds, referring especially to parental behaviour. A review. Anim. Behav., 8:153-181.

Ensor, D. M. 1975. Prolactin and adaptation. Symp. Zool. Soc. London, 35:129-148.

———. 1978. Comparative endocrinology of prolactin. Chapman Hall, London, ix+309 pp.

Ensor, D. M., and J. G. Phillips. 1970a. The effect of salt loading on the pituitary prolactin levels of the domestic duck (Anas platyrhynchos) and juvenile herring or lesser black-backed gulls (Larus argentatus and Larus fuscus). J. Endocrinol., 48:167-172.

———. 1970b. The effect of environmental stimuli on the circadian rhythm of prolactin production in the duck (Anas platyrhynchos). J. Endocrinol., 48:lxxi-lxxii.

———. 1972. The effect of dehydration on salt and water balance in gulls (Larus argentatus and L. fuscus). J. Zool., London, 168:127-137.

Ensor, D. M., I. M. Simons, and J. G. Phillips. 1973. The effect of hypophysectomy and prolactin replacement therapy on salt and water metabolism in Anas platyrhynchos. J. Endocrinol., 57:xi.

Ensor, D. M., J. G. Phillips, and M. J. O'Halloran. 1976. The effect of extreme cold stress on nasal gland function in the domestic duck Anas platyrhynchos. Gen. Comp. Endocrinol., 31:317-329.

Goldsmith, A. R., and D. M. Williams. 1980. Incubation in mallards (Anas platyrhynchos): changes in plasma levels of prolactin and luteinizing hormone. J. Endocrinol., 86:371-379.

Gourdji, D., and A. Tixier-Vidal. 1966. Variations due contenu hypophysaine en prolactin chez le canard Pekin male au cours du cycle sexuelle et de la photostimulation expérimentale. C. R. Acad. Sci., Paris, 262:1746-1749.

Harvey, S., C. G. Scanes, A. Chadwick, and N. J. Bolton. 1979. Growth hormone and prolactin secretion in growing domestic fowl: influence of sex and breed. Brit. Poult. Sci., 20:9-17.

Hirano, T., and S. Utida. 1968. Effects of ACTH and cortisol on water movement in isolated intestine of the eel, Anguilla japonica. Gen. Comp. Endocrinol., 11:1373-1380.

Holmes, W. N. 1978. Endocrine response in osmoregulation: hormonal adaptation in aquatic birds. Pp. 230-239, in Environmental endocrinology (I. Assenmacher and D. S. Farner, eds.), Springer-Verlag, Berlin, xv+334 pp.

Holmes, W. N., and D. Pierce. 1978. Hormones and osmoregulation in vertebrates. Pp. 413-533, in Mechanisms of osmoregulation in animals (R. Gilles, ed.), John Wiley and Sons, New York, xii+667 pp.

Holmes, W. N., and A. Wright. 1968. Some aspects of the control of osmoregulation and homeostasis in birds. Proc. Third Internat. Cong. Endocrinol., Mexico, July 1968, pp. 237-248.

Holmes, W. N., L. N. Lockwood, and E. L. Bradley. 1972. Adenohypophysial control of extrarenal excretion in the duck (Anas platyrhynchos). Gen. Comp. Endocrinol., 18:59-68.

Lockett, M. F., and B. Nail. 1965. A comparative study of the renal actions of growth hormone and lactogenic hormones in rats. J. Physiol. (London), 180:147-156.

McFarland, L. Z. 1964. Minimal salt load required to induce secretion from the nasal glands of sea gulls. Nature (London), 204:1202-1203.

McNeilly, A. S., R. J. Etches, and H. G. Friesen. 1978. A heterologous radioimmunoassay for avian prolactin: application to the measurement of prolactin in the turkey. Acta Endocrinol., 89:60-69.

Meier, A. H. 1975. Chronoendocrinology of vertebrates. Pp. 469-549, in Hormonal correlates of behaviour (B. E. Eleftheriou and R. L. Sprott, eds.), Plenum Press, New York, xi+439 pp.

MEIER, A. H., D. S. FARNER, AND J. R. KING. 1965. A possible endocrine basis for migratory behaviour in the white-crowned sparrow, *Zonotrichia leucophrys gambelii.* Anim. Behav., 13:453-465.

MORLEY, M. 1979. Osmoregulation in the fowl *Gallus domesticus* and the trout *Salmo trutta,* some effects of prolactin and aldosterone. Ph.D. thesis, Univ. Leeds, xi + 383 pp.

MORLEY, M., C. G. SCANES, AND A. CHADWICK. 1980. The effect of ovine prolactin on sodium and water transport across the intestine of the fowl (*Gallus domesticus*). Comp. Biochem. Physiol., 67A:695-697.

————. 1981. Water and sodium transport across the jejunum of normal and sodium loaded domestic fowl (*Gallus domesticus*). Comp. Biochem. Physiol., 68A:61-66.

NELSON, B. 1968. Galapagos. Longmans, London, xx+337 pp.

NICOLL, C. S. 1972. Some observations and speculations on the mechanism of 'depletion,' 'repletion' and release of adenohypophysial hormones. Gen. Comp. Endocr., Suppl., 3:85-96.

————. 1974. Problems in interpreting the physiological significance of results obtained with exogenous prolactin and with data on endogenous circulating levels of the hormone. Pp. 69-83, *in* Lactogenic hormones, fetal nutrition and lactation (J. B. Josimorich, ed.), John Wiley and Sons, New York, xi+483 pp.

————. 1975. Radioimmunoassay and radioreceptor assays for prolactin and growth hormone: a critical appraisal. Am. Zool. 15:881-903.

————. 1978. Comparative aspects of prolactin physiology: is prolactin the initial growth hormone in mammalian species also? Pp. 175-187, *in* Symposium on prolactin, 1977 (C. Robyn and M. Harter, eds.), Elsevier, North Holland Biomed. Press, Amsterdam, xi+425 pp.

NICOLL, C. S., AND H. A. BERN. 1972. On the actions of prolactin among the vertebrates: is there a common denominator? Pp. 299-374, *in* Lactogenic hormones (G. E. W. Wolstenholme and J. Knight, eds.), Churchill Livingston, Edinburgh, x+416 pp.

PEAKER, M., AND J. L. LINZELL. 1975. Salt glands in birds and reptiles. Cambridge Univ. Press, London, xi+307 pp.

PEAKER, M., AND J. G. PHILLIPS. 1969. The role of prolactin and other hormones in the adaptation of the duck (*Anas platyrhynchos*) to saline conditions. J. Endocrinol., 43:1x.

PEAKER, M., J. G. PHILLIPS, AND A. WRIGHT. 1970. The effect of prolactin on the secretory activity of the nasal salt-gland of the domestic duck (*Anas platyrhynchos*). J. Endocrinol., 47:123-127.

PHILLIPS, J. G., AND D. M. ENSOR. 1972. The significance of environmental factors in the hormone mediated changes of nasal (salt) gland activity in birds. Gen. Comp. Endocrinol., Suppl., 3:393-403.

PHILLIPS, J. G., AND S. HARVEY. 1980. Salt glands in birds. Pp. 517-532, *in* Avian endocrinology (A. Epple and M. T. Stetson, eds.), Academic Press, New York, xv+577 pp.

PICKFORD, G. E., AND J. G. PHILLIPS. 1959. Prolactin, a factor in promoting survival of hypophysectomised kill-fish in freshwater. Science, 130:454-455.

SCANES, C. G., A. CHADWICK, AND N. J. BOLTON. 1976. Radioimmunoassay of prolactin in the plasma of the domestic fowl. Gen. Comp. Endocrinol., 30:12-20.

SKADHAUGE, E. 1978. Hormonal regulation of salt and water balance in granivorous birds. Pp. 222-229, *in* Environmental endocrinology (I. Assenmacher and D. S. Farner, eds.), Springer Verlag, Berlin, xv+334 pp.

TAI, S-W., AND A. CHADWICK. 1976. The effect of salt loading on the prolactin cells of the chicken pituitary gland. IRCS Med. Sci., 4:509.

TIXIER-VIDAL, A., AND D. GOURDJI. 1972. Cellular aspects of the control of prolactin secretion in birds. Gen. Comp. Endocrinol., Suppl., 3:51-64.

THOMAS, D. H., AND J. G. PHILLIPS. 1978. The anatomy and physiology of the avian nasal salt glands. Pavo, 16:89-104.

THOMAS, D. H., AND E. SKADHAUGE. 1979. Chronic aldosteronetherapy and the control of transepthelial transport of ions and water by the colon and coprodeum of the domestic fowl (*Gallus domesticus*) *in vivo.* J. Endocrinol., 83:239-250.

Thomas, D. H., E. Skadhauge, and M. W. Read. 1975. Steroid effects on gut functions in birds. Biochem. Soc. Trans., 3:1164-1168.

———. 1979. Acute effects of aldosterone on water and electrolyte transport in the colon and coprodeum of the domestic fowl (*Gallus domesticus*) *in vivo*. J. Endocrinol., 83:229-257.

Utida, S., and T. Hirano. 1971. Effects of changes in environmental salinity on salt and water movement in the intestine and gills of the eel, *Anguilla japonica*. Pp. 240-269, *in* Response of fish to environmental changes (W. Chavin and N. Egami, eds.), C. C. Thomas, Springfield, x+459 pp.

Wright, A., J. G. Phillips, and D. P. Huang. 1966. The effect of adenohypophysectomy on the extrarenal and renal excretion of the saline-loaded duck (*Anas platyrhynchos*). J. Endocrinol., 36:249-256.

Wright, A. M., J. G. Phillips, M. Peaker, and S. J. Peaker. 1967. Some aspects of the endocrine control of water and salt-electrolytes in the duck (*Anas platyrhynchos*). Proc. Third Asia and Oceania Congress of Endocrinol., Manila, January, 1967, xvi+ 825 pp.

Reprinted from
ASPECTS OF AVIAN ENDOCRINOLOGY:
PRACTICAL AND THEORETICAL IMPLICATIONS (C. G. Scanes *et al.*, eds.)
Grad. Studies, Texas Tech Univ., 1982, 26:1-411.

PROLACTIN AND OSMOREGULATION IN LOWER VERTEBRATES: A PHYLOGENETIC PERSPECTIVE

D. M. ENSOR

Department of Zoology, University of Liverpool, Liverpool L69 3BX England

Prolactin is possibly more diverse in its actions than any other anterior pituitary hormone. Comparative endocrinologists have struggled for years to find some common denominator underlying the wide array of actions in the vertebrates. Although not necessarily a common denominator, one pattern which does emerge is the role of prolactin in controlling osmoregulation in vertebrates (Ensor, 1979; Clarke and Bern, 1980). In all classes of vertebrates, osmoregulatory or more strictly ionoregulatory effects of prolactin have been reported. This enables comparative endocrinologists to study changes in the action of prolactin within a phylogenetic framework. It is, however, rash to extrapolate from studies of living vertebrates to possible evolutionary changes which have occurred. Bearing this in mind, it is possible to compare the actions of prolactin within the various vertebrate classes and hence to draw some limited conclusions as to the underlying mechanisms and the changes that have occurred between groups. When prolactin exerts an osmoregulatory role, two features stand out: a) it often acts upon integumentery structures, and b) it acts together with steroid hormones.

Prolactin and Osmoregulation in Fish

Most reports of prolactin affecting osmoregulation in fish relate to studies in teleosts. Prolactin has been reported to increase muscle sodium concentrations in hagfish in 60 per cent seawater (Chester Jones *et al.*, 1962), although no alteration in pituitary cytology in freshwater fish exposed to prolactin has been observed (Larssen, 1969). In the elasmobranchs, prolactin appears to reduce water permeability across the gills (Payan and Maetz, 1971) and can counteract an elevation of plasma sodium observed in hypophysecto-mized *Dasyatis* in dilute seawater (De Vlaming *et al.*, 1975). In teleosts a wide range of species have been studied and only the overall trends will be reviewed here (for reviews, see Ensor and Ball, 1972; Bern, 1975; Clarke and Bern, 1980).

The main osmoregulatory organ in teleosts is undoubtedly the gill, if only because of the large surface area it presents to the environment. In most reports on teleosts, hypophysectomy in a freshwater environment leads to an increase in the sodium efflux across the gills, which in turn reduces plasma sodium levels (Potts and Evans, 1966; Maetz *et al.*, 1967; Ensor and Ball, 1968; Dharmamba and Maetz, 1972). In all cases prolactin has been shown to limit sodium loss by decreasing the passive efflux component. Nonphysiological treatment of seawater fish with prolactin again causes sodium retention (Clarke, 1973). This sodium retentive effect of prolactin in freshwater appears to occur in parallel with an ACTH or cortisol-mediated stimulation of the branchial sodium uptake.

Prolactin has also been reported to affect other osmoregulatory sites in teleosts. Both Stanley and Fleming (1967) and Lahlou and Giordan (1970) have reported a fall in urinary volume following hypophysectomy which can be restored by prolactin replacement. Prolactin has also been shown to restore renal $Na^+ K^+$ -ATPase activity in hypophysectomized *Fundulus heteroclitus* (Pickford *et al.*, 1970). Studies by Lam and Leatherland (1969) have shown that, as well as stimulating enzyme activity, prolactin can also cause increased glomerular recruitment in migrating sticklebacks. In addition, prolactin blocks water uptake from the urinary bladder of *Gillichthys mirabilis*, thus opposing cortisol stimulation (Doneen and Bern, 1974; Doneen, 1976). Other prolactin effects have been reported on fish skin (Marshall, 1978) and intestine (Hirano, 1975).

In summary, prolactin, at least in the teleosts, causes sodium retention and possible water loss in freshwater adapted fish. As prolactin levels are low in seawater adapted fish, reports of effects in such fish must, for the time being, be regarded as pharmacological (Ensor and Ball, 1972).

Prolactin and Osmoregulation in Amphibians

The function of prolactin in the Amphibia appears to be similar to that reported in freshwater teleosts; thus it tends to cause sodium retention and presumably increased water loss, although this latter point has not been studied extensively. Prolactin in *Rana temporaria* causes an increase in sodium influx in hypophysectomized frogs and also causes an increase in potential difference, short-circuit current, and sodium uptake across the isolated skin (Howard and Ensor, 1975). This is an effect which parallels the response of the isolated skin to arginine vasotocin (AVT). It can, however, be distinguished from the AVT response by the modifying effect of amiloride (see Table 1). The response is also modified by pre-treatment of the skin with either estradiol-17β or testosterone. These hormones alter the initial potential difference across the skin and therefore the response to prolactin (Howard and Ensor, 1975; Ensor, 1979). Sampietro and Vercelli (1968) have shown sodium retention caused by prolactin treatment of hypophysectomized *Triturus cristatus* as it also does in *Necturus maculosus* (Pang and Sawyer, 1974). These results suggest a sodium retentive/water excretory role for prolactin in the Amphibia. On the other hand, results from other work suggest that this interpretation of the data might prove to be an oversimplification. Crim (1972) failed to produce an effect of prolactin replacement in either *Rana pipiens* or *Taricha torosa*. Prolactin has also been shown to depress sodium transport across the newt skin (Lodi *et al.*, 1978). Goldenberg and Warburg (1977) have also shown water retention in larval *Rana ridibunda* and *Bufo*

TABLE 1.—*The effect of amiloride at varying doses on isolated frog skins pretreated with either prolactin or arginine vasopressin. The figures are expressed as mean percentage drop in potential difference* ± SE *(N=8, all males).*

Dose of amiloride (m)	Prolactin treated	Prolactin control	AVP treated	AVP control
5.0×10^{-8}	12.5 ± 1.75	13.2 ± 1.65	10.2 ± 1.28	13.4 ± 1.84
1.0×10^{-7}	24.1 ± 2.54	23.9 ± 1.69	17.8 ± 2.10*	24.0 ± 2.89
2.5×10^{-7}	42.7 ± 2.81	42.6 ± 2.29	34.3 ± 3.29	41.9 ± 3.59

*$P<0.05$ with respect to appropriate controls. Amiloride lowers AVP response but not prolactin effect.

viridis. One possible explanation for this apparently contrary evidence is the suggestion that prolactin may be of osmoregulatory importance only in gilled amphibia (pers. comm. by P. S. Brown and S. C. Brown, cited by Clarke and Bern, 1980). This would not, however, explain the changes seen in adult *Rana temporaria* (Howard and Ensor, 1975) nor the "water-drive" effect seen in certain newts (Bern and Nicoll, 1968). A possible extension of Brown and Brown's hypothesis might be to include amphibia during the reproductive phase; prolactin protects the adult from environmental change prior to and during reproduction. This might also explain its adaptive significance to birds and mammals (Ensor, 1975, 1979).

Prolactin and Osmoregulation in Reptiles

Reports of prolactin acting as an osmoregulatory hormone in the Reptilia are limited. In 1970, Chan and his colleagues found that hypophysectomy of the iguanid lizard *Dipsosaurus dorsalis* led to a rise in plasma water and sodium concentrations as well as a rise in total muscle water and intracellular sodium concentration. Prolactin replacement restored these parameters but only when it was administered together with corticosterone.

Recent work on the *Chelonia* has suggested that the role of prolactin might differ between aquatic and terrestrial reptiles. In the freshwater chelonian *Chrysemys picta*, prolactin appears to increase glomerular filtration rate in intact animals. Furthermore, hypophysectomy reduces glomerular filtration rate, and prolactin partially restores the normal situation (Brewer and Ensor, 1980a). In salt loaded animals, prolactin appears to cause sodium and magnesium retention. This effect is not shown in normal animals where magnesium excretion is increased by prolactin injection. Similar effects on sodium and magnesium excretion were also observed with aldosterone and corticosterone administration, thus raising the possibility that prolactin might act synergistically with or stimulate the release of these two hormones.

In the terrestrial chelonian *Testudo graeca*, prolactin caused a reduction in urine volume (the opposite effect to that observed in aquatic animals) and a rise in urine potassium and magnesium concentrations. Hypophysectomy led to a fall in plasma sodium levels, which could not be restored by injection of either prolactin or corticosterone but was exacerbated by injection of the two hormones together. This again raises the possibility that the two

TABLE 2.—*The effect of prolactin on urine flow in the domestic duck,* Anas platyrhynchos.
Mean $\pm$ SE, N = 6.

Dose of prolactin (i.u./kg)	Urine flow (ml/kg^{-1} h^{-1})
0.001	8.2±0.37
0.004	6.1±0.42***
0.04	3.6±0.21***
0.08	2.3±0.56
0.16	2.1±0.60

***$P<0.001$ with respect to previous dosage group. Mean $\pm$ SE.

hormones are acting synergistically although the adaptive significance of this response is not as yet clear (Brewer and Ensor, 1980*b*). Water loading of the tortoises caused the abolition of the antidiuretic response and prolactin then stimulated an increase in urine flow, suggesting that the response to prolactin may depend on either the internal environment of the animal or the circulating steroid levels (see below).

Prolactin and Osmoregulation in Birds

In aquatic vertebrates (fish, amphibians, and terrapins) which have been studied, prolactin appears to act as a sodium retentive/water excretory hormone. However, in the terrestrial vertebrates (tortoises, birds, and mammals), prolactin appears primarily to bring about water conservation, if it has any effect at all. The one exception to this rule is the salt loaded terrapin in which prolactin may cause water retention (Brewer and Ensor, 1980*a*, 1980*b*). Another fact does emerge from the literature, however. The effect caused by prolactin might depend upon the presence or absence of steroid hormones. This possibility becomes clearer if one looks at the osmoregulatory effects of prolactin in the domestic duck. In 1970, Peaker *et al.* investigated the role of, prolactin in salt excretion by the avian nasal gland and found that there was a rapid rise in salt excretion following prolactin administration. Maintenance of the domestic duck on hypertonic media caused an initial increase in pituitary prolactin levels followed by a sharp decline (Ensor and Phillips,

TABLE 3.—*The effect of 0.04 i.u./ml of ovine prolactin on the disappearance rate of* ^3H *water from isolated colonic sacs of* Anas platyrhynchos. *Values are expressed as a percentage of the zero time concentration of* ^3H *water. Mean $\pm$ SE, N=6.*

Time	Control (%)	Prolactin (%)
0	100	100
20	83.7±2.11	84.2±1.72
40	79.4±2.47	70.2±1.27
60	69.3±2.71	60.3±1.20*
80	57.2±3.62	48.7±1.37**
100	56.1±3.20	36.2±2.20***

*$P<0.05$ with respect to control, **$P<0.01$, ***$P<0.001$.

TABLE 4.—*Specific birding of I^{125} labelled ovine prolactin to membrane fractions derived from homogenates of colonic sacs of ducks undergoing various treatments. Values expressed as mean specific birding (percentage) ± SE, corrected for protein weight. Four assays per group (each assay consisted of colon from 3 birds).*

Treatment	% Specific birding
Saline loaded (birds maintained a 300 m M.S.W.)	13.7±2.16**
Control (tap water)	4.3±1.20
Control and prolactin (0.04 i.u./kg)	7.8±1.37**
Control and corticosterone (20 mg/kg)	11.2±0.46**

**$P<0.01$ with respect to control group

1970). Further work (Ensor *et al.*, 1972) showed that hypophysectomy resulted in a drop in nasal gland secretion rate in ducks which could be partially restored by injections of ovine or avian prolactin.

The probable site of action of prolactin in birds appears to be the cloaca, although a direct effect of the hormone on the function of the nasal gland or on water and food intake cannot be ignored. Urine flow in domestic ducks is proportional to the dose of prolactin administered (see Table 2); increased doses of prolactin causes a decrease in urine flow. Similar doses of prolactin cause an increase in tritiated water uptake from isolated colonic sacs of salt loaded ducks (see Table 3). The effect is greatly reduced in non-saline loaded birds. However, administration of corticosterone to nonsaline loaded birds causes an increase in prolactin stimulated water flow. These data can be compared with data obtained from prolactin binding experiments with membrane fractions from isolated colonic sacs. Although the experiment is a preliminary one, it appears that specific binding for prolactin is high in sacs from salt loaded birds, but reduced in non-salt loaded birds. Administration of prolactin to non-salt loaded birds causes an increase in specific binding, but a much greater effect is caused by the administration of corticosterone *in vivo* (Table 4). Thus it is known from the work of Phillips and others (Phillips and Ensor, 1972) that corticosterone levels rise in duck plasma following saline loading. It would appear that corticosterone in the presence of high levels of prolactin might induce prolactin receptors in the colon, thus stimulating sodium/water transport which further enhances the function of the nasal gland and hence providing the bird with osmotically free water (Ensor, 1975).

If one looks at the phylogenetic changes in the osmoregulatory actions of prolactin, the obvious reaction is to consider that prolactin is a primarily sodium retentive hormone acting in aquatic animals to maintain sodium and water balance (Bern, 1975), but has evolved as an antidiuretic hormone in birds and possibly mammals. However, in the future it may become necessary for comparative endocrinologists to consider the possibility that the prolactin molecule possesses a wide range of possible actions and that manifestation of these actions depends as much on the circulating levels of various steroid hormones as it does on the level of prolactin itself.

LITERATURE CITED

BERN, H. A. 1975. Prolactin and osmoregulation. Amer. Zool., 15:937-949.

BERN, H. A., AND C. S. NICOLL. 1968. The comparative endocrinology of prolactin. Recent Prog. Horm. Res., 24:681-720.

BREWER, K. J., AND D. M. ENSOR. 1980a. Hormonal control of osmoregulation in the Chelonia. I. The effects of prolactin and interrenal steroids in freshwater chelonians. Gen. Comp. Endocrinol., 42:304-309.

———. 1980b. Hormonal control of osmoregulation in the Chelonia. II. The effect of prolactin and corticosterone on Testudo graeca. Gen. Comp. Endocrinol., 42:310-314.

CHAN, D. K. O., I. P. CALLARD, AND I. CHESTER JONES. 1970. Observations on the water and electrolyte composition of the iguanid lizard Dipsosaurus dorsalis dorsalis (Baird and Girard) with special reference to the control by the pituitary gland and adrenal cortex. Gen. Comp. Endocrinol., 15:374-387.

CHESTER JONES, I., J. G. PHILLIPS, AND D. BELLAMY. 1962. Osmoregulation and prolactin in the hagfish. Gen. Comp. Endocrinol., Suppl., 1:36.

CLARKE, W. C. 1973. Sodium-retaining bioassay of prolactin in the intact teleost Tilapia mossambica in seawater. Gen. Comp. Endocrinol., 21:498-512.

CLARKE, W. C., AND H. A. BERN. 1980. Comparative endocrinology of prolactin. Pp. 106-197, in Hormonal proteins and peptides, Vol. IV (C.H.Li, ed.), Academic Press, New York, xiv+231 pp.

CRIM, J. W. 1972. Studies on the possible regulation of plasma sodium by prolactin in Amphibia. Comp. Biochem. Physiol., 43A:349-357.

DE VLAMING, V. L., M. SAGE, AND B. BIETZ. 1975. Pituitary, adrenal and thyroid influences on osmoregulation in the euryhaline elasmobranch, Dasyaks sabina. Comp. Biochem. Physiol., 52A:505-515.

DHARMAMBA, M., AND J. MAETZ. 1972. Effects of hypophysectomy and prolactin on sodium balance of Tilapia mossambica in freshwater. Gen. Comp. Endocrinol., 19:175-183.

DONEEN, B. A. 1976. Water and ion movements in the urinary bladder of the gobiid teleost Gillichthys mirabilis in response to prolactin and cortisol. Gen. Comp. Endocrinol., 28:33-41.

DONEEN, B. A. AND H. A. BERN. 1974. In vitro effects of prolactin and cortisol on water permeability of the urinary bladder of the teleost Gillichthys mirabilis. J. Exp. Zool., 187:173-179.

ENSOR, D. M. 1975. Prolactin and adaptation. Symp. Zool. Soc. London, 35:129-148.

———. 1979. Comparative endocrinology of prolactin. Chapman and Hall, London, ix+309 pp.

ENSOR, D. M., AND J. N. BALL. 1968. Prolactin and freshwater sodium fluxes in Poecilia latipinna (Teleosteii). J. Endocrinol., 41:xvi.

———. 1972. Prolactin and osmoregulation in fishes. Fed. Proc., 31:1615-1622.

ENSOR, D. M., AND J. G. PHILLIPS. 1970. The effect of salt loading on the pituitary prolactin levels of the duck (Anas platyrhynchos) and juvenile herring or lesser black backed gulls (Larus argentatus and Larus fuscus). J. Endocrinol., 48:167-172.

ENSOR, D. M., I. M. SIMONS, AND J. G. PHILLIPS. 1972. The effect of hypophysectomy and prolactin replacement therapy on salt and water metabolism in Anas platyrhynchos. J. Endocrinol., 57:xi.

GOLDENBERG, S., AND M. R. WARBURG. 1977. Osmoregulatory effect of prolactin during ontogenesis in two anurans. Comp. Biochem. Physiol., A56:137.

HIRANO, T. 1975. Effect of prolactin on water and electrolyte movements in the isolated urinary bladder of the flounder Platichthys flesus. Gen. Comp. Endocrinol., 27:88-94.

HOWARD, K., AND D. M. ENSOR. 1975. Effect of prolactin on sodium transport across frog skin in vitro. J. Endocrinol., 67:56P-57P.

———. 1978. The mechanism of action of prolactin on the isolated frog skin. In Comparative endocrinology (P. J. Guillard and D. Boer, eds.), Elsevier Press, Amsterdam, 450 pp.

LAM, T. J., AND J. F. LEATHERLAND. 1969. Effects of prolactin on the glomerulus of the marine Three Spine Stickleback *Gasterosteus aculeatus* L. from trachurus after transfer from seawater to freshwater in late autumn and early winter. Can. J. Zool., 47:245-250.

LAHLOU, B., AND A. GIORDAN. 1970. Le côntrole hormonal des êchanges et de la balance de l'eau chez le téléostean d'eau douce *Carassius auratus* intacte et hypophysectomisé. Gen. Comp. Endocrinol., 14:491-509.

LARRSEN, L. O. 1969. Effect of hypophysectomy before and during sexual maturation in the cyclostome *Lampetra fluviatilis*. Gen. Comp. Endocrinol., 12:200-208.

LODI, G., M. BICIOTTI, AND M. SACERDOTE. 1978. Osmoregulatory activity of prolactin in the skin of the crested newt. Gen. Comp. Endocrinol., 37:369-373.

MAETZ, J., W. H. SAWYER, G. E. PICKFORD, AND M. MAYER. 1967. Evolution de la balance minérale du sodium chez *Fundulus heteroclitus* au cours du transfert d'eau de mer en eau douce: effets de l'hypophysectomie et de la prolactine. Gen. Comp. Endocrinol., 8:163-176.

MARSHALL, W. S. 1978. On the involvement of mucous secretion in teleost osmoregulation. Can. J. Zool., 56:1088.

PANG, P. K. T., AND W. H. SAWYER. 1974. Effects of prolactin on hypophysectomised mud puppies *Necturus maculosus*. Amer. J. Physiol., 226:458-462.

PAYAN, P., AND J. MAETZ. 1971. Balance hydrique chez les elasmobranches: arguments en faveur d'un control endocrien. Gen. Comp. Endocrinol., 16:535-554.

PEAKER, M., J. G. PHILLIPS, AND A. WRIGHT. 1970. The effect of prolactin on the secretory activity of the nasal gland of the domestic duck (*Anas platyrhynchos*). J. Endocrinol., 47:123-127.

PHILLIPS, J. G., AND D. M. ENSOR. 1972. The significance of environmental factors in the hormone mediated changes of the nasal (salt) gland activity in birds. Gen. Comp. Endocrinol., Suppl., 3:393-404.

PICKFORD, G. E., R. W. GRIFFITH, J. TARETTE, E. HENDLER, AND F. EPSTEIN. 1970. Branchial reduction and renal stimulation of (Na^+ K^+) ATPase by prolactin in hypophysectomised killifish in freshwater. Nature (London), 228:378-379.

POTTS, W. T. W., AND D. H. EVANS. 1966. The effect of hypophysectomy and bovine prolactin on sodium fluxes in freshwater adapted *Fundulus heteroclitus*. Biol. Bull., 131:362.

SAMPIETRO, A., AND P. VERICELLI. 1968. Effetti della prolattina sul tasso ematico del sodiu nel *Tritone cristata* normale ed ipofisectomizzate. Boll. Zool., 35:419.

STANLEY, J. G. AND W. R. FLEMING. 1967. Effect of prolactin and ACTH on the serum and urine sodium levels of *Fundulus kansae*. Comp. Biochem. Physiol., 20:199-208.

Reprinted from
ASPECTS OF AVIAN ENDOCRINOLOGY:
PRACTICAL AND THEORETICAL IMPLICATIONS (C. G. Scanes *et al.*, eds.)
Grad. Studies, Texas Tech Univ., 1982, 26:1-411.

ONTOGENETIC ASPECTS OF STEROIDOGENESIS WITH SPECIAL REFERENCE TO CORTICOSTEROIDOGENESIS OF THE CHICKEN (*GALLUS DOMESTICUS*)

YUICHI TANABE

Department of Poultry and Animal Sciences, Faculty of Agriculture, Gifu University, Yanagido, Gifu 501-11, Japan

An important role of corticosterone, a major circulating corticoid hormone in avian species, is in the control of salt gland function involved in the extrarenal osmoregulatory mechanism in marine birds, including ducks (Bellamy and Phillips, 1966; Phillips and Ensor, 1972).

Ontogenetic aspects of steroidogenesis, especially that of corticosteroids of avian species, are not well established. Recently, the presence of cortisol in the plasma and adrenal glands of embryonic and young chickens (*Gallus domesticus*) has been demonstrated by both *in vivo* and *in vitro* studies (Kalliecharam and Hall, 1974, 1976, 1977; Idler *et al.*, 1976; Nakamura *et al.*, 1978). Furthermore, the amounts of sex-steroid hormones produced by embryonic and postembryonic chicken gonads have been reported (Guichard *et al.*, 1977, 1979; Tanabe *et al.*, 1979).

The present paper will describe and review recent results, mostly obtained in our laboratory, on the ontogenetic aspects of steroidogenesis and sex differentiation in the domestic fowl.

Ontogenetic Aspects of Corticosteroidogenesis in the Chicken

Phylogeny of corticosteroidogenesis in vertebrate species has been extensively reviewed by Idler and Truscott (1972, 1980) and Sandor (1972). It is well established that corticosterone is the principal adrenal steroid in avian blood, although aldosterone, cortisol, and cortisone are present in very limited amounts (see Sandor, 1972; Nakamura *et al.*, 1973).

Nakamura and Tanabe (1973), using *in vitro* techniques, showed that the main pathway for steroidogenesis in the adult chicken adrenal is: pregnenolone → progesterone → 11-deoxycorticosterone → corticosterone → aldosterone. Later, using the same technique, the effect of age on the pathway of steroidogenesis in the embryonic and postembryonic chicken adrenal was investigated (Nakamura *et al.*, 1978). They found that the activity of 17α-hydroxylase, which is responsible for the conversion of progesterone to 17α-hydroxyprogesterone and the subsequent formation of cortisol and cortisone, or androstenedione and testosterone, was present in 17- and 21-day-old embryos and 3- and 7-day-old chicks, but disappeared between 7 and 14 days posthatching (Nakamura *et al.*, 1978).

A map representing the pathways of corticosteroidogenesis in the chicken is illustrated in Fig. 1. The embryonic and very young chicken adrenal has two pathways for the production of two kinds of principal corticoids (corticosterone and cortisol), and another for the production of androgens such as androstenedione and testosterone. The activity of 17α-hydroxylase, which is responsible for the production of cortisol, cortisone, and androgens, decreases with advancing age, whereas 21-hydroxylase, which is responsible for corticosterone formation, increases with advancing age until 14 days posthatching, as illustrated in Fig. 2 (Nakamura et al., 1978). These results strongly suggest that the embryonic adrenal glands of the chicken are capable of producing several steroid hormones including corticosterone, aldosterone, cortisol, cortisone, androstenedione, and testosterone, whereas the adrenals of the adult chicken are capable of producing only two: corticosterone and aldosterone (Nakamura et al., 1978).

We (Y. Tanabe, T. Nakamura and K. Fujioka, unpubl. observ.) have measured corticosterone and cortisol (using radioimmunoassay) in the plasma and adrenals of 13-, 15-, 17-, and 20-day-old male and female chick embryos, and of 1-, 3-, 7-, 14-, and 150-day-old male and female chickens. Procedures for sampling of plasma and adrenal glands of embryonic and postembryonic chickens are the same as previously described (Tanabe et al., 1979). Briefly, this entailed taking all the samples at 13:00-15:00 during the day. After hatching, chicks were fed ad libitum and were maintained under 14 hours of light (from 5:00-19:00) per day.

Rabbit-anti-corticosterone-3-oxime-BSA serum and anti-cortisol-3-oxime-BSA serum were obtained from Teikokuzoki Pharmaceutical Company, Tokyo. The anti-corticosterone serum is highly specific and crossreacts with cortisol, cortisone, 11-deoxy-corticosterone, aldosterone, progesterone, and testosterone at 2.20, 0.23, 4,80, 0.17, 5.47, and 0.35 per cent levels, respectively. The anti-cortisol-serum, however, is not so highly specific and crossreacts with corticosterone, cortisone, progesterone, 11-deoxycorticosterone, aldosterone, and testosterone at 42.2, 13.7, 8.8, 3.5, 2.9, and 2.7 per cent levels, respectively. Sephadex LH-20-column chromatography was employed for the separation of corticosterone and cortisol of chicken plasma and adrenal gland extracts (Makino and Kanbegawa, 1973). After chromatography, [1,2,6,7-^{3}H] corticosterone (92 Ci/mmol) and [1,2,6,7-^{3}H] cortisol (80.6 Ci/mmol) were used as the tracers in the radioimmunoassays. The following procedures are the same as previously described (Shodono et al., 1975). The radioactivity was measured with a liquid scintillation spectrometer (LS 9000, Beckman).

The concentration of corticosterone and cortisol in the plasma and adrenal glands of the embryonic and postembryonic chickens at various ages are given in Table 1. The levels of corticosterone were always considerably higher than those of cortisol in both the plasma and adrenal glands studied at all the ages. Plasma corticosterone concentrations were low in 13- and 15-day-old embryos, elevated in 17-day-old embryos, and reached a maximum around hatching time (day 20 of incubation and 1 day posthatching). The peak values are approximately 20-fold higher than the levels in adult

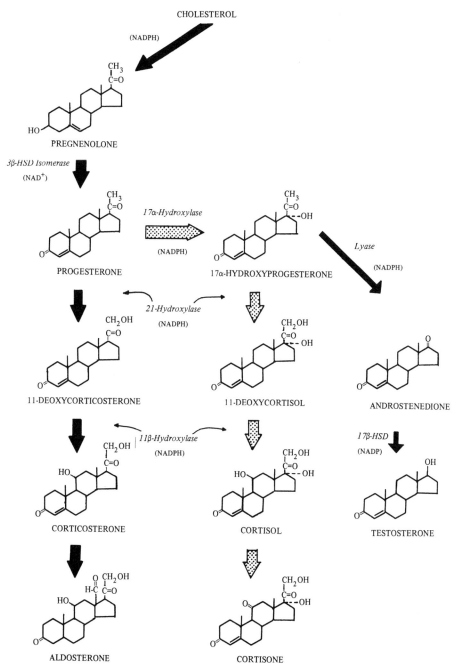

Fig. 1.—Probable pathways of steroidogenesis by the adrenal glands of embryonic and young chickens; based on Nakamura *et al.* (1978).

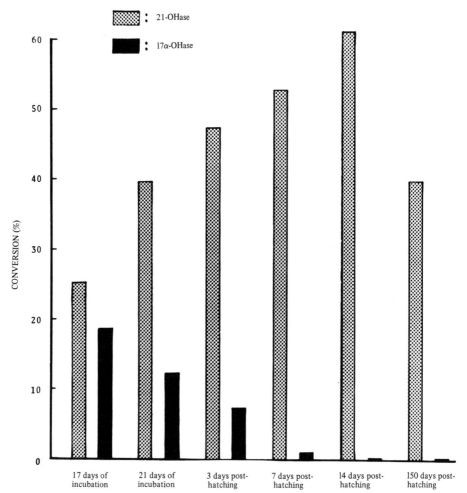

Fig. 2.—Comparison of the activities of 21-hydroxylase (based on the total conversions to 11-deoxycorticosterone and corticosterone from [4-^{14}C] progesterone), and 17α-hydroxylase (based on the total conversions to 17α-hydroxyprogesterone, 11-deoxycortisol, cortisol, androstenedione, and testosterone from [4-^{14}C] progesterone), by the adrenal glands of embryonic and postembryonic chicken (after Nakamura *et al.*, 1978).

chickens. The high plasma concentrations of corticosterone observed at hatching time might reflect its important role in the response of the body to changes in its environment.

The plasma levels of cortisol were highest in 17- and 20-day-old embryos and in 1-day-old chicks. This concentration decreased gradually with advancing age and reached a very low level of about 0.1 ng/ml in adult (150-day-old) male and female chickens.

Similar tendencies in adrenal corticosterone and cortisol concentrations were observed; peak values of the steroids occurred at one day posthatching.

TABLE 1.—*Concentrations of corticosterone and cortisol in the plasma and the adrenal glands of the embryonic and young chickens at various ages. Mean of 5 embryos or birds ± SE.*

Age	Days	Sex	Plasma (ng/ml)		Adrenal glands (ng/mg)	
			Corticosterone	Cortisol	Corticosterone	Cortisol
Embryo	13	M	0.61[a]	<0.10[b]	ND[c]	ND
		F	1.91[a]	<0.10[b]	ND	ND
	15	M	4.49[a]	<0.10[b]	ND	ND
		F	5.74± 0.24	<0.10[b]	ND	ND
	17	M	12.44± 1.32	5.21±0.73	2.09±0.20	0.49±0.03
		F	17.21± 6.50	6.33±1.13	2.32±0.24	0.35±0.02
	20	M	55.15±10.20	6.72±1.84	5.71±0.15	0.54±0.04
		F	52.13± 4.32	6.17±0.92	5.56±0.12	0.52±0.04
Chick	1	M	42.37± 3.43	6.05±1.42	17.55±2.41	0.80±0.11
		F	62.81±16.54	5.62±0.78	16.97±0.24	0.47±0.06
	3	M	17.42± 2.49	4.77±0.62	14.88±1.25	0.74±0.88
		F	37.06± 8.29	4.56±0.71	15.81±2.34	0.64±0.09
	7	M	19.74± 1.53	4.19±0.69	13.45±1.72	0.43±0.05
		F	36.05±14.32	4.03±1.34	10.20±0.74	0.50±0.06
	14	M	30.71± 1.40	4.72±0.64	12.56±1.62	0.36±0.04
		F	29.69± 5.92	4.77±0.77	13.56±1.59	0.30±0.02
Adult	150	M	2.53± 0.67	<0.10[b]	ND	0.10±0.02
		F	3.70± 0.25	0.18±0.03	ND	0.28±0.03

[a]Pooled sample of 20 embryos.
[b]Not detectable, level being less than 100 pg/mg.
[c]Not determined.

These *in vivo* data on the levels of corticosteroids roughly correlate with the *in vitro* data on corticosteroidogenesis (see above). Relatively high cortisol levels, however, were observed in the plasma from 7- and 14-day-old chicks in the *in vivo* studies, while the activity of 17α-hydroxylase (responsible for the formation of cortisol) was found to have almost disappeared in seven and 14-day-old chicks in the *in vitro* metabolism studies (see Fig. 2).

Ontogenetic Aspects of Sex Steroid Steroidogenesis in the Chicken

Production and secretion of progesterone, testosterone, and estradiol by the testes, the ovary, and the adrenal glands of embryonic and postembryonic chickens have been studied (Tanabe *et al.*, 1979). Eggs obtained from White Leghorn hens were incubated at 37°C and 55 per cent relative humidity. Blood samples, adrenal pairs, and testes or ovary (left) were collected from 17- and 20-day-old male and female chick embryos and from 1-, 3-, 7-, 14-, 21-, 28-, 35-, 42-, and 150-day-old male and female chickens. Plasma and tissue concentrations of sex steroid hormones were measured by radioimmunoassay following a procedure described elsewhere (Tanabe *et al.*, 1979). All samples were taken at between 13:00-15:00. The chickens received 14 hours of light (from 5:00-19:00) per day after hatching.

Testosterone concentrations in the plasma, testes, and adrenal pairs of the male embryonic and postembryonic chickens are shown in Fig. 3. Plasma testosterone concentrations in the embryonic chick were relatively high. In

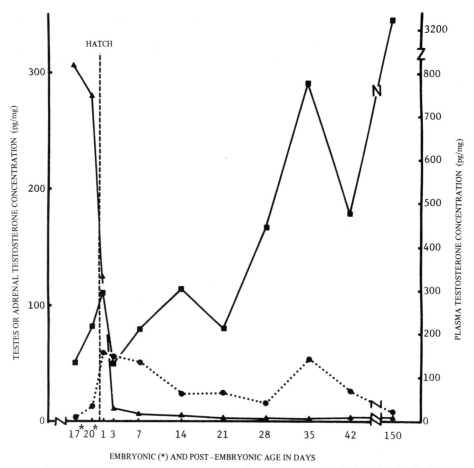

Fig. 3.—Changes in testosterone concentrations in plasma, testes and adrenal pairs during the prehatching and posthatching ages of male chickens. Closed squares, testosterone concentration in plasma; closed circles, that in testes; closed triangles, that in adrenal pairs. Each point represents an average of five embryos or birds (after Tanabe *et al.*, 1979).

addition, the plasma testosterone concentration started to increase at 21 days posthatching, and reached the first peak at 35 days posthatching. The concentration of testosterone in the testes of 17-day-old embryos was very low. This increased rapidly around the time of hatching to reach a first peak at one day posthatching. The level of testosterone in the testes than steadily decreased until 28 days posthatching. It then increased between 25 days and 35 days posthatching.

On the other hand, in the adrenal glands, testosterone concentration was at its highest in 17-day-old male embryos, decreased rapidly between one and three days posthatching, and remained at low levels thereafter. Testosterone concentrations in the adrenal glands at 17 days and at 20 days of incubation were 110- and 21-fold higher, respectively, than those in the testes. On the other hand, 35 days following hatching, the concentration of testosterone in the testes was 23-fold higher than in adrenals.

The plasma levels of estradiol in the male were very low (less than 7.5 pg/ml), and were mostly below the detection limit of the assay. Estradiol concentrations in the testes were also low, usually less than 2 pg/mg or 15 pg/testis pair. Estradiol concentrations in the male adrenal glands were also very low (less than 1.5 pg/mg or 15 pg/adrenal pairs), and mostly were below the detection limit except in the 17-day-old male embryo (9.7 pg/mg, a value fairly comparable to that of the female at 11.4 pg/mg).

Testosterone concentrations in the plasma, ovary (left), and adrenal pairs of the female embryonic and postembryonic chickens are illustrated in Fig. 4. Plasma testosterone concentrations in the female embryos were somewhat lower than the male, increased after hatching, and reached the first major peak at 28 days posthatching. Testosterone concentrations in the female gonad were much higher than in those of the male embryos, and decreased gradually after hatching. Testosterone concentrations in the female adrenal glands at embryonic age were very high; much higher (2- to 5-fold) than those of the left ovary. These levels decreased drastically between one and three days posthatching, and remained at low levels thereafter.

Estradiol concentrations in the plasma, left ovary, and adrenal pairs of the female embryonic and postembryonic chickens are illustrated in Fig. 5. Plasma estradiol concentrations in the female embryos were relatively high, and fairly constant levels were maintained until 42 days posthatching. The concentration of estradiol in the plasma of 150-day-old laying hens was about 5-fold higher than in the chicks. The estradiol concentration in the ovary from 17-day-old female embryos was higher (about 5-fold) than that in adrenal pairs. The plasma concentration of estradiol increased gradually after hatching, and reached a maximum in 150-day-old laying hens. The estradiol concentration in the adrenal glands was maintained at low, fairly constant levels in the females.

Progesterone concentration in the adrenal glands was always high throughout prehatching and posthatching periods. Progesterone concentrations either in testes or ovary were much lower than those in the adrenal glands throughout prehatching and posthatching periods.

These results show that the adrenal glands have a more important role in the production and secretion of testosterone than the testes or the ovary of the embryonic chick, at least in the later stage of development. Furthermore, the embryonic testes appear to be much less active than the embryonic ovary in production and secretion of testosterone and estradiol. After hatching, the testes or ovary produces and secretes more testosterone than does the adrenal gland. The concentrations of testosterone in the plasma of the male embryonic chicks were higher than those of the female embryos as reported by Tanabe et al. (1979) and Woods et al. (1975). This testosterone might have been derived from the adrenal glands and not from the testes. A drastic decrease in the testosterone concentration in adrenal pairs of both sexes

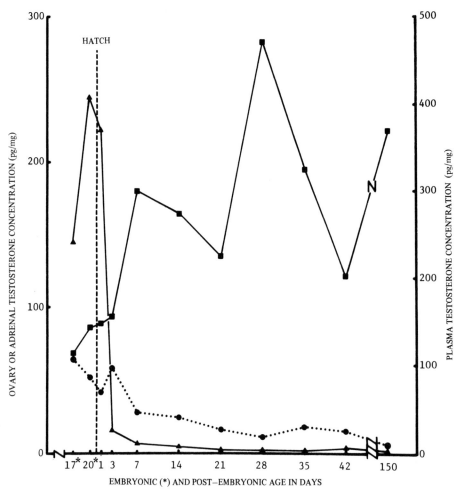

Fɪɢ. 4.—Changes in testosterone concentrations in plasma, ovary, and adrenal pairs during the prehatching and posthatching ages of female chickens. Closed squares, testosterone concentration in plasma; closed circles, that in ovary; closed triangles, that in adrenal pairs. Each point, represents an average of five embryos or birds (after Tanabe *et al.*, 1979).

between one and three days posthatching was observed. These data also indicate that embryonic adrenals are not totally differentiated organs, but have several functions so that the glands can produce various kinds of steroid hormones such as progesterone, androstenedione, testosterone, estradiol, corticosterone, and cortisol; after hatching, the glands become more differentiated and specialized organs produce mainly corticosterone.

Ontogenetic Aspects of Sex Differentiation in the Chicken

The control of the development of secondary sex organs in birds is well established. The male secondary sex organs (including the development of

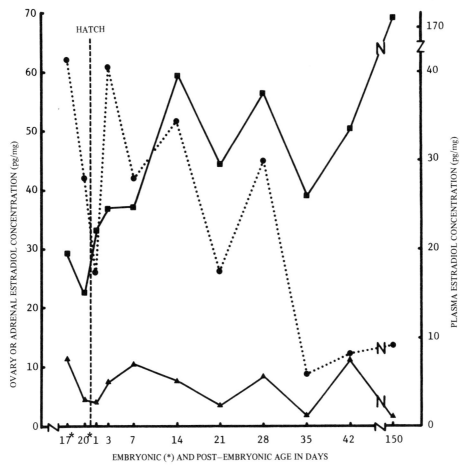

Fɪɢ. 5.—Changes in estradiol concentrations in plasma, ovary, and adrenal pairs during the prehatching and posthatching ages of female chickens. Closed squares, estradiol concentration in plasma; closed circles, that in ovary; closed triangles, that in adrenal pairs. Each point represents an average of five embryos or birds (after Tanabe *et al.*, 1979).

phallus, the laryngeal extension in ducks, *Anas platyrhynchos domestica*, and the development of phallus and the male type pointed plumage in the chicken), appear in castrates by self-differentiation. The female type depends on the presence of an ovary or that of estrogenic hormones (reviews: Wolff, 1979; Masui, 1967; Witschi, 1967).

This situation is the reverse of that found in mammals (Witschi, 1967; Ohno, 1979). It may be pointed out that sex-steroid applications to a species with male heterozygosity (XX/XY) are likely to respond by masculinization of the female (XX), while those with female heterozygosity (ZW/ZZ) are apt to react with feminization of the male (ZZ); this is especially true in amphibians, reptiles, and birds (Witschi, 1967; Wolff, 1979). Ohno (1979) suggests that the major sex-determining gene in mammals is present on the Y chromosome. This produces the H-Y (histocompatibility-Y) antigen that has testis organizing action. Wachtel *et al.* (1975) performed an extensive phylogenic study on the H-Y antigen, and found that it is present in XY male cells in mammals and amphibians with male heterozygosity (XX/XY) such as *Homo sapiens*, *Mus musculus*, and *Rana pipiens*. However, it is present in the ZW female cells (possibly on the W chromosome) in birds and amphibians with female heterozygosity (ZW/ZZ) such as *Gallus domesticus* and *Zenopus laevis*. Ohno (1979) concluded that the "original sex" of the mammals, with male heterozygosity (XX/XY), should be female which has homozygosity of sex chromosomes (XX).

The present results (see also Tanabe *et al.*, 1979; Guichard *et al.*, 1977, 1979) demonstrate that the chick female embryonic gonad (left ovary) is much more active in secreting sex steroid hormones than the male embryonic gonads (testes). These results strongly suggest that the "original sex" of the chicken (probably of all bird species) should be male (ZZ), and estrogenic hormones secreted from the embryonic ovary have a very important role in sexual development (feminizing) of the female. It is plausible that the major sex-determining (inducing) gene in the chicken is present on the W chromosome.

CONCLUSIONS

Ontogenetic aspects of corticosteroidogenesis and steroidogenesis of the chicken were studied. In *in vitro* studies the major products from progesterone were 11-deoxycorticosterone and 17α-hydroxyprogesterone. The yield of 17α-hydroxyprogesterone decreased with advancing age and disappeared at seven days posthatching. 11-Deoxycortisol and cortisol were produced from progesterone by the homogenates from 17- and 21-day-old embryos and 3-day-old chicks, but neither was produced by those from 7-day-old chicks or those from 150-day-old hens. Radioactive 17α-hydroxyprogesterone was converted to 11-deoxycortisol and cortisol in large amounts and to cortisone in small amounts. Androstenedione and testosterone were detected in the adrenal homogenate from 17 days of incubation to seven days posthatching, but not in the tissue from 14 days posthatching. The activity of 17α-hydroxylase was high at 17 days of incubation, decreased with advancing age, and disappeared between seven and 14 days posthatching. These results represent evidence of cortisol and testosterone formation *in vitro* by embryonic and very young chick adrenals. Radioimmunoassay data showed the presence of high levels of corticosterone and somewhat lower levels of cortisol in both the plasma and the adrenals in the late embryonic and postembryonic chickens. Testosterone concentration in adrenal pairs from 17- and 20-day-old male and female embryos were high, but drastically decreased between one and three days posthatching, and remained low thereafter. Testosterone concentration in the testes of 17-day-old male embryo was very low, but it increased rapidly in pre- and posthatching time and reached the first peak at seven days posthatching. Plasma level of testosterone at embryonic ages was relatively high. Similar results were obtained with the female embryos and female chicks, except for relatively high testosterone concentration in the embryonic ovary. Estradiol concentration in the ovary from the 17-day-old female embryo was 5-fold higher than that of the adrenals. Estradiol concentration in the adrenals and the testes from the male embryos was very low and below the limits of detection. It seems that the adrenal glands have a more important role in the production and secretion of testosterone than the testes or the ovary in the embryonic chicks. Furthermore, the embryonic testes are less active than the embryonic ovary in the production of both testosterone and estradiol. After hatching, the testis or ovary produces and secretes more sex steroid hormones than does the adrenal gland. These results, along with information obtained by other researchers, suggest that the original sex of the chicken should be male (ZZ), and estrogens secreted from the embryonic left ovary have an important role in sexual development (feminizing) of the female chicken.

ACKNOWLEDGMENTS

I thank Dr. Takao Nakamura, Mr. Hirokazu Hirano, and Mr. Koro Fujioka, Gifu University, for their collaboration, and Mr. Osamu Doi, Gifu University, for his assistance in preparing the manuscript.

LITERATURE CITED

BELLAMY, D., AND J. G. PHILLIPS. 1966. Effects of the administration of sodium chloride solutions on the concentration of radioactivity in the nasal gland of ducks (*Anas platyrhynchos*) injected with (^{3}H) corticosterone. J. Endocrinol., 36:97-98.

GUICHARD, A., L. CEDARD, Th. M. MIGNOT, D. SCHEIB, AND K. HAFFEN. 1977. Radioimmunoassay of steroids produced by cultured chick embryonic glands: differences according to age, sex and size. Gen. Comp. Endocrinol., 32:255-265.

———. 1979. Radioimmunoassay of steroids produced by chick embryo gonads cultured in the presence of some exogenous steroid precursors. Gen. Comp. Endocrinol., 39:9-19.

IDLER, D. R., AND B. TRUSCOTT. 1972. Corticosteroids in fish. Pp. 126-252, *in* Steroids in nonmammalian vertebrates (D. R. Idler ed.), Academic Press, New York, xii+504 pp.

———. 1980. Phylogeny of vertebrate adrenal corticosteroids. Pp. 357-372, *in* Evolution of vertebrate endocrine systems (P. K. T. Pang and A. Epple, eds.), Texas Tech Press, Lubbock, 404 pp.

IDLER, D. R., J. M. WALSH, R. KALLIECHARAN, AND B. K. HALL. 1976. Identification of cortisol in plasma of the embryonic chick. Gen. Comp. Endocrinol., 30:539-540.

KALLIECHARAN, R., AND B. K. HALL. 1974. A developmental study of the levels of progesterone, cortisol, and cortisone circulating in plasma of chick embryo. Gen. Comp. Endocrinol., 24:364-372.

———. 1976. A developmental study of progesterone, corticosterone and cortisone in the adrenal glands of the embryonic chick. Gen. Comp. Endocrinol., 30:404-409.

———. 1977. The *in vitro* biosynthesis of steroids from pregnenolone and cholesterol and the effects of bovine ACTH on corticoid production by adrenal glands of embryonic chicks. Gen. Comp. Endocrinol., 33:147-159.

MASUI, K. 1967. Sex determination and sex differentiation in the fowl. Univ. Tokyo Press, Tokyo, x+225 pp.

NAKAMURA, T., AND Y. TANABE. 1973. *In vitro* corticosteroidogenesis by the adrenal gland of the chicken (*Gallus domesticus*). Gen. Comp. Endocrinol., 21:99-107.

NAKAMURA, T., Y. TANABE, AND H. HIRANO.   1978.   Evidence of the *in vitro* formation of cortisol by the adrenal gland of embryonic and young chickens (*Gallus domesticus*). Gen. Comp. Endocrinol., 35:302-308.

OHNO, S.   1979.   Major sex-determining genes.   Springer-Verlag, Berlin, vii+140 pp.

PHILLIPS, J. G., AND D. M. ENSOR.   1972.   The significance of environmental factors in the hormone mediated changes of nasal (salt) gland activity in birds.   Gen. Comp. Endocrinol., Suppl., 3:393-404.

SANDOR, T.   1972.   Corticosteroids in Amphibia, Reptilia and Aves.   Pp. 253-327, *in* Steroids in nonmammalian vertebrates (D. R. Idler ed.), Academic Press, New York, xii+504 pp.

SHODONO, M., T. NAKAMURA, Y. TANABE, AND K. WAKABAYASHI.   1975.   Simultaneous determinations of oestradiol-17β, progesterone and luteinizing hormone in the plasma during the ovulatory cycle of the hen.   Acta. Endocrinol., 78:565-573.

TANABE, Y., T. NAKAMURA, K. FUJIOKA, AND O. DOI.   1979.   Production and secretion of sex steroid hormones by the testes, the ovary, and the adrenal glands of embryonic and young chickens (*Gallus domesticus*).   Gen. Comp. Endocrinol., 39:26-33.

WACHTEL, S. S., G. C. KOO, AND E. A. BOYSE.   1975.   Evolutionary conservation of H-Y male antigen.   Nature (London), 254:270-272.

WITSCHI, E.   1967.   Biochemistry of sex differentiation in vertebrate embryos.   Pp. 193-225, *in* The biochemistry of animal development, Vol. II (R. Weber, ed.), Academic Press, New York, xiv + 481 pp.

WOLFF, E.   1979.   Old experiments and new trends in avian sex differentiation.   *In vitro*, 15:6-10.

WOODS, J. E., R. M. SIMPSON, AND P. L. MOORE.   1975.   Plasma testosterone levels in the chick embryo.   Gen. Comp. Endocrinol., 27:543-547.

SECTION VI

EFFECTS OF ENVIRONMENTAL POLLUTANTS ON AVIAN ENDOCRINE SYSTEMS

INTRODUCTION: POLLUTANTS AND ENDOCRINE SYSTEMS

JAMES CRONSHAW

*Department of Biological Sciences, University of California,
Santa Barbara, California 93106 USA*

On January 24, 1969, a blowout occurred at an offshore oil drilling operation in the Santa Barbara Channel, California. During the next ten days, approximately 11,000 tons of crude oil were released into the surrounding water, and as a result many seabirds died (Table 1). This incident and similar ones around the world have raised the consciousness of scientists as well as the public to the problems associated with increasing petroleum contamination in the environment.

The acute decimation of large numbers of marine birds in regions of accidental petroleum spillage is due mainly to physical factors such as oiling of the feathers and the associated impairment of flight, loss of waterproofing, insulation, and buoyancy (Tuck, 1961; Hartung and Hunt, 1966; Hartung, 1967; Mc Ewan and Koelink, 1973; Holmes and Cronshaw, 1977). For instance, Hartung (1967) has shown that the application of small amounts of oil to the breast feathers of birds can cause significant increases in thermal conductivity, and results in an increased basal metabolism to compensate for the high rate of heat loss. At an environmental temperature of 15°C, contamination of a 900 gram black duck with only 20 grams of oil will cause almost a two-fold increase in metabolic rate; this rate of energy consumption would be equivalent to that needed to maintain the normal body temperature of an uncontaminated bird living at a temperature below -10°C.

The acute effects of heavy petroleum spillage on local seabird populations are perhaps more rapid and more conspicuous than they are on other forms of wildlife. As a result, the spectacle of oil-soaked birds lying on the beach has become an extremely emotional public issue. An examination of the effects resulting from many major spillage accidents, however, indicates that seabird mortality is often surprisingly low in relation to the amount of the oil spilled (Table 1). As the world-wide consumption of petroleum has increased during the past 50 years, the frequency of accidental spillage has likewise increased dramatically. At the same time, extensive reductions have occurred in the size of many seabird colonies, suggesting that factors other than physical contamination with oil might have contributed to their demise. For example, colonies of puffins living on the islands of the western approaches were exceedingly large at the beginning of the twentieth century, but populations now have been reduced drastically. There is little doubt that chronic low-level persistent contamination of their environment with oil has been a major factor in their decline.

TABLE 1.—*The estimated mortalities sustained by seabird populations following some of the major oil spills that have occurred since 1937 (adapted from Holmes and Cronshaw, 1977).*

Incident	Spillage	Mortality	Species
March 1937 San Francisco Bay Calif., USA	Crude oil 9,000 tons	10,000 (1.1 birds/ton)	Murre, grebe, scoter
January 1953 Howacht Bay Baltic Sea	Oil residues 500 tons	10,000 (20 birds/ton)	Eider, merganser, scoter
January 1955 Gerd Maersk Elbe River, Germany	Crude oil 8,000 tons	275,000 (34 birds/ton)	Scoter
March 1967 Torrey Canyon SW England	Crude oil 117,000 tons	30,000 (0.3 birds/ton)	Guillemot, razorbill
March 1969 Santa Barbara Calif., USA	Crude oil 11,000 tons	3,600 (0.3 birds/ton)	Western grebe, loon, scoter, cormorant
April 1969 Hamilton Trader Irish Sea	Heavy fuel oil 600-700 tons	6,000 (9 birds/ton)	Gillemot, razorbill
May 1969 Palva, Kokar Finland	Crude oil 150 tons	3,000-3,500 (22 birds/ton)	Eider, long-tail duck
January 1970 NE Britain	Fuel oil 1,000 tons	50,000 (50 birds/ton)	Sea duck, auk
February 1970 Delian Apollon, Tampa Florida, USA	Bunker C fuel oil 80-100 tons	9,000 (90 birds/ton)	No record
Feb.-April 1970 Arrow & Irving Whale Newfoundland	BunkerC fuel oil 10,000 tons	12,800 (0.8 birds/ton)	Sea duck, auk, alcid, eider, ducks
January 1971 San Francisco Bay Calif., USA	Bunker C fuel oil 300-350 tons	7,000 (22 birds/ton)	Grebe, guillemot, scoter

Experimental studies have shown that the ingestion of sublethal doses of petroleum can have a variety of detrimental physiological and behavioral effects on birds (Giles and Livingston, 1960; Hawkes, 1961; Hunt, 1961; Erickson, 1963; Hartung, 1963; Hartung and Hunt, 1966; Hartung, 1967; Snyder *et al.*, 1973; Crocker *et al.*, 1974, 1975; Szaro and Albers, 1977; Holmes and Cronshaw, 1977; Szaro et al., 1978; Holmes et al., 1978a, 1978b, 1979; Miller et al., 1978; Holmes and Gorsline, 1980; Patton and Dieter, 1980). Developing embryos as well as juveniles and mature birds appear to be sensitive to petroleum, and each of the effects, both physiological and behavioral, is controlled either directly or indirectly by endocrine mechanisms associated with the pituitary-adrenal and the pituitary-gonadal axes.

Early evidence for the involvement of the endocrine system in petroleum contamination of seabirds came from studies dealing with increasing solute-linked water transfer through intestinal mucosa cells in juvenile birds adapting to seawater. During adaption of ducklings to seawater, rates of sodium and water transfer across the intestinal mucosa are increased; this development is stimulated through the action of an adrenocortical steroid hormone, probably corticosterone (Crocker and Holmes, 1971, 1976). In studies where small volumes of crude oil were fed to ducklings also receiving seawater, the ducklings failed to develop increased solute-linked mucosal water transfer rates sufficient to ensure their survival (Crocker et al., 1974, 1975). These inhibitory effects of the ingested crude oil, however, could be prevented by treating the birds with corticosterone prior to giving them crude oil (Crocker and Holmes, 1976). Thus, ingestion of sublethal doses of crude oil seems to interfere with a hormonally regulated mechanism in ducklings, and probably involves the pituitary-adrenal axis; as a result, the birds are unable to maintain homeostasis and adapt to life in a simulated marine environment.

Although adult mallard ducks seem to be more tolerant of petroleum-contaminated food than the juvenile birds, some adults show very low tolerances to food conaining high concentrations of petroleum. Even when consuming low concentrations of crude oil, some birds show signs of extreme sensitivity to cold stress. This is particularly apparent among birds that must sustain high levels of adrenocortical activity, such as those individuals that have been adapted to a simulated seawater environment (Holmes et al., 1978a, 1979). Under these conditions, sensitivity to cold stress is manifested among groups of contaminated birds as mortality episodes that often start earlier, last longer, and involve a greater number of birds than among groups consuming uncontaminated food (Fig. 1). Also, the pattern of each mortality episode sometimes is quantitatively related to the concentration of petroleum in the food. Additionally, a wide range of relative toxicities is found among crude oils from different geographic regions (Fig. 1; Holmes et al., 1978a, 1979). These observations suggest that the consumption of oil-contaminated food might interfere with the ability of the birds to respond to nonspecific stressors. Thus, contaminated birds already exposed to stressors, such as hyperosmotic drinking water and persistent cold, tend to succumb as a result of an inability of the adrenal cortex to respond adequately.

In mammals it is known that both the oral and the parenteral doses of DDD cause profound cytotoxic adrenocortical atrophy and that this atrophy is similar to that seen following adenohypophysectomy (Nichols, 1961; Hart et al., 1973). In ducks that have been consuming food contaminated with

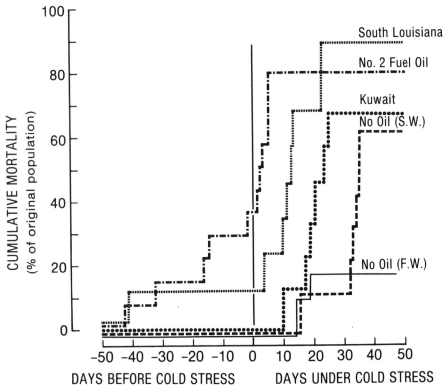

Fig. 1.—Cumulative mortalities among groups of freshwater- and seawater-adapted mallard ducks (*Anas platyrhynchos*) consuming petroleum-contaminated food. The levels of contamination were such that each bird consumed each day either 3 ml of No. 2 fuel oil or 6 ml of crude oil from either Kuwait or South Louisiana. During the first 50 days the birds were maintained at 27°C and for the next 50 days the room temperature was lowered to 3°C (from Holmes *et al.*, 1978a).

petroleum, a structural zonation develops among the adrenocortical cells, and a prominent subcapsular zone can be distinguished from the central inner zone (Fig. 2; Pearce *et al.*, 1979). The cells of the inner zone show structural changes in the mitochondria similar to those that have been described in the dog adrenal cortex following intravenous injection of DDD (Hart *et al.*, 1973). The extensive mitochondrial degeneration in cells from

→

Fig. 2 (top).—A section through the adrenal gland of a seawater-maintained duck that had consumed food contaminated with Santa Barbara crude oil for 50 days. The gland is clearly zonated. The inner zone contains numerous basophilic polygonal adrenocortical cells with densely staining nuclei. The interrenal cells of the subcapsular zone appear to be more normal than those of the inner zone and are arranged in cords two cells wide. 160×.

Fig. 3 (bottom left).—Mitochondria in an inner zone adrenocortical cell from a seawater-maintained duck consuming uncontaminated food. 31,000×.

Fig. 4 (bottom right).—Mitochondria in an inner zone adrenocortical cell from a duck that had consumed food containing 3% South Louisiana crude oil for 7 days. 31,000×.

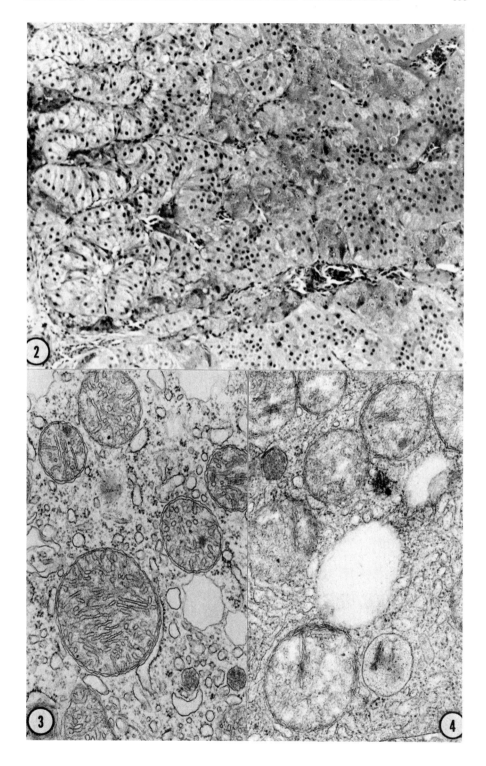

the inner zone of the duck adrenal gland is characterized by swelling, a reduction in the number of cristae and a loss of electron opacity of the mitochondrial matrix (Figs. 3 and 4). The few intact cristae remaining in the mitochondria of these cells are in the tubular configuration. This configuration is very labile and occurs only in cells that are subject to corticotropic stimulation (Pearce *et al.*, 1979). Because corticosterone is synthesized primarily in the adrenocortical cells of the inner zone of the adrenal gland, these degenerative changes probably indicate a reduced capacity for hormone synthesis.

Other effects of ingested petroleum on adrenocortical and gonadal function in birds will be examined in the following presentations in which the effects of crude oils from different parts of the world will be compared. Some of these studies have focused primarily on the effects of ingestion of food contaminated with crude oils from North America and the Middle East. The effects of these crude oils on adrenal and hepatic function will be introduced by J. Gorsline, followed by discussions of their effects on reproduction (Holmes, Cavanaugh). Other studies have been concerned with the effects of the recently discovered crude oil from the North Sea. These have also dealt primarily with effects on adrenocortical and gonadal function and, in the joint presentation by Harvey, Phillips and Sharp, the results will be discussed and compared to those from experiments using North American and Middle East oils. Most of the results have been obtained from laboratory studies using as the experimental model either the mallard duck (*Anas platyrhynchos*) that has been periodically outbred with wild stock or some domesticated variant of this species. The crude oils used in these studies were obtained either from offshore drilling areas or from oilfields where they are transported in large quantities throughout the world by tanker. In both instances, the probability for spillage and contamination of coastal waters, harbors and shipping lanes is extremely high.

LITERATURE CITED

CROCKER, A. D., AND W. N. HOLMES. 1971. Intestinal absorption in ducklings (*Anas platyrhynchos*) maintained on freshwater and hypertonic saline. Comp. Biochem. Physiol., 40A:203-211.

———. 1976. Factors affecting intestinal absorption in ducklings (*Anas platyrhynchos*). Proc. Soc. Endocrinol., 71:88-89.

CROCKER, A. D., J. CRONSHAW, AND W. N. HOLMES. 1974. The effect of a crude oil on intestinal absorption in ducklings (*Anas platyrhynchos*). Environ. Poll., 7:165-178.

———. 1975. The effect of several crude oils and some petroleum distillation fractions on intestinal absorption in ducklings (*Anas platyrhynchos*). Environ. Physiol. Biochem., 5:92-106.

ERICKSON, R. C. 1963. Oil pollution and migratory birds. Atlantic Nat., 18:5-14.

GILES, L. A., AND J. LIVINGSTON. 1960. Oil pollution on the seas. Trans. N. Amer. Wildl. Nat. Resource Conf., 25:297-302.

HART, M. M., R. L. REAGAN, AND R. H. ADAMSON. 1973. The effects of isomers of DDD on the ACTH-induced steroid output, histology and ultrastructure of the dog adrenal cortex. Toxicol. Appl. Pharmacol., 24:101-113.

HARTUNG, R. 1963. (Wildlife) Ingestion of oil by waterfowl. Papers Michigan Acad. Sci., Arts and Letters, 48:49-55.

———. 1967. Energy metabolism in oil-covered ducks. J. Wildl. Mgmt., 31:789-804.

HARTUNG, R., AND G. S. HUNT. 1966. Toxicity of some oils to waterfowl. J. Wildl. Mgmt., 30:564-570.

HAWKES, A. L. 1961. A review of the nature and extent of damage caused by oil pollution at sea. Trans. N. Amer. Wildl. Nat. Resource Conf., 26:343-355.

HOLMES, W. N., AND J. CRONSHAW. 1977. Biological effects of petroleum on marine birds. Pp. 359-398, *in* Effects of petroleum on Arctic and subarctic marine environments and organisms. Vol. II. Biological effects (D. C. Malins, ed.), Academic Press, New York, xx+500 pp.

HOLMES, W. N., AND J. GORSLINE. 1980. Effects of some environmental pollutants on the adrenal cortex. Pp. 311-314, *in* Symposium on adrenal steroid biosynthesis. Proc. Sixth Internat. Cong. Endocrinol. (I. A. Cumming, J. W. Funder, and F. A. O. Mendelsohn, eds.), Australian Acad. Sci., Canberra, xii+737 pp.

HOLMES, W. N., J. CRONSHAW, AND J. GORSLINE. 1978a. Some effects on ingested petroleum on seawater-adapted ducks (*Anas platyrhynchos*). Environ. Res., 17:177-190.

HOLMES, W. N., K. P. CAVANAUGH, AND J. CRONSHAW. 1978b. The effects of ingested petroleum on oviposition and some aspects of reproduction in experimental colonies of mallard ducks (*Anas platyrhynchos*). J. Reprod. Fert., 54:335-347.

HOLMES, W. N., J. GORSLINE, AND J. CRONSHAW. 1979. Effects of mild cold stress on the survival of seawater-adapted mallard ducks (*Anas platyrhynchos*) maintained on food contaminated with petroleum. Environ. Res., 20:424-444.

HUNT, G. S. 1961. Waterfowl losses on the lower Detroit River due to oil pollution. Pp. 10-26, *in* Proc. Fourth Conf., Internat. Assoc. Great Lakes Res. Dir., Inst. Sci. Technol. Publ., 000+000 pp.

McEWAN, E. H., AND A. F. C. KOELINK. 1973. The heat production of oiled mallards and scaup. Can. J. Zool., 51:27-31.

MILLER, D. S., D. B. PEAKALL, AND W. B. KINTER. 1978. Ingestion of crude oil; sublethal effects in herring gull chicks. Science, 199:315-317.

NICHOLS, J. 1961. Studies on an adrenal cortical inhibitor. Pp. 84-107, *in* The adrenal cortex (H. D. Moon, ed.), Internat. Acad. Pathol. Monograph, Harper, New York, xi+315 pp.

PATTON, J. F., AND M. P. DIETER. 1980. Effects of petroleum hydrocarbons on hepatic function in the duck. Comp. Biochem. Physiol., 650:33-36.

PEARCE, R. B., J. CRONSHAW, AND W. N. HOLMES. 1979. Structural changes occurring in interrenal tissue of the duck (*Anas platyrhynchos*) following adenohypophysectomy and treatment *in vivo* and *in vitro* with corticotropin. Cell Tissue Res., 196:429-447.

SNYDER, S. B., J. G. FOX, AND O. A. SOAVE. 1973. Mortalities in waterfowl following Bunker C fuel exposure: an examination of the pathological, microbiological, and oil hydrocarbon residue findings in birds that died after the San Francisco Bay oil spill, January 18, 1971. Mimeographed report on work supported by a grant from the Standard Oil Corp. and by the National Institutes of Health, Research and Resources, grant no. RR00282-06.

SZARO, R. C., AND P. H. ALBERS. 1977. Effects of external applications of No. 2 fuel oil on common eider eggs. Pp. 164-167, *in* Fate and effects of petroleum hydrocarbons in marine ecosystems and organisms (D. A. Wolfe, ed.), Pergamon Press, New York, xviii+478 pp.

SZARO, R. C., M. P. DIETER, G. H. HEINZ, AND J. F. FERRELL. 1978. Effects of chronic ingestion of South Louisiana crude oil on mallard ducklings. Environ. Res., 17:426-436.

TUCK, L. M. 1961. The murres: their distribution, populations and biology, a study of the genus *Uria*. Can. Wildl. Ser. 1, Monogr. Nat. Parks Branch, Ottawa, 260 pp.

Reprinted from
ASPECTS OF AVIAN ENDOCRINOLOGY:
PRACTICAL AND THEORETICAL IMPLICATIONS (C. G. Scanes *et al.*, eds.)
Grad. Studies, Texas Tech Univ., 1982, 26:1-411.

THE EFFECTS OF SOUTH LOUISIANA CRUDE OIL
ON ADRENOCORTICAL FUNCTION

J. GORSLINE

Department of Biological Sciences, University of California,
Santa Barbara, California 93106 USA

The observation that severe adrenocortical atrophy developed in dogs fed DDD (dichlorodiphenyldichloroethane) provided the first indication that ingested hydrocarbon pollutants can interfere with the adrenal function of organisms inhabiting contaminated environments (Nelson and Woodward, 1949). These changes, which involved a breakdown in the organization of the mitochondria in the cells of the zonae fasciculata and reticularis, were associated with decreases in the secretion of glucocorticoids following corticotropic stimulation (Vilar and Tullner, 1959; Nichols, 1961; Cazorla and Moncloa, 1962; Kaminsky *et al.*, 1962; Hart *et al.*, 1973).

There is now clear evidence that mallard ducks (*Anas platyrhynchos*) consuming petroleum-contaminated food also develop some structural damage to the mitochondria of inner zone cells in the adrenal cortex. Because these cells are the principal site of corticosterone synthesis (Klingbeil *et al.*, 1979), any disruption of their mitochondrial organization could adversely affect the synthesis of corticosteroids in a way similar to that seen in mammals exposed to chlorinated hydrocarbon insecticides.

An examination of the effects of ingested petroleum hydrocarbons on the diurnal variations in plasma corticosterone concentrations of mallard ducks has confirmed this possibility (Gorsline and Holmes, 1981). Uncontaminated birds maintained on a short photoperiod (6 hours light and 18 hours darkness) show two daily maxima: one during the first hour of the light phase and a second early in the night. In contrast, birds consuming food contaminated with three per cent South Louisiana crude oil (3 ml per 100 g dry food) for one week show a significant dampening of the first peak, an almost complete obliteration of the second peak and a mean daily concentration that is only one-sixth of that found in the uncontaminated birds (1.2±0.37 vs 7.7±0.84 ng corticosterone per ml plasma). Furthermore, the consumption of food containing lower concentrations of this crude oil causes a similar suppression of the plasma concentration during the interval corresponding to the first diurnal peak; the extent to which the plasma corticosterone concentration is suppressed is inversely proportionate to the amount of crude oil consumed (Fig. 1).

The concentration of corticosterone in plasma, however, reflects not only the rate of synthesis and the distribution of hormone within the organism, but also the rate at which it is metabolized and removed from the circula-

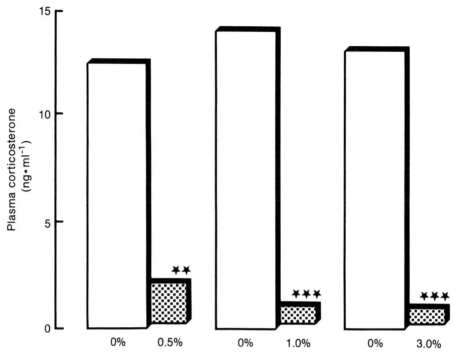

Fig. 1.—Changes in the mean plasma corticosterone concentration in groups of mallard ducks (*Anas platyrhynchos*) following their exposure for 1 week to food contaminated with different concentrations of South Louisiana crude oil. All plasma samples were taken during the first hour of the daily photoperiod. **$P<0.01$ and ***$P<0.001$ with respect to corresponding control values (adapted from Gorsline and Holmes, 1981).

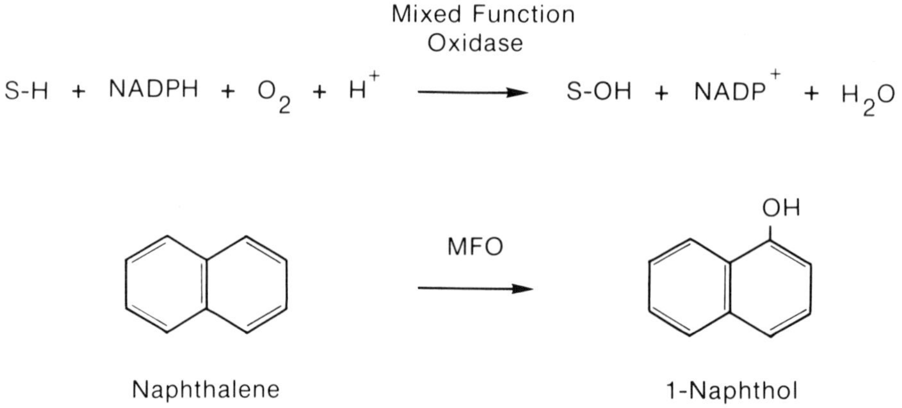

$$\text{Mixed Function Oxidase}$$

$$\text{S-H} + \text{NADPH} + O_2 + H^+ \longrightarrow \text{S-OH} + \text{NADP}^+ + H_2O$$

Fig. 2.—The general equation representing the conversion of a substrate (S) to hydroxylated derivatives (S-OH) through the action of the hepatic mixed function oxidase (MFO) system. In this manner, naphthalene is transformed to metabolites such as 1-naphthol.

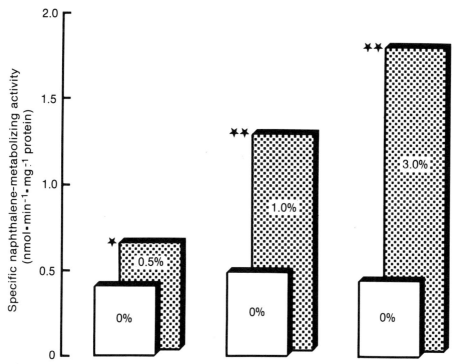

Fig. 3.—Induced increases in the hepatic mixed function oxidase activity in mallard ducks (*Anas platyrhynchos*) following their exposure for 1 week to food contaminated with different concentrations of South Louisiana crude oil. Enzyme activities were estimated in terms of microsomal naphthalene-metabolism *in vitro*. *$P < 0.05$ and **$P < 0.01$ with respect to corresponding control values (adapted from Gorsline and Holmes, 1981).

tion. Studies in mammals have suggested that the rate of degradation of endogenous corticosteroid might be affected by hydrocarbon-induced changes in liver function (see Conney, 1967; Kupfer and Bulger, 1976). These induced changes might involve increases in the hydrocarbon-metabolizing activities of enzymes associated with the mixed function oxidase (MFO) system in the endoplasmic reticulum. It is believed that, through the action of these enzymes, contaminants are rendered more water soluble and excretion is facilitated; thus their accumulation in the tissues is minimized. Changes of this type can be identified by comparing the hydrocarbon-metabolizing properties of hepatic microsomes prepared from both contaminated and uncontaminated organisms. This can be done by measuring the rate at which hydroxylated hydrocarbon derivatives are formed when liver microsomes are incubated in the presence of an appropriate hydrocarbon substrate such as naphthalene, together with the necessary cofactors (Fig. 2). Using this type of assay, we have shown that hydrocarbons derived from food contaminated with any one of a variety of different crude oils are potent inducers of this enzyme system in mallard ducks (Gorsline *et al.*, 1981). For example, the hydrocarbon-metabolizing activity associated with the hepatic MFO

Tetrahydrocorticosterone

CH_2OH
$C=O$

HO

H

Corticosterone

CH_2OH
$C=O$

HO

O

Reductase Pathway

MFO Pathway

6-Hydroxycorticosterone

CH_2OH
$C=O$

HO

O

OH

Fɪɢ. 4.—Metabolic pathways for corticosterone that may occur in the liver of birds exposed to petroleum-contaminated food.

system increased considerably during the first week of exposure to food contaminated with South Louisiana crude oil, and the magnitude of the response was roughly proportionate to the amount of crude oil consumed (Fig. 3).

Although the depuration of contaminated tissues is probably the primary function of the induced hepatic MFO activity, this enzyme system might

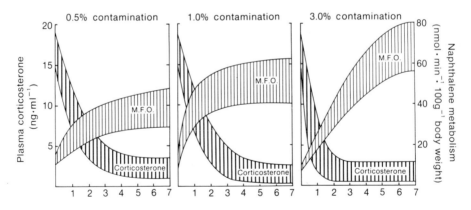

Time of exposure to petroleum-contaminated food (days)

Fɪɢ. 5.—The simultaneous changes in the plasma corticosterone concentrations and the hepatic MFO activity in groups of mallard ducks (*Anas platyrhynchos*) during their first week of exposure to food contaminated with different concentrations of South Louisiana crude oil (adapted from Gorsline and Holmes, 1981).

also, as we have stated earlier, accelerate the turnover of endogenous corticosteroid hormones (Kuntzman *et al.*, 1964; Bledsoe *et al.*, 1964; Kupfer *et al.*, 1964; Balazs and Kupfer, 1966; Kupfer and Peets, 1966; Bradlow *et al.*, 1973). For example, guinea pigs treated with o,p'DDD (an isomer of DDD) not only show high rates of hydrocarbon metabolism, but they also secrete less cortisol and show simultaneous increases in the urinary excretion of some polar derivatives of cortisol (Kupfer *et al.*, 1964). Similar effects occurring in birds exposed to petroleum hydrocarbons could also result in a change in the hepatic metabolism of corticosterone so that an increased hydroxylation of the molecule at the carbon-6 position might occur either to replace or supplement the reductive pathway (Fig. 4). It is therefore particularly interesting to note that the patterns of change in plasma corticosterone concentration and hepatic MFO activity are inversely related during the first week that mallard ducks are exposed to petroleum-contaminated food (Fig. 5). Thus, the decline in plasma corticosterone concentration could be due not only to the disruptive effects of ingested hydrocarbons on adrenocortical mitochondria, but also to their indirect effects on corticosterone metabolism. We have only circumstantial evidence to support this hypothesis, but a rigorous examination of this possible relationship would seem to be worthy of further attention.

Literature Cited

Balazs, T., and D. Kupfer. 1966. Adrenocortical function test: estimation of cortisol production rate in the guinea pig. Toxicol. Appl. Pharmacol., 8:152-158.

Bledsoe, T., D. P. Island, R. L. Ney, and G. W. Liddle. 1964. An effect of o,p'DDD on the extra-renal metabolism of cortisol. Man. J. Clin. Endocr., 24:1303-1311.

Bradlow, H. L., B. Zumoff, D. K. Fukushima, L. Hellman, D. R. Bickers, A. P. Alvares, and A. Kappas. 1973. Drug-induced alterations of steroid hormone metabolism in man. Ann. New York Acad. Sci., 212:148-155.

Cazorla, A., and F. Moncloa. 1962. Action of 1,1,Dichloro-2-p-chlorophenyl-2-o-chlorophenylethane on dog adrenal cortex. Science, 136:47.

Conney, A. H. 1967. Pharmacological implications of microsomal enzyme induction. Pharm. Rev., 19:317-366.

Gorsline, J., and W. N. Holmes. 1981. Effects of petroleum on adrenocortical activity and on hepatic naphthalene-metabolizing activity in mallard ducks (*Anas platyrhynchos*). Environ. Contam. Toxicol., 10:765-777.

Gorsline, J., W. N. Holmes, and J. Cronshaw. 1981. The effects of ingested petroleum on the naphthalene-metabolizing properties of liver tissue in seawater-adapted mallard ducks (*Anas platyrhynchos*). Environ. Sci., 24:377-390.

Hart, M. M., R. L. Reagan, and R. H. Adamson. 1973. The effects of isomers of DDD on the ACTH-induced steroid output, histology and ultrastructure of the dog adrenal cortex. Toxicol. Appl. Pharmacol., 24:101-113.

Kaminsky, N., S. Luse, and P. Hartroft. 1962. Ultrastructure of the adrenal cortex of the dog during treatment with DDD. J. Nat. Cancer Inst., 29:127-159.

Klingbeil, C. K., W. N. Holmes, R. B. Pearce, and J. Cronshaw. 1979. Functional significance of interrenal cell zonation in the duck (*Anas platyrhynchos*) adrenal gland. Cell Tissue Res., 201:23-36.

Kuntzman, R., M. Jacobson, K. Schneidman, and A. H. Conney. 1964. Similarities between oxidative drug-metabolizing enzymes and steroid hydroxylases in liver microsomes. J. Pharm. Exp. Therap., 146:280-285.

KUPFER, D., AND W. H. BULGER. 1976. Interaction of chlorinated hydrocarbons with steroid hormones. Fed. Proc., 35:2603-2608.

KUPFER, D., AND L. PEETS. 1966. The effect of o,p'DDD on cortisol and hexobarbital metabolism. Biochem. Pharm., 15:573-581.

KUPFER, D., T. BALAZS, AND D. A. BUYSKE. 1964. Stimulation of o,p'DDD of cortisol metabolism in the guinea pig. Life Sci., 3:959-964.

NELSON, A. A., AND G. WOODWARD. 1949. Severe adrenal cortical atrophy (cytotoxic) and hepatic damage produced in dogs by feeding 2,2-bis(parachlorophenyl)-1,1-dichloroethane (DDD or TDE). Arch. Path., 48:387-394.

NICHOLS, J. 1961. Studies on an adrenal cortical inhibitor. Pp. 84-107, in The adrenal cortex (H. D. Moon, ed.), Internat. Acad. Pathol. Monograph, Harper, New York, xi+315 pp.

VILAR, O., AND W. W. TULLNER. 1959. Effects of o,p'DDD on histology and 17-hydroxycorticosteroid output of the dog adrenal cortex. Endocrinology, 65:80-86.

Reprinted from
Aspects of Avian Endocrinology:
Practical and Theoretical Implications (C. G. Scanes *et al.*, eds.)
Grad. Studies, Texas Tech Univ., 1982, 26:1-411.

SOME COMMON POLLUTANTS AND THEIR EFFECTS ON STEROID HORMONE-REGULATED MECHANISMS

W. N. Holmes

*Department of Biological Sciences, University of California,
Santa Barbara, California 93106 USA*

The list of pesticides and pollutants known to affect adrenocortical function is long and includes a diverse range of compounds including the chlorinated hydrocarbon, the organophosphate, the phosphothioate, and the carbamate pesticides, as well as the polychlorinated biphenyls, aromatic amines, and aflatoxins (Toth *et al.*, 1971; Bruckner *et al.*, 1974; Civen and Brown, 1974; Sanders and Kirkpatrick, 1975, 1977; Singh and Venkitasubramanian, 1975; Civen *et al.*, 1977a, 1977b). In mammals, all these compounds seem to act primarily on the corticotropin-sensitive cells in the zona fasciculata to cause a decrease in the secretion of glucocorticoids (Kovacs *et al.*, 1970; Horvath *et al.*, 1971).

Although their overall effects may be similar, these compounds might not always interfere with the same step in the steroidogenic pathway. The strikingly consistent pattern of effects on corticosteroidogenesis, however, might not be entirely coincidental. All tend to be hydrophobic, and this property causes them to accumulate primarily in the fatty tissues of contaminated organisms. Clearly, the lipid-rich cells of the adrenal cortex are well-suited to such an accumulation of contaminants and this could account for the fairly uniform pattern of cytotoxic effects seen in adrenocortical cells. Studies on the effects of hydrocarbon pollutants in birds suggest that some of them, particularly those derived from petroleum might also act to disrupt normal steroidogenesis in the adrenal cortex (Holmes and Gorsline, 1980; Gorsline and Holmes, 1981; Holmes *et al.*, 1981).

The pathways whereby any vertebrate organism can become contaminated with hydrocarbon pollutants from the environment, with the development of a hypoadrenocortical state through the action of these compounds on the adrenal cortex, is summarized in Fig. 1.

Many of the pesticides and pollutants that affect adrenocortical function, however, can also either impair or modify reproduction. Again, much of this information has been derived from studies on the effects of chlorinated hydrocarbons, and studies in both birds and mammals suggest that these compounds have two distinct types of effect (see reviews, Moriarty, 1975; Thomas, 1975).

The first type of effect involves an apparent competition for receptor binding sites between the circulating hydrocarbon and the endogenous gonadal steroid hormones (Bitman *et al.*, 1968; Levin *et al.*, 1968a; Welch *et al.*, 1969;

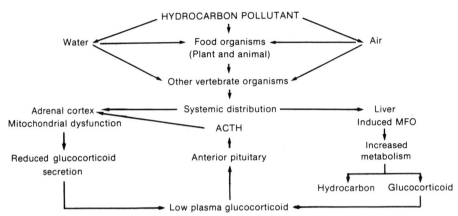

Fig. 1.—The pathways whereby hydrocarbons may contaminate organisms to affect adrenal-cortical activity (adapted from Holmes *et al.*, 1981).

Bitman and Cecil, 1970; Blend and Schmidt, 1971; Smith *et al.*, 1972; Wakeling and Visek, 1973; Wakeling *et al.*, 1973; Kupfer and Bulger, 1976; Gellert, 1978; Gellert and Wilson, 1979). The competitive binding of these compounds to estrogen receptors is frequently followed, especially in birds, by a spectrum of apparently normal estrogenic effects. Dichlorodiphenyltrichloroethane (DDT), dichlorodiphenyldichloroethane (DDD), dichlorodiphenylethylene (DDE), methoxychlor, decachlorotetracyclodecanone (Kepone), and polychlorinated biphenyls have each been shown to possess these properties to a greater or lesser degree, and in some instances their effects are quite profound. For example, in birds, several isomers of DDT, as well as Kepone, have each been shown to bind to estrogen receptors in the oviduct and to cause structural changes resembling those induced by the natural estrogens. These changes involve increases in oviduct weight, the cellular differentiation of the tissues in the oviduct, the generation of cilia, and the formation of tubular glands. (Eroschenko and Wilson, 1974, 1975). Furthermore, the pollutant-receptor complex has even been shown to interact with the nucleus in cells of the oviduct and to induce the production of the specific mRNA necessary for ovalbumin and conalbumin synthesis (Palmiter and Mulvihill, 1978; Hammond *et al.*, 1979).

Although many hydrocarbon pollutants can induce reproductive change similar to that normally evoked by the endogenous estrogens, there is no reason to conclude that these compounds can successfully sustain the complete reproductive cycle. Indeed, many of them seem to disturb normal reproductive cyclicity, and it seems reasonable to presume that these changes reflect impaired gonadal steroidogenesis resulting in abnormally low circulating concentrations of hormone during critical phases of the cycle. This type of change, therefore, might constitute a second category of effects attributable to hydrocarbon pollutants. There have been only a few observations, however, that directly support this hypothesis. Reports on studies in mammals have indicated that a suppression of the cyclical peaks in the plasma progesterone concentration might occur in monkeys given polybrominated biphenyls, and a similar reduction in plasma progesterone concentration has been observed in rats given polychlorinated biphenyls (Jonsson et al., 1976; Allen et al., 1978; Lambrecht et al., 1978). In one study on birds, Ring doves exposed to p,p'DDT have been shown to suffer significant declines in plasma estradiol concentrations (Peakall, 1970). Nevertheless, in spite of the dearth of relevant data on this topic, the frequent impairment of normal reproduction found in birds exposed to pesticides provides strong evidence, albeit circumstantial, that low plasma concentrations of gonadal steroid hormones must occur following their contamination. Furthermore, many of these pesticides are also potent inducers of hepatic mixed function oxidase (MFO) activity and this effect alone might indirectly contribute to a disturbance of normal reproductive cyclicity by increasing the turnover of endogenous steroid hormones (Kuntzman et al., 1966; Peakall, 1970; Levin et al., 1968b; Welch et al., 1971; Nowicki and Norman, 1972).

Studies on birds provide quite convincing evidence that circulating hydrocarbons, particularly those derived from ingested petroleum, can impair normal reproduction by reducing plasma concentrations of ovarian steroid hormones (Holmes et al., 1978, 1981; Cavanaugh and Holmes, 1982). The first such study indicating that petroleum hydrocarbons can suppress an ovarian steroid hormone-dependent mechanism in birds was reported by Hartung in 1965. In this study, he found that egg-laying stopped completely for approximately two weeks when mallard ducks were given a single oral dose of lubricating oil. The evidence was inconclusive, however, because the lubricating oil used in the experiments was a commercial product containing additives of unknown toxicity. The first unequivocal evidence that ingested petroleum hydrocarbons could adversely affect reproduction in birds was provided by Grau et al. (1977) in experiments on the Japanese quail. These investigators found that a single oral dose of Bunker C. fuel oil, containing no additives, caused a significant reduction in the frequency of laying for several days. Also, when incubated, the eggs laid during the period immediately following the administration of the fuel oil showed unusually high incidences of embryo mortality. Later, it was shown that the frequency of oviposition was reduced significantly by the hydrocarbons contained in both ether- and chloroform-extracts of the fuel oil (Fig. 2a), whereas the chloroform extract of this oil contained more of the material affecting hatchability (Fig. 2b).

Considering the lipophilic nature of these hydrocarbon contaminants and their effects on cells of the adrenal cortex, it is tempting to speculate that they might also accumulate in the gonads and interfere directly with steroid hormone synthesis. Although there is no direct evidence to support this hypothesis, a persuasive body of data is accumulating which suggests its feasibility. Should future investigations provide convincing evidence for this type of effect, then the pathways outlined in Fig. 1 could be extended to

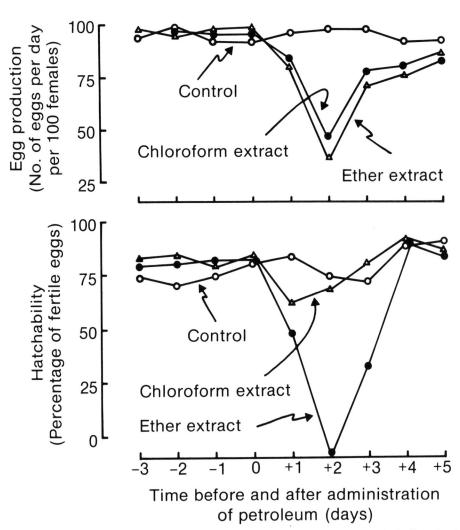

FIG. 2.—The effects of petroleum ether and chloroform extracts of Bunker C fuel oil on (top) egg production and on (bottom) the hatchability of fertile eggs in Japanese quail. Each bird was given a single oral dose of extract derived from 400 mg of fuel oil (adapted from Wootton *et al.*, 1979).

include interference with gonadotropic as well as corticotropic stimulation of steroidogenic targets.

LITERATURE CITED

ALLEN, J. R., L. K. LAMBRECHT, AND D. A. BARSOTTIE. 1978. Effects of polybrominated biphenyls in nonhuman primates. J. Amer. Vet. Med. Assoc., 173:1485-1489.

BITMAN, J., AND H. C. CECIL. 1970. Estrogenic activity of DDT and polychlorinated biphenyls. J. Agric. Food Chem., 18:1108-1112.

BITMAN, J., H. C. CECIL, S. HARRIS, AND G. F. FRIES. 1968. Estrogenic activity of o,p'DDT in the mammalian uterus and avian oviduct. Science, 162:371-372.

BLEND, M. J., AND T. J. SCHMIDT. 1971. *In vitro* uptake of steroid hormones and dieldrin by rat and canine prostatic tissue. Fed. Proc., 30:577.

BRUCKNER, J. V., K. L. KHANNA, AND H. H. CORNISH. 1974. Polychlorinated biphenyl-induced alteration of biologic parameters in the rat. Toxicol. Appl. Pharmacol., 28:189-199.

CAVANAUGH, K. P., AND W. N. HOLMES. 1982. Effect of injected petroleum on plasma levels of ovarian steroid hormones in photostimulated mallard ducks. Arch. Environ. Contam. Toxicol., in press.

CIVEN, M., AND C. B. BROWN. 1974. The effect of organophosphate insecticides on adrenal corticosterone formation. Pest. Biochem. Physiol., 4:254-259.

CIVEN, M., C. B. BROWN, AND R. J. MORIN. 1977a. Effects of organophosphate insecticides on adrenal cholesteryl ester and steroid metabolism. Biochem. Pharmacol., 26:1901-1907.

CIVEN, M., E. LIFRAK, AND C. B. BROWN. 1977b. Studies on the mechanism of inhibition of adrenal steroidogenesis by organophosphate and carbamate compounds. Pest. Biochem. Physiol., 7:169-182.

EROSCHENKO, V. P., AND W. O. WILSON. 1974. Photoperiods and age as factors modifying the effects of Kepone in Japanese quail. Toxicol. Appl. Pharmacol., 29:329-339.

———. 1975. Cellular changes in the gonads, livers and adrenal glands of Japanese quail as affected by the insecticide Kepone. Toxicol. Appl. Pharmacol., 31:491-504.

GELLERT, R. J. 1978. Kepone, mirex, dieldrin, and aldrin: estrogenic activity and the induction of persistent vaginal estrus and anovulation in rats following neonatal treatment. Environ. Res., 16:131-138.

GELLERT, R. J., AND C. WILSON. 1979. Reproductive function in rats exposed prenatally to pesticides and polychlorinated biphenyls (PCB). Environ. Res., 18:437-443.

GORSLINE, J., AND HOLMES, W. N. 1981. Effects of petroleum on adrenocortical activity and on hepatic naphthalene-metabolizing activity in mallard ducks (*Anas platyrhynchos*). Environ. Contam. Toxicol., 10:765-777.

GRAU, C. R., T. E. ROUDYBUSH, J. C. DOBBS, AND J. WATHEN. 1977. Altered yolk structure and reduced hatchability of eggs from birds fed single doses of petroleum oils. Science, 195:779-781.

HAMMOND, B., B. S. KATZENELLENBOGEN, N. KRAUTHAMMER, AND J. McCONNELL. 1979. Estrogenic activity of the insecticide chlordecone (Kepone) and interaction with uterine estrogen receptors. Proc. Nat. Acad. Sci., U.S.A., 76:6641-6645.

HARTUNG, R. 1965. Some effects of oiling on reproduction of ducks. J. Wildl. Mgmt., 29:872-874.

HOLMES, W. N., AND J. GORSLINE. 1980. Effects of some environmental pollutants on the adrenal cortex. Pp. 311-314, *in* Symposium on adrenal steroid biosynthesis, Proc. Sixth Internat. Cong. Endocrinol. (I. A. Cumming, J. W. Funder, and F. A. O. Mendelsohn, eds.), Australian Acad. Sci., Canberra, xii+736 pp.

HOLMES, W. N., K. P. CAVANAUGH, AND J. CRONSHAW. 1978. The effects of ingested petroleum on oviposition and some aspects of reproduction in experimental colonies of mallard ducks (*Anas platyrhynchos*). J. Reprod. Fert., 54:335-347.

HOLMES, W. N., J. GORSLINE, AND K. P. CAVANAUGH. 1981. Some effects of environmental pollutants on endocrine regulatory mechanisms. Pp. 1-11, *in* Recent progress of avian endocrinology, XXVII Internat. Cong. Physiol. Sci., Budapest, Hungary, Akademiai Kiado, Hungary, Pergamon Press, London, xvii+469 pp.

HORVATH, E., K. KOVACS, AND E. YEGHIAYAN. 1971. Histochemical study of the "adrenocortical lipid hyperplasia" induced in rats by aniline. Acta Histochem. Bd., 39:154-161.

JONSSON, H. T. JR., J. E. KEIL, R. G. GADDY, C. B. LOADHOLT, G. R. HENNIGAR, AND E. M. WALKER, JR. 1976. Prolonged ingestion of commercial DDT and PCB; effects on progesterone levels and reproduction in the mature female rat. Arch. Environ. Contam. Toxicol., 3:479-490.

KOVACS, K., E. YEGIAYAN, S. HATAKEYANA, AND H. SELYE. 1970. Adrenocortical lipoid hyperplasia induced in rats by aniline. Experientia, 26:1014-1015.

KUNTZMAN, R., R. M. WELCH, AND A. H. CONNEY. 1966. Factors influencing steroid hydroxylases in liver microsomes. Adv. Enz. Reg., 4:149-160.

KUPFER, D., AND W. H. BULGER. 1976. Interactions of chlorinated hydrocarbons with steroid hormones. Fed. Proc., 35:2603-2608.

LAMBRECHT, L. K., D. A. BARSOTTI, AND J. R. ALLEN. 1978. Responses of nonhuman primates to a polybrominated biphenyl mixture. Environ. Health Perspectives, 23:139-145.

LEVIN, W., R. M. WELCH, AND A. H. CONNEY. 1968a. Estrogenic action of DDT and its analogs. Fed. Proc., 27:649.

——. 1968b. Effect of phenobarbitol and other drugs on the metabolism and uterotropic action of estradiol-17β and estrone. J. Pharm. Exp. Therap., 159:362-371.

MORIARTY, F. 1975. Exposure and residues. Pp. 1-302, in Organochlorine insecticides: persistant organic pollutants (F. Moriarty, ed.) Academic Press, London, xii+302 pp.

NOWICKI, H. G., AND A. W. NORMAN. 1972. Enhanced hepatic metabolism of testosterone, 4-androstene-3,17-dione and estradiol-17β in chickens pretreated with DDT or PCB. Steroids, 19:85-99.

PALMITER, R. D., AND E. R. MULVIHILL. 1978. Estrogenic activity of the insecticide Kepone on the chicken oviduct. Science, 201:356-358.

PEAKALL, D. B. 1970. p,p'-DDT: effect on calcium metabolism and concentration of estradiol in the blood. Science, 168:592-594.

SANDERS, O. T., AND R. L. KIRKPATRICK. 1975. Effects of a polychlorinated biphenyl (PCB) on sleeping times, plasma corticosteroids, and testicular activity of white-footed mice. Environ. Physiol. Biochem., 5:308-313.

——. 1977. Reproductive characteristics and corticoid levels of female white-footed mice fed ad libitum and restricted diets containing a polychlorinated biphenyl. Environ. Res., 13:358-363.

SINGH, N., AND T. A. VENKITASUBRAMANIAN. 1975. Effects of aflatoxin B_1 on lipids of rat tissues. Environ. Physiol. Biochem., 5:147-157.

SMITH, M. T., J. A. THOMAS, C. G. SMITH, M. G. MAWHINNEY, AND J. W. LLOYD. 1972. Effects of DDT on radioactive uptake from testosterone-1,2-^{3}H by mouse prostate glands. Toxicol. Appl. Pharmacol., 23:159-164.

THOMAS J. A. 1975. Effects of pesticides on reproduction. Pp. 205-223, in Molecular mechanisms of gonadal hormone action (J. A. Thomas and R. L. Singhal, eds.), Univ. Park Press, Baltimore, x+399 pp.

TOTH, S., E. YEGHIAYAN, AND K. KOVACS. 1971. Effect of aniline on plasma corticosterone in rats. Can. J. Physiol. Pharmacol., 49:433-435.

WAKELING, A. E., AND VISEK. W. J. 1973. Insecticide inhibition of 5α-dihydrotestosterone binding in the rat ventral prostate. Science, 181:659-660.

WAKELING, A. E., T. J. SCHMIDT, AND W. J. VISEK. 1973. Effects of Dieldrin on 5α-dihydrotestosterone binding in the cytosol and nucleus of the rat ventral prostate. Toxic. Appl. Pharmacol., 25:267-275.

WELCH, R. M., W. LEVIN, AND A. H. CONNEY. 1969. Effect of chlorinated insecticides on steroid metabolism. Pp. 390-407, in Chemical fallout (M. W. Miller and G. C. Berg, eds.), Charles C. Thomas, Publ., Springfield, xxii+531 pp.

WELCH, R. M., W. LEVIN, R. KUNTZMAN, M. JACOBSON, AND A. H. CONNEY. 1971. Effects of halogenated hydrocarbon insecticides on the metabolism and uterotropic action of estrogens in rats and mice. Toxicol. Appl. Pharmacol., 19:234-246.

WOOTTON, T. A., C. R. GRAU, T. E. ROUDYBUSH, M. E. HAHS, AND K. V. HIRSCH. 1979. Reproductive responses of quail to Bunker C oil fractions. Arch. Environ. Contam. Toxicol., 8:457-463.

Reprinted from
ASPECTS OF AVIAN ENDOCRINOLOGY:
PRACTICAL AND THEORETICAL IMPLICATIONS (C. G. Scanes *et al.*, eds.)
Grad. Studies, Texas Tech Univ., 1982, 26:1-411.

THE EFFECTS OF SOUTH LOUISIANA AND KUWAIT CRUDE OILS ON REPRODUCTION

K. P. CAVANAUGH

*Department of Biological Sciences, University of California,
Santa Barbara, California 93106 USA*

Recent studies, both in the field and in the laboratory, have provided much evidence indicating that the demise of some seabird populations could have been caused by the presence of petroleum hydrocarbon contaminants in their environments. The list of effects attributable to the ingestion of petroleum and the presumed systemic distribution of hydrocarbons is long, and many of them could well impinge directly on the reproductive success of birds living in polluted environments. These effects include a delay in gonadal maturation, reduced frequency of laying, diminished frequency of fertilization, imperfect patterns of prenuptial behavior, and high frequency of unsuccessful incubations (Hartung, 1965; Grau *et al.*, 1977; Holmes *et al.*, 1978; Goldsmith *et al.*, 1981). In addition, high incidences of teratogenic defects and prenatal mortality are found among the embryos developing in eggs with petroleum-contaminated shells (Hartung, 1965; Kopischke, 1972; Birkhead *et al.*, 1973; Albers, 1977; Szaro and Albers, 1977; Albers, 1978; Hoffman, 1978; Hoffman, 1979*a*, 1979*b*, 1979*c*; Szaro, 1979; Macko and King, 1980). Furthermore, prenatal exposure of ducklings to petroleum contaminants can impair growth during the first weeks of postnatal life and diminish their abilities to survive adverse conditions (Vangilder and Peterle, 1980). Clearly, the slow but persistent declines that have occurred among some bird populations might well have been due to either some or all of these direct and indirect effects of petroleum contaminants on reproduction (Bourne, 1968; Parslow, 1976*a*, 1976*b*; Holmes and Cronshaw, 1977). In this presentation I will attempt to outline some of the effects that have been observed in the laboratory when sexually mature mallard ducks (*Anas platyrhynchos*) are given food contaminated with crude oil.

In early studies, small standard-sized breeding colonies of mallard ducks were exposed to a stimulatory photoperiod (18 hours light, 6 hours dark) and given food contaminated with either South Louisiana or Kuwait crude oil. When compared to similar groups given uncontaminated food, the onset of oviposition among these birds was delayed, and during the ensuing weeks they frequently laid fewer eggs (Fig. 1). Also, the incidence of fertilized eggs was much lower than normal and, when incubated, they often yielded fewer live ducklings than fertilized eggs laid by birds consuming uncontaminated food (Fig. 1; Holmes *et al.*, 1978). In some instances, these adverse effects could be eliminated by removing petroleum from the diet. For example, when food containing South Louisiana crude oil was replaced with uncontaminated food, the rate of oviposition, the incidence of fertilization, and the hatchability of the fertilized eggs were each restored to normal. Some of the adverse effects, however, could not always be diminished by reducing the level of contamination. This was particularly evident among birds given Kuwait crude oil where, although a normal rate of oviposition was restored when the concentration was reduced from three to one per cent, the incidence of fertilization remained low and, when incubated, none of the fertilized eggs yielded live ducklings (Fig. 1; Holmes *et al.*, 1978).

Some, but not all, crude oils can also affect postovulatory mechanisms. For instance, the eggs laid by females consuming food contaminated with South Louisiana crude oil have significantly thinner shells than eggs laid at the same time by birds consuming uncontaminated food. In contrast, food contaminated with Kuwait crude oil has no effect on shell thickness (Fig. 2).

The influence of crude oil on the ovarian cycle is more clearly defined when the patterns of oviposition of individual females are examined (Cavanaugh and Holmes, 1982). As was the case among the females maintained in groups, solitary unmated birds consuming uncontaminated food start to lay on about day 14 of photostimulation. Thereafter, the patterns of oviposition consist of long sequences of daily egg-laying and only occasional pauses lasting not more than four days. Although the addition of crude oil to the diet of these solitary females always causes delays in ovarian maturation and the onset of laying, their individual patterns of response to the contaminant vary considerably. In most birds the onset of laying is delayed only a few days, but in others it might not start for several weeks and, in some, no eggs will be laid as long as the bird continues to consume contaminated food. Once laying has been initiated, however, individual patterns of oviposition usually consist of short sequences of egg-laying frequently interrupted by long pauses sometimes lasting more than two weeks. Exposure to petroleum can also alter the synchronization of ovulation and oviposition during the daily light-phase. For example, under the conditions of these experiments, the birds consuming uncontaminated food laid most of their eggs between the second and fifth hours of the light phase whereas those consuming

→

FIG. 1 (top).—The effects of consuming food contaminated with different concentrations of either South Louisiana or Kuwait crude oil on oviposition in mallard ducks. All birds were kept in small experimental groups consisting of six females and two males. The dotted and striped areas represent respectively the total number of fertilized and unfertilized eggs laid by each group during a 50-day period of photostimulation. The arrows along the horizontal axes indicate the days on which the first unfertilized and fertilized eggs were laid. Adapted from Holmes *et al.* (1978).

FIG. 2 (bottom).—Changes in the shell thicknesses of eggs laid by mallard ducks consuming food contaminated with different concentrations of either Kuwait or South Louisiana crude oil. The birds given food containing 3% Kuwait crude oil laid no eggs from which measurements could be derived. Levels of significance relate to a comparison of values derived from birds consuming uncontaminated food and those consuming food contaminated with crude oil. Adapted from Holmes *et al.* (1978).

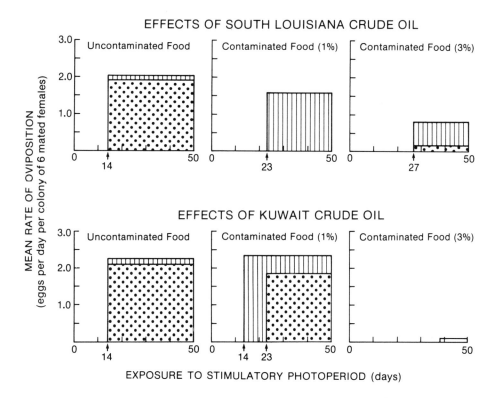

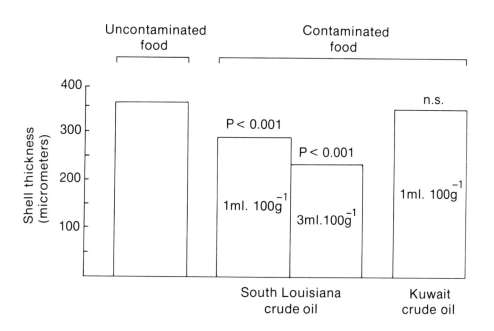

petroleum-contaminated food tended to lay more eggs during a much longer interval of the light phase (Cavanaugh and Holmes, 1982).

These petroleum-induced alterations in ovarian development and ovulatory cyclicity have many of the characteristics of endocrine dysfunction. The impaired ovarian development and abnormal patterns of oviposition might well reflect alterations in the normal patterns of gonadotropin release and ovarian steroid hormone synthesis. Evidence in support of this notion has been obtained recently by comparing the plasma estrogen and progesterone concentrations in uncontaminated and contaminated birds during the 24-hour period preceding ovulation (Cavanaugh and Holmes, 1982). In these experiments, we have found that the mean daily plasma concentrations of estradiol and estrone (but not progesterone) are reduced significantly in ducks consuming petroleum-contaminated food. Furthermore, these reduced concentrations of estrogen reflect a distinct suppression of the peaks in hormone concentration, particularly estradiol, that occur during the early and late phases of the ovulatory cycle (Fig. 3). Also, although the mean daily concentration of progesterone in contaminated birds is not significantly different than in contaminated birds, the single daily peak that usually occurs during the few hours preceding ovulation seems to occupy a longer than normal interval (Fig. 4).

On a pause day, when no preovulatory peaks in the plasma concentrations of either estrogen or progesterone occur, the mean daily concentrations in both contaminated and uncontaminated ducks are significantly lower than they are during a day preceding ovulation. The birds consuming petroleum-contaminated food, however, have more and longer sequences of pause days than do the uncontaminated birds and therefore have a higher frequency of days when characteristically low mean daily concentrations of ovarian steroid hormones occur.

The reduced plasma estrogen that occurs in birds exposed to petroleum, and perhaps also in birds exposed to other hydrocarbon pollutants, is probably due primarily to a suppression of ovarian synthesis. Since petroleum hydrocarbons are fat-soluble, they could accumulate in the lipid-rich cells of the differentiating ovarian follicles, and thus directly interfere with steroidogenesis in much the same way that these compounds affect the synthesis of glucocorticoids in cells of the adrenal cortex. The altered yolk structure found in eggs laid by quail during the days following the ingestion of petroleum and the large number of atretic follicles present in the ovaries of birds consuming petroleum certainly indicate a direct effect of these contaminants on ovarian tissue (Fig. 5; Grau et al., 1977). These effects would be consistent with such a hypothesis. Some other hydrocarbon contaminants such as hexachlorobenzene (HCB), DDT, polychlorinated biphenyls (PCB), and polycyclic aromatic hydrocarbons (PAH) have also been shown to impair reproduction in mammals by causing degenerative changes in the ovary and by altering normal patterns of androgen synthesis in the testis (Platonow et al., 1972; Iatropoulos et al., 1976; Jonsson et al., 1976; Freeman and Sangalang,

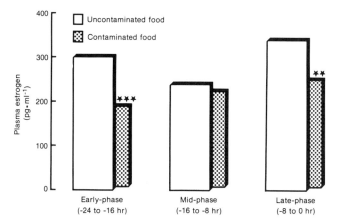

FIG. 3.—A comparison of the mean plasma concentrations of total estrogen (estradiol plus estrone) found in plasma during the three successive eight-hour intervals preceding ovulation in birds consuming uncontaminated food or petroleum-contaminated food. The petroleum-contaminated food contained 3 ml South Louisiana crude oil per 100 grams dry weight. Indicated levels of significance, **$P<0.01$, ***$P<0.001$, relate to the comparison of the corresponding value found in birds consuming uncontaminated food. Also, whereas the concentrations found in the uncontaminated birds during the early and late phases were each significantly higher than that occurring during the mid-phase, no such differences were observed among birds consuming the petroleum-contaminated food. Adapted from Cavanaugh and Holmes (1982).

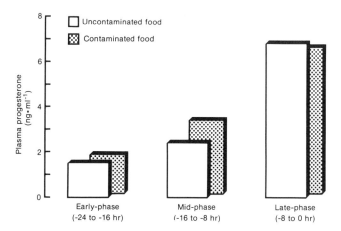

FIG. 4.—The mean plasma concentrations of progesterone in uncontaminated and contaminated birds during each phase of the ovulatory cycle. Although there were no significant differences between the uncontaminated and contaminated birds, the apparently higher concentration during mid-phase of birds consuming contaminated food reflected a broadening of the single daily peak into midcycle, which in the uncontaminated birds was restricted to the late phase. Adapted from Cavanaugh and Holmes (1982).

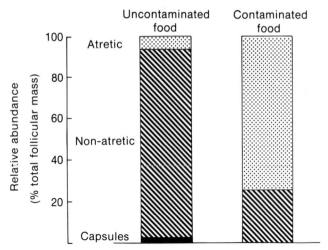

Fig. 5.—A comparison of the incidences of atretic and nonatretic follicles in ovaries taken from mallard ducks that have been maintained on uncontaminated food ($N=6$) or food contaminated with 3% South Louisiana crude oil ($N=6$). The capsules represent the remnants remaining in the ovary following the release of a mature follicle on the day preceding autopsy. Adapted from Holmes *et al.* (1978).

1977; Fuller *et al.*, 1980; Mattison, 1980). In this regard, the gonadal responses of birds to petroleum hydrocarbon contaminants might be similar.

LITERATURE CITED

ALBERS, P. H. 1977. Effects of external applications of fuel oil on hatchability of mallard eggs. Pp. 158-163, *in* Fate and effects of petroleum hydrocarbons in marine ecosystems and organisms (D. A. Wolfe, ed.), Pergamon Press, New York, xix+478 pp.

———. 1978. The effects of petroleum on different stages of incubation in bird eggs. Bull. Environ. Contam. Toxicol., 19:624-630.

BIRKHEAD, T. R., C. LLOYD, AND P. CORKHILL. 1973. Oiled seabirds successfully cleaning their plumage. Brit. Birds, 66:535-537.

BOURNE, W. R. P. 1968. Oil pollution and bird populations. Pp. 99-121, *in* The biological effects of oil pollution on littoral communities (J. D. Carthy and D.R. Arthur, eds.), E. W. Classey, Ltd., Hampton, England, vii+998 pp.

CAVANAUGH, K. P., AND W. N. HOLMES. 1982. Effects of ingested petroleum on plasma levels of ovarian hormones in photostimulated mallard ducks. Arch. Environ. Contam. Toxicol., in press.

FREEMAN, H. C., AND G. B. SANGALANG. 1977. A study of the effects of methyl mercury, cadmium, arsenic, selenium, and a PCB (Aroclor 1254) on adrenal and testicular steroidogeneses *in vitro*, by the gray seal *Halichoerus grypus*. Arch. Environ. Contam. Toxicol., 5:369-383.

FULLER, G. B., V. KNAUF, W. MUELLER, AND W. C. HOBSON. 1980. PCB augments LH-induced progesterone synthesis. Bull. Environ. Contam. Toxicol., 25:65-68.

GOLDSMITH, A. R., K. P. CAVANAUGH, B. K. FOLLETT, AND W. N. HOLMES. 1983. Oviposition and incubation in mallards: effects of crude oil on prolactin and LH secretion. J. Endocrinol., in press.

GRAU, C. R., T. E. ROUDYBUSH, J. C. DOBBS, AND J. WATHEN. 1977. Altered yolk structure and reduced hatchability of eggs from birds fed single doses of petroleum oils. Science, 195:779-781.

HARTUNG, R. 1965. Some effects of oiling on reproduction of ducks. J. Wildl. Mgmt., 29:872-874.

HOFFMAN, D. 1978. Embryotoxic effects of crude oil in mallard ducks and chicks. Toxicol. Appl. Pharmacol., 46:183-190.

———. 1979a. Embryotoxic and teratogenic effects of petroleum hydrocarbons in mallards (Anas platyrhynchos). J. Toxicol. Environ. Health., 5:835-844.

———. 1979b. Embryotoxic and teratogenic effects of crude oil on mallard embryos on day one of development. Bull. Environ. Contam. Toxicol., 22:632-637.

———. 1979c. Embryotoxic effects of crude oil containing nickel and vanadium in mallards. Bull. Environ. Contam. Toxicol., 23:203-206.

HOLMES, W. N., AND J. CRONSHAW. 1977. Biological effects of petroleum on marine birds. Pp. 359-398, in Effects of petroleum on Arctic and subarctic marine environments and organisms, Vol. II (D. C. Malins, ed.), Academic Press, New York, xx+500 pp.

HOLMES, W. N., K. P. CAVANAUGH, AND J. CRONSHAW. 1978. The effects of ingested petroleum on oviposition and some aspects of reproduction in experimental colonies of mallard ducks (Anas platyrhynchos). J. Reprod. Fert., 54:335-347.

IATROPOULOS, M. J., W. HOBSON, V. KNAUF, AND H. P. ADAMS. 1976. Morphological effects of hexachlorobenzene toxicity in female rhesus monkeys. Toxicol. Appl. Pharmacol., 37:433-444.

JONSSON, H. T., JR., J. E. KEIL, R. G. GADDY, C. B. LOADHOLT, G. R. HENNIGAR, AND E. M. WALKER, JR. 1976. Prolonged ingestion of commercial DDT and PCB; effects on progesterone levels and reproduction in the mature female rat. Arch. Environ. Contam. Toxicol., 3:479-490.

KOPISCHKE, E. D. 1972. The effect of 2,4-D and diesel fuel on egg hatchability. J. Wildl. Mgmt., 36:1353-1356.

MACKO, S. A., AND S. M. KING. 1980. Weathered oil: effect on hatchability of heron and gull eggs. Bull Environ. Contam. Toxicol., 25:316-320.

MATTISON, D. R. 1980. Morphology of oocyte and follicle destruction by polycyclic aromatic hydrocarbons in mice. Toxicol. Appl. Pharmacol., 53:249-259.

PARSLOW, J. L. F. 1976a. A census of auks. Brit. Trust Ornithol. News, 23:8-9.

———. 1976b. Changes in status among breeding birds in Britain and Ireland. Brit. Birds, 60:2-47, 97-122, 177-202.

PLATONOW, N. S., R. M. LIPTRAP, AND H. D. GEISSINGER. 1972. The distribution and excretion of polychlorinated biphenyls (Aroclor 1254) and their effect on urinary gonadal steroid levels in the boar. Bull. Environ. Contam. Toxicol., 7:358-365.

SZARO, R. C. 1979. Bunker C fuel oil reduces mallard egg hatchability. Bull. Environ. Contam. Toxicol., 22:731-732.

SZARO, R. C. AND P. H. ALBERS. 1977. Effects of external applications of No. 2 fuel oil on common eider eggs. Pp. 164-167, in Fate and effects of petroleum hydrocarbons in marine ecosystems and organisms (D. A. Wolfe, ed.), Pergamon Press, New York, xix+478 pp.

VANGILDER, L. D. AND T. J. PETERLE. 1980. South Louisiana crude oil and DDE in the diet of mallard hens: effects of reproduction and duckling survival. Bull. Environ. Contam. Toxicol., 25:23-28.

Reprinted from
ASPECTS OF AVIAN ENDOCRINOLOGY:
PRACTICAL AND THEORETICAL IMPLICATIONS (C. G. Scanes *et al.*, eds.)
Grad. Studies, Texas Tech Univ., 1982, 26:1-411.

REPRODUCTIVE PERFORMANCE AND ENDOCRINE RESPONSES
TO INGESTED NORTH SEA OIL

S. HARVEY[1], J. G. PHILLIPS[1], AND P. J. SHARP[2]

[1]*The Wolfson Institute, University of Hull, Hull HU6 7RX England; and* [2]*Agricultural Research Councils, Poultry Research Centre, Roslin, Scotland*

The decline of many populations of marine birds has been causally related to the persistent spillage of petroleum (see Cronshaw, this vol.). In addition to deaths resulting from plumage contamination and ingestion of lethal amounts of oil, ingestion of sublethal doses of petroleum can seriously endanger a colony's survival by impairing reproductive activity and thus reducing the rate of annual recruitment.

The adverse effect of ingested petroleum on avian fecundity might partly be due to impairment of pre-nuptial or nesting behavior (Hartung, 1965; Holmes *et al.*, 1978*b*; Goldsmith *et al.*, 1980). Fecundity might also be impaired as a result of delay in the onset of the reproductive cycle. For instance, we have found that chronic ingestion of food containing one and five per cent North Sea Oil causes a delay in the onset of lay in Khaki Campbell ducks by at least seven and 60 days, respectively; in fact, some birds do not lay at all while consuming contaminated diets (Harvey *et al.*, 1981*a*). Similar observations previously have been made in Mallard ducks (Holmes *et al.*, 1978*b*; Cavanaugh, this vol.). In the natural environment, a delay in the onset of reproductive activity might result in the fledglings being hatched under unfavorable environmental conditions or, in the case of migratory species, the fledglings being too immature to migrate to the wintering grounds.

Under favorable environmental conditions, the acute or chronic ingestion of oil also reduces the rate of oviposition (Hartung 1965; Grau *et al.*, 1977; Holmes *et al.*, 1978*b*; Eastin and Hoffman, 1978). These findings agree with our observations of Khaki Campbell ducks in which the rate of oviposition was greatly inhibited during a five month period of feeding two and five per cent crude oil and remained suppressed for at least six weeks after the oil-fed birds were refed an uncontaminated diet (Fig. 1). It has also been found that birds that chronically ingested sublethal doses of petroleum laid smaller and lighter eggs than did birds fed uncontaminated diets. This effect persisted even after oil-fed birds were returned to a pristine diet (Fig. 1). Moreover, the shell weight of the eggs laid by these birds before and after oil treatment was less than the shell weight of eggs laid by the control birds, even after making a correction for the reduced egg weight (Fig. 1). The thinning of the egg shell makes the egg more susceptible to accidental cracking and could also

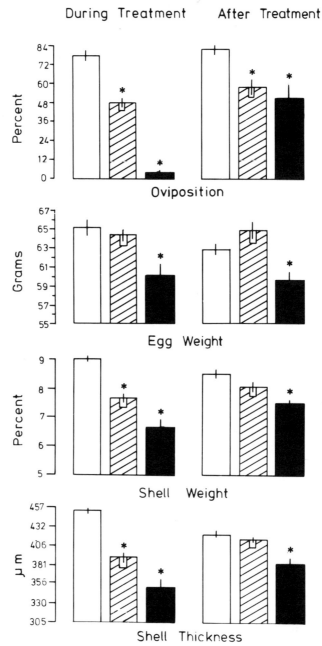

FIG. 1.—Mean percentage oviposition rate (100% represents 1 egg/bird/day) and weight (g), shell weight (%), and shell thickness (μm) of eggs laid by groups ($N=$ 5 or 6) or laying Khaki Campbell ducks fed 0% (open), 2% (shaded), or 5% (solid histograms) North Sea crude oil in the diet during a 20-week period of oil feeding and a 6-week period when all birds were fed an uncontaminated diet. All the birds were maintained under a long-day photoperiod of 16L:8D. Means±se. Parameters which are significantly ($P<0.05$) different from the controls are indicated by the asterisks. Adapted from Harvey et al. (1981a).

make them more prone to deliberate destruction by the parents (Ratcliffe, 1970).

Egg shell thinning also impairs the integrity of the physical barrier required to protect the developing embryo from dessication and from infection. It is therefore not surprising that the hatchability of fertilized eggs from ducks, chickens, and Japanese quail ingesting oil is low (Hartung, 1965; Grau et al., 1977; Holmes et al., 1977; Holmes et al., 1978b).

The hatchability of eggs laid by birds fed petroleum might also be reduced as a result of smaller size, a reduction in the yolk size, or abnormal structure of the yolk (Grau et al., 1977). In addition, eggs could be infertile because of impaired spermatogenesis in the male or because of defects in the oviducal environment that affect storage and viability of spermatozoa in the female. Application of crude oil to the egg shell also reduces the hatchability of the embryos. Thus, oiling of incubated eggs by the contaminated plumage of Cabot's terns and other shore birds has been reported to prevent the eggs from hatching even after 50 days of incubation (Rittinghaus, 1956; Birkhead et al., 1973). Similarly, the hatchability of eggs laid by Mallards with oil-contaminated plumage has also been reported to be reduced (Hartung, 1965). As a consequence, oiled birds continue to incubate infertile clutches longer than normal, thus reducing the likelihood of renesting. Application of very small amounts of crude oil and fuel oil to the external surface of fertile eggs from Mallard and Common Eider ducks and chickens results in decreased embryonic growth (Hoffman, 1978) and hatchability (Gross, 1950; Hartung, 1965; Kopischke, 1972; Albers, 1977, 1978; Szaro and Albers, 1977). These embryotoxic effects of petroleum are not due to the blockage of shell pores and a consequent interference with the supply of oxygen because they occur when as little as 10 per cent of the egg is covered with oil. These toxic effects of petroleum are believed to result from the presence of aromatic compounds, since paraffin has no comparable effect. The ingestion of sublethal doses of crude oil by neonatal birds following hatch also results in a retardation of growth (Miller et al., 1978) and can delay or prevent subsequent sexual maturation.

The mechanisms by which oil exerts these effects on avian reproduction are, however, unknown. Nevertheless, because reproductive activity is controlled by the pituitary-gonadal axis, alterations in the secretion of gonadotropins and prolactin could be involved. Such an endocrine dysfunction might result in impaired ovarian activity, leading to reduced egg production and a decrease in the secretion of gonadal steroids. These steroids are important in the induction of the preovulatory release of luteinizing hormone (LH) (Sharp, 1980) and in the development and maintenance of the functions of the oviduct which include egg shell formation (Gilbert, 1979). A similar endocrine dysfunction is thought to be partly responsible for the poor rates of oviposition in birds ingesting organochlorine pollutants. For instance, Peakall (1970) has demonstrated that in Ring doves ingestion of p, p[1]-DDT decreased the blood levels of oestradial as well as medullary bone deposition early in the breeding cycle. These effects were associated with delayed ovulation and decreased egg shell weight. Peakall (1970) suggested that the levels of plasma oestradial were reduced because the pesticide caused an increase in the activity of hepatic enzymes that regulate steroid catabolism. Cooke (1973) subsequently argued, however, that increased steroid catabolism would be expected to result in compensatory adjustments in the negative feedback system that regulates gonadotropin secretion and therefore should not lead to a reduction in levels of plasma steroids. He suggested that reduced levels of plasma steroids in birds affected by pesticides might be due to an inhibition of gonadotropin secretion caused by a direct effect of the pollutant on the pituitary gland or on the central nervous system.

It has been found, however (Fig. 2), that in ducks exposed to long days and fed North Sea oil, the resulting decrease in egg production is not associated with a decrease in the levels of plasma LH. Indeed, in control ducks fed an uncontaminated diet, the mean LH level was lower than that in birds fed five per cent North Sea oil. In the latter birds, LH levels fell to those found in the controls only after the oil was removed from the diet (Fig. 2). Moreover, the chronic ingestion of oil did not impair the photoperiodic response to increasing day length (Follett and Davies, 1975; Balthazart et al., 1977; Jallageas et al., 1978; Follett and Robinson, 1980). Thus the effects of petroleum on reproduction are unlikely to be mediated by impaired gonadotropin secretion.

The depressed rate of lay in ducks fed oil (Fig. 1) could be due to a loss of response of the mature ovarian follicle to LH, which stimulates steroidogenesis and ovulation. This effect could result from an interference by the petroleum hydrocarbons with the ovarian receptors for gonadotropins. Such a possibility should be considered since it has been found that some chlorinated hydrocarbons mimic the action of some hormones by competing for their receptors or binding sites (See Holmes, this vol.). It is not known whether or not petroleum hydrocarbons have similar effects in birds, although indirect (Miller et al., 1978; Holmes and Gorsline, 1980) and direct (Lawler et al., 1978a, 1978b, 1978c; Gay et al., 1980) evidence has shown that they are taken up and retained in body tissues after ingestion.

In birds, oviducal growth and many aspects of oviducal function are controlled by the gonadal steroids; both oestrogen and progesterone are necessary for the synthesis of albumin, the transport of the egg along the oviduct, and egg shell formation (Gilbert, 1971a, 1971b, 1971c, 1979). A decrease in the secretion of gonadal steroids might therefore be responsible for the effects of petroleum on avian reproduction. The finding that, in birds fed five per cent North Sea oil, plasma progesterone concentrations were significantly less than those in the controls throughout a five month period of photostimulation (Fig. 3) supports the view that gonadal steroid secretion could be suppressed by ingestion of petroleum. The progesterone concentration in these birds also remained low even after the oil had been removed from the diet, and might in some way be connected with the associated low rate of oviposition and poor egg shell quality (Fig. 1).

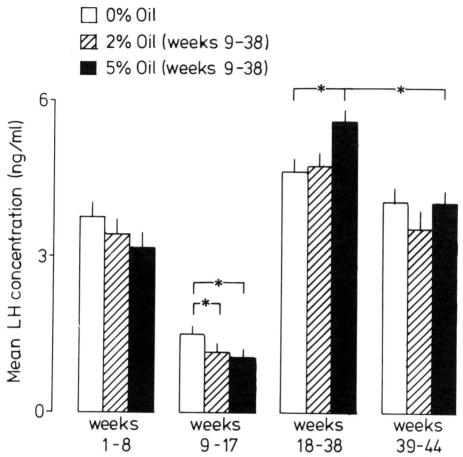

FIG. 2.—Mean concentrations of plasma LH (ng/ml) in blood samples taken at weekly intervals from laying Khaki Campbell ducks ($N = 5$ or 6) before (weeks 1-8), during (weeks 9-38), and after (weeks 39-44) the feeding of 0% (open), 2% (shaded), and 5% (solid histograms) North Sea crude oil in the diet. The birds were maintained under a long photoperiod (16L:8D) between weeks 1 and 8 and between weeks 18 and 44 and under a short photoperiod (8L:16D) between weeks 9 and 17. Means± SE. Concentrations significantly ($P<0.05$) different from the controls are indicated by the asterisk. Adapted from Harvey et al. (1981a).

In laying hens, progesterone is mainly secreted by the three largest preovulatory follicles (Wells et al., 1980). The observation that the levels of plasma progesterone in the ducks fed five per cent oil and exposed to long days were lower than in the control birds suggests that in the former, the yellow-yolky follicles in the ovary were only partially developed. In support of this view, Holmes et al. (1978b) found that the ovaries of ducks fed oil contained many atretic yellow-yolky follicles. The depressed levels of plasma progesterone in the ducks fed five per cent oil could therefore have resulted in an impairment of oviduct function. The low progesterone levels might have also been responsible for the increased LH levels in these birds, because

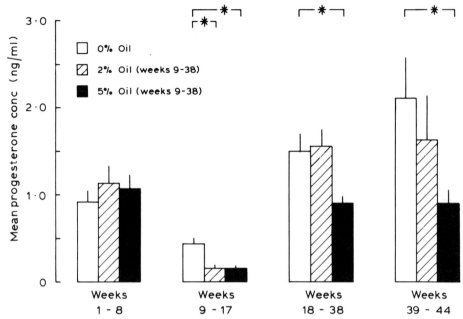

F<small>IG</small>. 3.—Mean concentrations of plasma progesterone (ng/ml) in blood samples taken at weekly intervals from laying Khaki Campbell ducks ($N = 5$ or 6) before (weeks 1-8), during (weeks 9-38), and after (weeks 39-44) the feeding of 0% (open), 2% (shaded), and 5% (solid histograms) North Sea crude oil in the diet. Other details as in Fig. 2. Adapted from Harvey *et al.* (1981*a*).

they would have reduced the negative feedback effect of ovarian steroids (Sharp, 1980).

However, in view of the fact that two per cent North Sea oil had no effect on progesterone secretion during photostimuation (Fig. 3), but did impair oviposition and egg quality (Fig. 1), decreased gonadal steroid secretion cannot be solely responsible for the effects of petroleum on reproductive performance.

Prolactin is also believed to play a role in the regulation of ovarian activity in birds because injections of this hormone cause gonadal regression (for example, Bates *et al.*, 1935; Opel, 1971), block LH induced ovulation (Tanaka *et al.*, 1971), inhibit FSH and LH induced secretion of oestrogen and progesterone (Camper and Burke, 1977), and induce and maintain broodiness (Eisner, 1960). However, although these observations might lead one to expect an increase in prolactin secretion after ingestion of petroleum, plasma prolactin levels in birds fed five per cent North Sea oil were invariably lower than those in birds fed uncontaminated diets (Fig. 4). This observation is consistent with the suggestion that prolactin could have progonadal effects (Meier, 1969); indeed, photo-induced gonadal activity in control and petroleum fed birds was accompanied by increased plasma prolactin levels (Fig. 4), as has been previously observed in other species (for example, Etches *et*

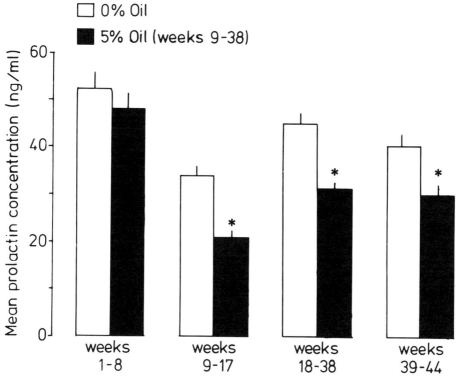

FIG. 4.—Mean concentrations of plasma prolactin (ng/ml) in blood samples taken at weekly intervals from laying Khaki Campbell ducks (N = 5 or 6) before (weeks 1-8), during (weeks 9-38), and after (weeks 39-44) the feeding of 0% (open histograms), and 5% (solid histograms), North Sea crude oil in the diet. Other details as in Fig. 2. Adapted from Harvey *et al.* (1981*b*).

al., 1979; Scanes *et al.*, 1979). These results also agree with the finding that prolactin secretion in the fowl is stimulated by positive gonadal steroid feedback (Bolton *et al.*, 1975; Bolton, 1976; Hall and Chadwick, 1978) and suggest that impaired gonadal function in birds fed petroleum is not due to increased prolactin secretion.

Impaired ovarian function in birds and mammals also has been frequently associated with stress-induced increases in adrenocortical activity. Thus, the administration of adrenal corticosteroids to laying birds induces shell thinning (Urist and Deutsch, 1960) and alters shell composition (reviewed by Cooke, 1973). A loss of reproductive potential in birds fed petroleum might, therefore, be secondary to chronic stress and the associated high levels of plasma corticotropin (Christian, 1975). The ingestion of oil has been shown to induce a general stress response in birds, as reflected by adrenal hypertrophy, involution of lymphoid tissue, liver enlargement, and retardation of growth (Hartung and Hunt, 1966; Holmes and Cronshaw, 1977; Miller *et al.*, 1978; Holmes *et al.*, 1978*a*; 1979; Szaro *et al.*, 1978; Patton and Dieter, 1980). However, the exposure of birds to food contaminated with petroleum results

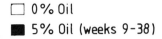

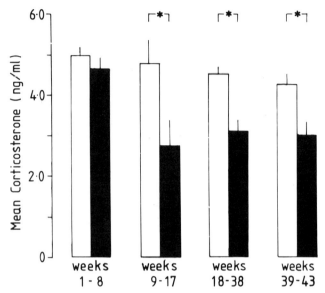

FIG. 5.—Mean concentrations of plasma corticosterone (ng/ml) in blood samples taken at weekly intervals from laying Khaki Campbell ducks ($N = 5$ or 6) before (weeks 1-8), during (weeks 9-38), and after (weeks 39-44) the feeding of 0% (open histograms) and 5% (solid histograms) North Sea crude oil in the diet. Other details as in Fig. 2. Adapted from Harvey *et al.* (1981*b*).

in an extensive degeneration of the mitochondria in the interrenal cells that secrete corticosterone and an increase in its hepatic metabolism (Holmes and Gorsline, 1980; Gorsline, this vol.). Not surprisingly therefore, the plasma corticosterone levels in laying ducks chronically exposed to petroleum are consistently lower than those in birds receiving an uncontaminated diet (Fig. 5; Gorsline, this vol.). This suggests that impaired ovarian function is not due to increased adrenocortical activity.

Hypoadrenocorticalism has also been thought to occur in chickens following chronic exposure to technical grade DDT or p, p^1-DDT, as judged by low adrenal and plasma corticosteroid levels and low liver glycogen concentrations (Srebocan *et al.*, 1971). The adrenals of pigeons dosed with p, p^1-DDT and p, p^1-DDB also increase in weight (Jefferies *et al.*, 1971; Jefferies and French, 1972), possibly as a result of changes in thyroid function, since hypo- and hyperthyroidism cause adrenal hypertrophy in the chicken (Jefferies, 1975).

In birds, the normal production of thyroid hormones is essential for gonadal growth and the development of secondary sexual characteristics, as well

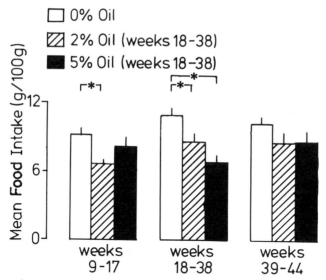

Fig. 6.—Mean levels of food intake (g/100 g body weight) by laying Khaki Campbell ducks (N = 5 or 6) during (weeks 9-17) and after (weeks 39-44) the feeding of 0% (open), 2% (shaded), and 5% (solid histograms) North Sea crude oil in the diet. Other details as in Fig. 2. Adapted from Harvey *et al.* (1981a).

as for successful reproduction (reviewed by Jefferies, 1975). Egg weight, egg production, shell weight, yolk weight, time of ovulation, and male fertility all can be altered by changes in thyroid activity because both hypothyroidism and severe hyperthyroidism reduce reproductive success. For instance, Jefferies and French (1971) and Jefferies *et al.* (1971) observed that pigeons, after being fed low doses of DDT, became hyperthyroid and laid fewer eggs. On the other hand, birds fed higher doses of DDT were hypothyroid, which also resulted in reduced fecundity and egg shell thinning. Thus it has been suggested (Jefferies, 1969; Jefferies and French, 1971) that most of the sublethal effects of environmental pollutants on avian reproduction could be due to a lesion of the thyroid; considerable evidence supporting this hypothesis has been the subject of a detailed review (Jefferies, 1975). However, in contrast to the effects of organochlorines, petroleum ingestion by laying ducks was not characterized by major differences in the circulating concentrations of either triiodothyronine (T_3) or thyroxine (T_4), which suggests that an endocrine dysfunction of the thyroid is not causally involved in mediating the effects of petroleum on avian reproduction (Harvey *et al.*, 1981b).

The reduction in the rate of lay and quality of eggs laid following petroleum ingestion might, however, also be due to a reduction in food consumption. In birds fed two and five per cent North Sea oil, the level of food intake was reduced by more than five per cent during the first two weeks of oil feeding, and in the case of the birds fed two per cent, the weekly level of food intake remained less than that by birds consuming uncontaminated food (Fig. 6). Although increasing the photoperiod stimulated food intake in control birds and in birds fed two per cent oil, birds fed two and five per cent oil consumed much less than the controls. This effect was especially marked in the birds fed five per cent oil, in which the level of weekly food intake was consistently 40-60 per cent of that in the controls (Fig. 6). The removal of oil from the diet had no effect on the amount of food eaten by the group which had been fed two per cent oil, but it significantly increased food intake by the group fed five per cent oil (Fig. 6). The reduction in the rate of lay and egg quality in ducks fed oil could, therefore, have been due to the reduction in food consumption, since in the domestic hen the rate of egg production can be depressed by starvation (Hosoda et al., 1955; Morris and Nalbandov, 1961). Such a reduction could be caused by a depression in the levels of plasma LH or prolactin because the secretion of these hormones is depressed in starved chickens (Scanes et al., 1976; Harvey et al., 1978). Additionally, a reduction in food consumption could lead to a reduction in calcium intake and thereby reduce the availability of minerals for egg shell formation. The observed concentrations of plasma LH and prolactin in ducks fed oil (Figs. 2 and 4) are, however, not entirely consistent with this view. Although the levels of prolactin were depressed, the level of plasma LH was elevated in the ducks fed five per cent oil and held on long days. Moreover, the adverse effects of oil on the formation of yolk are known to be different from those which result from food or water deprivation (Grau et al., 1977). Furthermore, although food consumption tended to be depressed in ducks fed oil and exposed to long days, there were no consistent corresponding reductions in body weight.

The body weight of the birds that had been fed five per cent oil did fall significantly after oil had been fed for one week and remained low while the birds were held on short days. However, after transfer to long days, body weight increased to pretreatment values (Fig. 7) despite the reduction in food intake (Fig. 6). Similarly, although food intake was reduced, the body weight of the birds fed two per cent oil was not reduced by petroleum ingestion (Fig. 7). Thus the birds fed oil might have had a better food conversion efficiency (FCE) or might have tended to eat less than the control groups because they did not have to meet increased demands for nutrient intake necessary for a high rate of egg production. In other studies, Patton and Dieter (1980) found that ducks fed 4000 ppm aromatic hydrocarbons also lost body weight during the first two months of oil feeding and then regained it within five months.

Although reduced food intake could partly be responsible for the effects of petroleum on oviposition and egg quality, other studies provide evidence that this factor is not solely responsible; the degree of food intake is independent of the detrimental effects of oil on reproductive performances (Holmes et al., 1978b; 1979). Moreover, in other studies, pair feeding control birds to the reduced level of food intake induced by insecticide ingestion does not impair reproductive activity in the controls (Cooke, 1973).

The effect of oil feeding on food intake might have been due to the food being unpalatable, to its noxious odor, or possibly, as a result of low prolac-

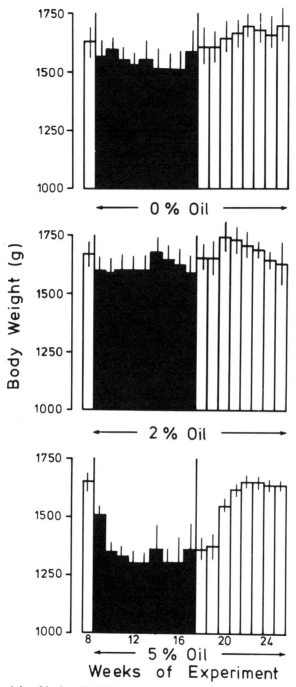

FIG. 7.—Body weight of laying Khaki Campbell ducks (N = 5 or 6) during the first 17 weeks (weeks 9-25) of being fed 0%, 2%, or 5% North Sea crude oil under short day (8L:16D, solid histograms) and long day (16L:8D, open histograms) photoperiods. Other details as in Fig. 2. Adapted from Harvey *et al.* (1981*a*).

tin and corticosterone concentrations (Holmes and Phillips, 1976; Ensor, 1978), to a depression of the appetite center of the brain. Contamination of the diet with crude oil could irritate the gastro-intestinal tract, because a number of fuel and diesel oils have been found to cause hyperemia of the intestinal wall, mucous and blood loss, and diarrhea with doses as low as 1 ml/kg (Hartung and Hunt, 1966). It is possible, therefore, that petroleum could have a direct effect on the absorptive properties of the mucosal cells in the small intestine, which might indirectly lead to a reduction in egg production and egg quality. *In vitro* studies have shown that a significant inhibition of the mechanisms responsible for the intestinal uptake of essential amino acids and possibly glucose might occur in herring gull chicks given sublethal doses of South Louisiana oil (Miller *et al.*, 1978). Moreover, oil fed birds that were affected in this way failed to grow even though they ate 15 per cent more food than birds fed uncontaminated diets. Petroleum and petroleum products also inhibit sodium and water absorption by the intestinal mucosa of ducklings given hypertonic saline solutions (Crocker *et al.*, 1974, 1975). An imbalance in plasma electrolyte levels could therefore be partly responsible for the alterations in yolk structure and shell thickness that result from petroleum ingestion (Cooke, 1973; Grau *et al.*, 1977). Indeed, as calcium restriction can put birds out of lay and induce egg shell thinning, a similar interference with calcium uptake or the effect of an electrolyte imbalance on the shell gland could be a major factor responsible for some of the effects of petroleum on reproductive activity.

Evidence that calcium uptake could be impaired by petroleum ingestion comes from the observation that the concentrations of total plasma calcium in birds fed two and five per cent North Sea oil were consistently less than those in control birds (Fig. 8). In the control birds, the level of plasma calcium was directly related to the rate of lay and to the length of the photoperiod, although qualitatively similar patterns in calcium levels were seen in the birds that were fed oil. After the oil had been removed from the food, the concentration of plasma calcium remained lower than that in the controls, despite the increase in food consumption and rate of lay (Fig. 8). This suggests that reduced food intake was not responsible for the low plasma calcium levels and, although this reduced intestinal uptake could be responsible, it should be emphasized that the concentrations of plasma calcium could also be low as a result of low estrogen levels (Simkiss, 1967; Cooke, 1973). It should be noted, however, that total plasma calcium might not be a true reflection of calcium status (Ladenson and Bowers, 1973; Luck and Scanes, 1979a) and, although it appeared to reflect petroleum induced changes in lay and egg quality, low blood calcium levels are not always associated with pesticide-induced egg shell thinning (Bitman and Cecil, 1970). Moreover, the fact that some of the control and oil fed birds laid normal eggs while exposed to short days when the plasma calcium level was at its lowest (Fig. 8), suggests that the level of plasma calcium in the oil-fed

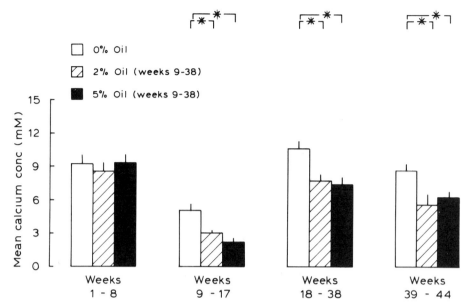

Fig. 8.—Mean concentrations of plasma total calcium (mM) in blood samples taken at weekly intervals from laying Khaki Campbell ducks ($N = 5$ or 6) before (weeks 1-8), during (weeks 9-38), and after (weeks 39-44) the feeding of 0% (open), 2% (shaded), and 5% (solid histograms) North Sea crude oil in the diet. Other details as in Fig. 2. Adapted from Harvey et al. (1981a).

birds held on long days was higher than the threshold required for normal egg shell formation and reproductive activity (Taylor et al., 1962; Taylor, 1965). In addition, low plasma LH levels are a characteristic feature of calcium deficient poultry, and could be due to a "cut-out" mechanism that operates when calcium levels are low to inhibit gonadotropin secretion (Taylor et al., 1962; Taylor, 1965; Bonney and Cunningham, 1977; Luck and Scanes, 1978, 1979b). The observation that the levels of plasma LH in the birds fed five per cent oil were greater than those in the controls (Fig. 2) suggests that calcium levels were not low enough to exert an inhibitory effect on gonadotropin secretion. Thus, as the effects of petroleum on reproductive activity cannot be satisfactorily explained simply by changes in plasma LH, prolactin, T_3, T_4, corticosterone, and calcium concentrations, it is possible that the effects of petroleum could be mediated by direct or indirect effects on the shell gland of the oviduct to impair egg and egg shell formation. This possibility awaits investigation.

ACKNOWLEDGMENTS

The authors would like to express their thanks to Mr. R. Troake (British Petroleum) for the generous gift of North Sea (Forties Field) crude oil. This work was supported in part by grants from the Science Research Council (GR A/29869 and GR B/47782).

LITERATURE CITED

ALBERS, P. H. 1977. Effects of external application of fuel oil on hatchability of mallard eggs. Pp. 158-163, *in* Fate and effects of petroleum hydrocarbons in marine ecosystems and organisms (D. A. Wolfe, ed.), Pergamon Press, New York, xix+478 pp.

———. 1978. The effects of petroleum on different stages of incubation in bird eggs. Bull. Environ. Contam. Toxicol., 19:624-630.

BALTHAZART, J , J. C. HENDRICK, AND P. DEVICHE. 1977. Diurnal variations of plasma gonadotrophins in male domestic ducks during the sexual cycle. Gen. Comp. Endocrinol., 32:376-389.

BATES, R. W., E. L. LAHR, AND O. RIDDLE. 1935. The gross action of prolactin and FSH on the mature ovary and sex accessories of the fowl. Amer. J. Physiol., 111:361-368.

BIRKHEAD, T. R., C. LLOYD, AND P. COCKHILL. 1973. Oiled seabirds successfully cleaning their plumage. Brit. Birds, 66:535-537.

BITMAN, J., AND H. C. CECIL. 1970. Estrogenic activity of DDT analogs and polychlorinated biphenyls. J. Agric. Food. Chem., 18:1108-1112.

BOLTON, N. J. 1976. Prolactin physiology in the fowl, *Gallus domesticus*. An investigation by radioimmunoassay of blood and pituitary gland hormone concentrations. Ph.D. thesis, Univ. Leeds, viii+218 pp.

BOLTON, N. J., C. G. SCANES, AND A. CHADWICK. 1975. A radioimmunoassay for prolactin in the circulation of birds. J. Endocrinol., 67:51P.

BONNEY, R. C., AND F. J. CUNNINGHAM. 1977. Effect of ionic environment on the release of LH from chicken anterior pituitary cells. Mol. Cell. Endocrinol., 7:245-251.

CAMPER, P. M., AND W. B. BURKE. 1977. The effect of prolactin on the gonadotrophin induced rise in serum estradiol and progesterone of the laying turkey. Gen. Comp. Endocrinol., 32:72-77.

CHRISTIAN, J. J. 1975. Hormonal control of population growth. Pp. 205-274, *in* Hormonal correlates of behaviour (B. E. Eleftheriou and R. L. Sprott, eds.), Plenum Press, New York, xi+439+xii-xvii pp.

COOKE, A. S. 1973. Shell thinning in avian eggs by environmental pollutants. Environ. Poll., 4:85-152.

CROCKER, A. D., J. CRONSHAW AND W. N. HOLMES. 1974. The effect of a crude oil on intestinal absorption in ducklings (*Anas platyrhynchos*). Environ. Poll., 7:165-178.

———. 1975. The effect of several crude oils and some petroleum distillation fractions on intestinal absorption in ducklings (*Anas platyrhynchos*). Environ. Physiol. Biochem., 5:92-106.

EASTIN, W. C., AND D. J. HOFFMAN. 1978. Biological effects of petroleum on aquatic birds, Pp. 561-582, *in* AIBS conference on assessment of ecological impacts of oil spills. Keystone, Colorado, vii+936 pp.

EISNER, E. 1960. The relationship of hormones to the reproductive behaviour of birds, referring especially to parental behaviour: a review. Anim. Behav., 8:155-179.

ENSOR, D. M. 1978. Comparative endocrinology of prolactin. Chapman and Hall, London, ix+ 309 pp.

ETCHES, R. J., A. S. MCNEILLY, AND C. E. DUKE. 1979. Plasma concentrations of prolactin during the reproductive cycle of the domestic turkey (*Meleagris gallopavo*). Poultry Sci., 58:963-970.

FOLLETT, B. K., AND D. T. DAVIES. 1975. Photoperiodicity and the neuroendocrine control of reproduction in birds. Pp. 199-224, *in* Advances in avian physiology (M. Peaker, ed.), Academic Press, London, xiv+377 pp.

FOLLETT, B. K., AND J. E. ROBINSON. 1980. Photoperiod and gonadotrophin secretion in birds. Prog. Reprod. Biol., 5:39-61.

GAY, M. L., H. A. BELISLE, AND J. F. PATTON. 1980. Quantification of petroleum type hydrocarbons in avian tissue. J. Chromotag., 187:153-160.

GILBERT, A. B. 1971a. The female reproductive effort. Pp. 1153-1162, *in* Physiology and biochemistry of the domestic fowl (D. J. Bell and B. M. Freeman, eds.), Academic Press, New York, xxii+1428+xiii-cxlv pp.

———. 1971b. The endocrine ovary in reproduction. Pp. 1449-1468, *in* Physiology and biochemistry of the domestic fowl (D. J. Bell and B. M. Freeman, eds.), Academic Press, New York, xxii+1428+xiii-cxlv pp.

———. 1971c. Transport of the egg through the oviduct and oviposition. Pp. 1345-1352, *in* Physiology and biochemistry of the domestic fowl, (D. J. Bell and B. M. Freeman, eds.), Academic Press, New York, xxii+1428+xxiii-cxlv pp.

———. 1979. Female genital organs. Pp. 238-260, *in* Form and function in birds (A. S. King and J. Mclelland, eds.), Academic Press, New York, xi+459 pp.

GOLDSMITH, A. C., K. P. CAVANAUGH, B. K. FOLLETT, AND W. N. HOLMES. 1980. Oviposition and incubation in mallards: effects of crude oil on prolactin and LH secretion. Proc. Soc. Endocrinol.

GRAU, C. R., T. ROUDYBUSH, J. DOBBS, AND J. WATHEN. 1977. Altered yolk structure and reduced hatchability of eggs from birds fed single doses of petroleum oils. Science, 195:779-781.

GROSS, A. O. 1950. The herring gull-cormorant control project. Proc. Internat. Ornith. Conf., 10:532-536.

HALL, T. R., AND A. CHADWICK. 1978. Effects of oestradiol 17β and testosterone on prolactin secretion by the pituitary of the domestic fowl *in vitro*. IRCS Med. Sci., 6:327.

HARTUNG, R. 1965. Some effects of oiling on reproduction of ducks. J. Wildl. Mgmt., 29:872-874.

HARTUNG, R., AND G. S. HUNT. 1966. Toxicity of some oils to waterfowl. J. Wildl. Mgmt., 30:564-570.

HARVEY, S., P. J. SHARP, AND J. G. PHILLIPS. 1981a. Influence of sublethal doses of ingested petroleum on reproductive performance and the pituitary-gonadal axis in domestic ducks (*Anas platyrhynchos*). Comp. Biochem. Physiol., in press.

HARVEY, S., H. KLANDORF AND J. G. PHILLIPS. 1981b. Reproductive performance and endocrine responses to ingested petroleum in domestic ducks (*Anas platyrhynchos*). Gen. Comp. Endocrinol., 45:372-380.

HARVEY, S., C. G. SCANES, A. CHADWICK, AND N. J. BOLTON. 1978. Influence of fasting, glucose and insulin on the levels of growth hormone and prolactin in the plasma of the domestic fowl. (*Gallus domesticus*). J. Endocrinol., 76:501-506.

HOFFMAN, D. J. 1978. Embryotoxic effects of crude oil in mallard ducks and chicks. Toxicol., Appl. Pharmacol., 183:190.

HOLMES, W. N., AND J. CRONSHAW. 1977. Biological effects of petroleum on marine birds. Pp. 359-398, *in* Effects of petroleum on Arctic and subarctic marine environments and organisms (D. C. Malins, ed.), Academic Press, New York, xx+500 pp.

HOLMES, W. N., AND J. GORSLINE. 1980. Effects of some environmental pollutants on the adrenal cortex. Pp. 311-314, *in* Proc. Third Internat. Cong. Endocrinol., Melbourne (I. A. Cummings, J. W. Funder, and S. A. O. Mendelsohn, eds.), xii+736 pp.

HOLMES, W. N., AND J. G. PHILLIPS. 1976. The adrenal cortex of birds. Pp. 293-420, *in* General comparative and clinical endocrinology of the adrenal cortex (I. Chester Jones and I. W. Henderson, eds.), Academic Press, New York, xv+461 pp.

HOLMES, W. N., J. CRONSHAW, AND J. GORSLINE. 1978a. Some effects of ingested petroleum on sea water-adapted ducks (*Anas platyrhynchos*). Environ. Res., 17:177-190.

HOLMES, W. N., K. P. CAVANAUGH, AND J. CRONSHAW. 1978b. The effects of ingested petroleum on oviposition and some aspects of reproduction in experimental colonies of mallard ducks (*Anas platyrhynchos*). J. Reprod. Fert., 54:335-347.

HOLMES, W. N., J. GORSLINE, AND J. CRONSHAW. 1979. Effects of mild cold stress on the survival of sea water-adapted mallard ducks (*Anas platyrhynchos*) maintained on food contaminated with petroleum. Environ. Res., 20:425-444.

HOSODA, T., T. KANEKO, K. MOGI, AND T. ABE. 1955. Effect of gonodotrophic hormone on ovarian follicles and serum vitallin of fasting hens. Proc. Soc. Exp. Biol. Med., 88:502-504.

JALLAGEAS, M., A. TASMISIER, AND I. ASSENMACHER. 1978. A comparative study of the annual cycles in sexual and thyroid function in male peking ducks (*Anas platyrhynchos*) and teal (*Anas crecca*). Gen. Comp. Endocrinol., 36:201-210.

JEFFERIES, D. J. 1969. Induction of apparent hyperthyroidism in birds fed DDT. Nature, 222:578-579.

———. 1975. The role of the thyroid in the production of sublethal effects by organochlorine insecticides and polychlorinated biphenyls. Pp. 131-230, *in* Organochlorine insecticides: persistent organic pollutants (F. Moriarty, ed.), Academic Press, London, xii+302 pp.

———. 1975. The role of the thyroid in the production of sublethal effects by organochlorine insecticides and polychlorinated biphenyls. Pp. 131-230, *in* Organochlorine insecticides: persistent organic pollutants (F. Moriarty, ed.), Academic Press, London, xii+302 pp.

JEFFERIES, D. J., AND M. C. FRENCH. 1971. Hyper- and hypothyroidism in pigeons fed DDT: an explanation for the "thin eggshell phenomenon." Environ. Poll. 1:235-242.

———. 1972. Changes induced in the pigeon thyroid by p, p'-DDE and dieldrin. J. Wildl. Mgmt., 36:24-30.

JEFFERIES, D. J., M. C. FRENCH, AND B. E. OSBORNE. 1971. The effect of p, p'-DDT on the rate, amplitude and weight of the heart of the pigeon and Bengalese finch. Brit. Poult. Sci., 12:387-399.

KOPISCHKE, E. D. 1972. The effect of 2,4-D and diesel fuel on egg hatchability. J. Wildl. Mgmt., 36:1353-1356.

LADENSON, J. H., AND G. N. BOWERS, JR. 1973. Free calcium in serum II. Rigor of homeostatic control, correlations with total serum calcium and review of data on patients with disturbed calcium metabolism. Clin. Chem., 19:575-582.

LAWLER, G. C., W.-A. LOONG, AND J. L. LASETER. 1978a. Accumulation of saturated hydrocarbons in tissues of petroleum-exposed mallard ducks. Environ. Sci. Technol., 12:47-51.

———. 1978b. Accumulation of aromatic hydrocarbons in tissues of petroleum-exposed mallard ducks. Environ. Sci. Technol., 12:51-54.

LAWLER, G. C., J. P. HOLMES, B. FIORITO, J. L. LASETER, AND R. C. SZARO. 1978c. Quantification of petroleum hydrocarbons in selected tissues of male mallard ducklings chronically exposed to South Louisiana Crude oil. Pp. 583-612, *in* AIBS Conference on assessment of ecological impacts of oil spills, Keystone, Colorado, vii+936 pp.

LUCK, M. R., AND C. G. SCANES. 1978. Gonadotrophin secretion in the domestic fowl during calcium deficiency. Gen. Comp. Endocrinol., 34:80-81.

———. 1979a. Plasma levels of ionized calcium in the laying hen (*Gallus domesticus*). Comp. Biochem. Physiol., 63A:177-181.

———. 1979b. The relationship between reproductive activity and blood calcium in calcium-deficient hen. Brit. Poult. Sci., 20:559-564.

MEIER, A. H. 1969. Antigonadal effects of prolactin in the white-throated sparrow, *Zonotrichia albicollis*. Gen. Comp. Endocrinol., 13:222-225.

MILLER, D. S., D. B. PEAKALL, AND W. B. KINTER. 1978. Ingestion of crude oil; sublethal effects in herring gull chicks. Science, 199:315-317.

MORRIS, T. R., AND A. V. NALBANDOV. 1961. The induction of ovulation in starving pullets using mammalian and avian gonadotrophins. Endocrinology, 68:687-697.

OPEL, H. 1971. Induction of incubation behaviour in the hen by brain implants of prolactin. Poultry Sci., 50:1613.

PATTON, J. F., AND M. P. DIETER. 1980. Effects of petroleum hydrocarbons on hepatic function in the duck. Comp. Biochem. Physiol., 650:33-36.

PEAKALL, D. B. 1970. p, p^1-DDT: effect on calcium metabolism and concentration of estradiol in the blood. Science, 168:592-594.

RATCLIFFE, D. A. 1970. Changes attributable to pesticides in egg breakage frequency and egg shell thickness in some British birds. J. Appl. Ecol., 7:67-115.

RITTINGHAUS, H. 1956. Etwasüber die "indirekte" verbreitung der Ölpest in einem seevagels-schutzgebeit. Orn. Mitt., 8:43-46.

SCANES, C. G., S. HARVEY, AND A. CHADWICK. 1976. Plasma luteinizing hormone and follicle-stimulating hormone concentrations in fasted immature male chickens. IRCS Med. Sci., 4:371.

SCANES, C. G., P. J. SHARP, S. HARVEY, P. M. M. GODDEN, A. CHADWICK, AND W. S. NEWCOMER. 1979. Variations in plasma prolactin, thyroid hormones, gonadal steroids and growth hormone in turkeys during the induction of egg laying and moult by different photoperiods. Brit. Poult. Sci., 20:143-148.

SHARP, P. J. 1980. Female reproduction. Pp. 435-454, in Avian endocrinology (A. Epple and M. T. Stetson, eds.), Academic Press, New York, xv+577 pp.

SIMKISS, K. 1967. Calcium in reproductive physiology. Chapman and Hall, London, xiv+264 pp.

SREBOCAN, V., J. P. GOTAL, V. ADAMOVIC, B. SOKIC, AND M. DELAK. 1971. Effect of technical grade DDT and p,p^1-DDT on adrenocortical functions in chicks. Poultry Sci., 50:1271-1279.

SZARO, R. C., AND P. H. ALBERS. 1977. Effects of external applications of No. 2 fuel oil on common eider eggs. Pp. 164-167, in Fate and effects of petroleum hydrocarbons in marine ecosystems and organisms (D. A. Wolfe, ed.), Pergamon Press, New York, xix+478 pp.

TANAKA, K., M. KAMIYOSHI, AND Y. TANABE. 1971. Inhibition of premature ovulation by prolactin in the hen. Poultry Sci., 50:63-66.

TAYLOR, T. G. 1965. Calcium-endocrine relationships in the laying hen. Proc. Nutr. Soc., 24:49-54.

TAYLOR, T. G., T. R. MORRIS, AND F. HERTELENDY. 1962. The effect of pituitary hormones on ovulation in calcium deficient pullets. Vet. Rec., 74:123-125.

URIST, M. R., AND N. W. DEUTSCH. 1960. Effects of cortisone upon blood, adrenal cortex, gonads and the development of osteoporosis in birds. Endocrinology, 66:805-818.

WELLS, J. W., A. B. GILBERT, AND J. CULBERT. 1980. The effect of luteinizing hormone on progesterone secretion in vitro by the granulosa cells of the domestic fowl (Gallus domesticus). J. Endocrinol., 34:249-254.

INDEX